KINETIC CONTROL IN SYNTHESIS AND SELF-ASSEMBLY

Index

KINETIC CONTROL IN SYNTHESIS AND SELF-ASSEMBLY

Edited by

MUNENORI NUMATA
Department of Biomolecular Chemistry
Graduate School of Life and Environmental Sciences
Kyoto Prefectural University, Shimogamo
Sakyo-ku, Kyoto, Japan

SHIKI YAGAI
Graduate School of Engineering, and
Institute for Global Prominent Research (IGPR)
Chiba University
Chiba, Japan

TOSHIYUKI HAMURA
Department of Applied Chemistry for Environment
Kwansei Gakuin University
Sanda, Japan

ELSEVIER

ACADEMIC PRESS
An imprint of Elsevier

Academic Press is an imprint of Elsevier
125 London Wall, London EC2Y 5AS, United Kingdom
525 B Street, Suite 1650, San Diego, CA 92101, United States
50 Hampshire Street, 5th Floor, Cambridge, MA 02139, United States
The Boulevard, Langford Lane, Kidlington, Oxford OX5 1GB, United Kingdom

Library of Congress Cataloging-in-Publication Data
A catalog record for this book is available from the Library of Congress

British Library Cataloguing-in-Publication Data
A catalogue record for this book is available from the British Library

ISBN: 978-0-12-812126-9

For information on all Academic Press publications visit our website at https://www.elsevier.com/books-and-journals

Publisher: Susan Dennis
Acquisition Editor: Emily M. McCloskey
Editorial Project Manager: Katerina Zaliva
Production Project Manager: Bharatwaj Varatharajan
Cover Designer: Christian J. Bilbow

Typeset by TNQ Technologies

CONTENTS

CONTRIBUTORS

Manabu Abe
Department of Chemistry, Graduate School of Science, Hiroshima University, Higashi-Hiroshima, Japan
Hiroshima University Research Center for Photo-Drug-Delivery Systems (HiU-P-DDS), Hiroshima University, Higashi-Hiroshima, Japan

Bimalendu Adhikari
Department of Chemical Sciences, Indian Institute of Science Education and Research (IISER) Mohali, Punjab, India

Shuji Akai
Graduate School of Pharmaceutical Sciences, Osaka University, Osaka, Japan

Rikuo Akisaka
Department of Chemistry, Graduate School of Science, Hiroshima University, Higashi-Hiroshima, Japan

Toshiyuki Hamura
Department of Applied Chemistry for Environment, Kwansei Gakuin University, Sanda, Japan

Shuhei Higashibayashi
Faculty of Pharmacy, Keio University, Tokyo, Japan

Sung Ho Jung
National Institute for Materials Science (NIMS), Tsukuba, Japan

Ryoichi Kuwano
Department of Chemistry, Faculty of Science, Kyushu University, Fukuoka, Japan

Yusuke Makida
Department of Chemistry, Faculty of Science, Kyushu University, Fukuoka, Japan

Mihoko Maruyama
Graduate School of Engineering, Osaka University, Suita, Osaka, Japan
Graduate School of Life and Environmental Sciences, Kyoto Prefectural University, Kyoto, Japan

Yusuke Mori
Graduate School of Engineering, Osaka University, Suita, Osaka, Japan
Osaka University, Suita, Osaka, Japan

Takashi Morii
Institute of Advanced Energy, Kyoto University, Uji, Kyoto, Japan

Gamal A.I. Moustafa
Graduate School of Pharmaceutical Sciences, Osaka University, Osaka, Japan
Department of Medicinal Chemistry, Faculty of Pharmacy, Minia University, Minia, Egypt

Shun Nakano
Institute of Advanced Energy, Kyoto University, Uji, Kyoto, Japan

Eiji Nakata
Institute of Advanced Energy, Kyoto University, Uji, Kyoto, Japan

Kosuke Namba
Graduate School of Pharmaceutical Science, Tokushima University, Tokushima, Japan

Munenori Numata
Department of Biomolecular Chemistry, Graduate School of Life and Environmental Sciences, Kyoto Prefectural University, Kyoto, Japan

Eisaku Ohashi
Graduate School of Pharmaceutical Science, Tokushima University, Tokushima, Japan

Arivazhagan Rajendran
Institute of Advanced Energy, Kyoto University, Uji, Kyoto, Japan

Kazunori Sugiyasu
National Institute for Materials Science (NIMS), Tsukuba, Japan

Kazufumi Takano
Graduate School of Life and Environmental Sciences, Kyoto Prefectural University, Kyoto, Japan

Masayuki Takeuchi
National Institute for Materials Science (NIMS), Tsukuba, Japan

Katsunori Tanaka
Biofunctional Synthetic Chemistry Laboratory, RIKEN, Saitama, Japan
GlycoTargeting Research Team, RIKEN, Saitama, Japan
Max Planck-RIKEN Joint Center for Chemical Biology Research, Saitama, Japan
Biofunctional Chemistry Laboratory, A. Butlerov Institute of Chemistry, Kazan Federal University, Kazan, Russia
JST PRESTO, Saitama, Japan

Takashi Uemura
Department of Advanced Materials Science, Graduate School of Frontier Sciences, The University of Tokyo, Kashiwa, Japan
CREST, Japan Science and Technology Agency (JST), Saitama, Japan

Kenward Vong
Biofunctional Synthetic Chemistry Laboratory, RIKEN, Saitama, Japan

Shiki Yagai
Department of Applied Chemistry and Biotechnology, Graduate School of Engineering, Chiba University, Chiba, Japan
Institute for Global Prominent Research (IGPR), Chiba University, Chiba, Japan

Mitsuaki Yamauchi
Department of Applied Chemistry and Biotechnology, Graduate School of Engineering, Chiba University, Chiba, Japan
Department of Applied Chemistry for Environment, School of Science and Technology, Kwansei Gakuin University, Sanda, Japan

Masashi Yoshimura
Graduate School of Engineering, Osaka University, Suita, Osaka, Japan
Institute of Laser Engineering, Osaka University, Suita, Osaka, Japan

PREFACE

Kinetic control is essentially related to the time required for the transformation of molecules to occur. Kinetic control in a synthetic reaction is related to ways of shortening the time required to reach this transition state. The main concern of attempting to kinetically control the reactive species is how to prolong their lifetime in addition to developing a strategy to obtain the desired reactive species through an appropriate kinetic pathway. In a controlled self-assembly, the required time for numerous number of molecules or polymers to fully self-assemble should be taken into consideration. When using thermodynamic control, the time related to molecular transformation is not controlled. The concept of controlling the reaction time is not present under thermodynamic conditions, whereas it is directly linked with molecular functions under kinetic conditions. Introducing appropriate substituents attached to molecules, changing the molecular environments (e.g., temperature, solvent polarities, and viscosities), compartmentalizing molecules into restricted spaces, supplying energy via heating/cooling, photoirradiation, and sonication, and the combination of these techniques results in positive control of the time required for each molecular transformation step. Elaborate and specialized strategies are therefore required when using kinetic control depending on the system being examined, in contrast to the use of thermodynamic control. Many strategies for kinetic control have been practiced in different areas of chemistry; these have led to the development of new frontiers in molecular sciences. Although these subfields of molecular sciences have examined the use of kinetic control, the strategies that have been used across these fields have been regarded as irrelevant to one another and discussed independently. Integrating these kinetic control strategies, however, can allow innovation in each area and give rise to new concepts that may allow for more precise regulation of kinetic processes in a versatile manner. The molecular design required for a selective reaction, the kinetic trapping of reactive species, the controlled self-assembly of these species, and how the molecular environments surrounding the molecules influence the kinetic processes are all factors that should be explored. This book, entitled "Kinetic Control in Synthesis and Self-assembly," highlights techniques that may be used to control kinetic processes across different fields of chemistry. An interdisciplinary book will attract the interest of a multidisciplinary community of chemistry, especially of researchers active in molecular sciences such as organic synthesis, supramolecular chemistry, polymer chemistry, crystal engineering, and materials sciences.

The editorial team, Numata, Yagai, and Hamura, have assembled a diverse collection of chapters from leading world experts in chemistry across diverse disciplines, including synthetic, material, and biological chemistry. In order to account for even more fields of chemistry, the editors have also contributed to chapters in this book. This book is composed of 13 chapters. The first half of this book (Chapters 1–6) focuses on controlled reactions and reactivity under kinetically controlled conditions. The effects of the molecular structures, peripheral substitutions, and auxiliary units on reactivity are discussed from the viewpoint of kinetic control. Chapters 7 and 8 discuss the kinetics related to biological processes. In these chapters, biological molecules and polymers such as oligosaccharides, proteins, and DNA participate in kinetic reactions and in interpolymer interactions. These chapters also focus on the concept of self-assembly of biological components through precise molecular recognition. Kinetic control in the self-assembly of synthetic molecules or polymers has developed rapidly, but it may still lack versatile strategies in comparison with those of organic and biological reactions. The editors arranged chapters related to the self-assembly of synthetic components in the latter part of this book (from Chapters 9–13) using the latest research.

The editors express our heartfelt thanks to all chapter contributors who agreed on the concept of this book and participated in this project. We would like to thank the editorial team of Elsevier for their kind support and continuous encouragement for this project.

SUMMARY OF CONTENTS

Chapter 1

Reactive species are characterized by the low-valence and weak chemical bonds, such as singlet diradicals. They are important intermediates not only for developing new chemical reactions but also for designing novel functional materials. In this chapter, Akisaka and Abe take a detailed look at the kinetic stabilization of singlet diradicals, a new type of π-bonding system. They describe how to isolate new molecules with π-single bonds.

Chapter 2

Developing an effective methodology for obtaining optically pure compounds at a quantitative yield has been the main concern in the field of organic chemistry. In this chapter, Akai and Moustafa review the concept of dynamic kinetic resolution as a potential solution for overcoming this fundamental issue. The advantage of the use of an enzyme (such as lipase) and its combination with metal catalysts is demonstrated in this chapter.

Chapter 3

Exploiting versatile synthetic routes to produce polycyclic aromatic compounds with desired substituents at desired positions has been important in organic chemistry. The development of new synthetic routes using highly reactive intermediates as a reactive platform is introduced in this chapter. The controlled reactions of highly reactive intermediates under both kinetic and thermodynamic conditions are discussed by Hamura.

Chapter 4

Bowl-shaped aromatic compounds that have a partial, spherical fullerene structure, known as "buckybowls," have been an attractive synthetic target because of their unique shape and fundamental properties arising from their curved π-electron systems. In this chapter, Higashibayashi focuses on the chirality of buckybowls, the so-called "bowl chirality," and reviews the dynamics of the bowl inversion of many kinds of buckybowls in terms of the effect of molecular structure on the inversion barrier.

Chapter 5

The asymmetric hydrogenation of aromatic compounds results in the synthesis of optically active six-membered rings. Because of the high stability of aromatic compounds, higher temperatures are generally required for the hydrogenation process, but this is unfavorable for stereoselective reactions. Overcoming these difficulties relies on the use of fundamental asymmetric hydrogenation; a solution for controlling the kinetic pathway required to achieve the desired chirality is demonstrated in this chapter by Makida and Kuwano.

Chapter 6

In this chapter, Namba and Ohashi focus on the total synthesis of natural compounds, for which kinetically controlled reactions play a crucial role. In the field of total synthesis, an appropriate combination of kinetic and thermodynamic control is required to achieve the construction of a desired molecular skeleton with a desired selectivity. For the synthesis of palau'-amine, a complex natural product, kinetically controlled reactions were used with competing thermodynamic reactions.

Chapter 7

Oligosaccharides play an important role in physiological processes. Vong and Tanaka survey the development of the latest strategies for selective glycosylation and the biological functions of the resulting *N*-glycosylated proteins. On the basis of kinetics, a comprehensive description of the biological stabilities of *N*-glycosylated proteins (oligosaccharides) correlated with their biological distributions is provided in this chapter.

Chapter 8

Morii and coworkers describe the recent development of DNA-based assembly systems. The ability to locate different components (functions) at desired positions on a regular mesh grid remains a challenge in these systems. A detailed discussion of the kinetic strategies used for loading different proteins onto a DNA scaffold and the application of the resultant protein–DNA conjugation to cascade reactions are provided in this chapter.

Chapter 9

In polymer chemistry, controlling kinetic processes is indispensable for the preparation of polymer materials with desired functions. Unique methodologies for the kinetically controlled polymerization of synthetic polymers and their self-assemblies utilizing metal–organic frameworks as templates are introduced by Uemura in this chapter.

Chapter 10

Molecular self-assemblies potentially comprise both thermodynamic and kinetic processes. Stable and closed self-assembled structures, such as capsules, form under thermodynamic processes, whereas the incorporation of kinetic processes allows the formation of larger structures. A comprehensive description of the use of kinetic control in supramolecular chemistry is described by Sugiyasu and Takeuchi in this chapter.

Chapter 11

Yagai and coworkers describe the strategy toward creation of higher-order structures, such as ring and helix, from one-dimensional supramolecular chains. The topological control in noncovalent chains has still been challenging task in supramolecular chemistry. The creation of supramolecular polymers with well-defined topologies has been successfully achieved by focusing on the kinetic aspect of the polymerization.

Chapter 12

Self-assemblies coupled with energy consumption provide a kinetic pathway toward forming metastable supramolecules. The precise regulation of the chemical and physical properties of a self-assembly space at a micrometer scale is indispensable for achieving such molecular self-assemblies that are seen in nature. Numata describes the use of microflow as an ideal self-assembly space for kinetic control.

Chapter 13

Maruyama's chapter provides an impressive overview of crystal polymorphism. Controlling the polymorphism of pharmaceutical compounds has been a central subject in the pharmaceutical industry. The issue posed by polymorphism may be overcome by the kinetically controlled self-assembly of the pharmaceutical compounds. Unique strategies coupled with various external stimuli are discussed in this chapter.

Munenori Numata
Shimogamo, Kyoto
August 30, 2018

CHAPTER 1

Kinetic Stabilization of Localized Singlet Diradicaloids With π Single Bonding Character

Rikuo Akisaka[1], **Manabu Abe**[1,2]
[1]Department of Chemistry, Graduate School of Science, Hiroshima University, Higashi-Hiroshima, Japan; [2]Hiroshima University Research Center for Photo-Drug-Delivery Systems (HiU-P-DDS), Hiroshima University, Higashi-Hiroshima, Japan

Contents

1. INTRODUCTION

Reactive species with high energy, such as low-valent chemical species, electronically excited states, and compounds with weak bonds, play important roles in the chemical and material-science fields, although they have, in general, short lives. Studies on the structure and reactivity of such elusive species contribute much to understanding the reaction mechanisms and developing new chemical reactions and functional materials. For example, the isolation and crystallization of singlet carbene was achieved by

Kinetic Control in Synthesis and Self-Assembly
ISBN 978-0-12-812126-9
https://doi.org/10.1016/B978-0-12-812126-9.00001-8

Arduengo and coworkers (Fig. 1.1) *(1)*. Subsequent studies showed that such stable singlet carbenes, the so-called NHCs, are useful as ligands for metal-mediated coupling reactions *(2)*. Chauvin et al. clarified the intervention of the metalacyclic intermediate in the synthetically important olefin metathesis reactions *(3)*. Spectroscopic analyses, including low-temperature NMR, UV-vis, and IR, largely contributed to the elucidation of the structure and reactivity of reactive intermediates *(4)*.

When such high-energy molecules are isolated and handled like stable molecules, rapid and great progresses are possible in chemistry. A kinetic stabilization effect, which increases the lifetimes of reactive species without changing the original electronic character, has emerged to elucidate the structure and reactivity of the highly reactive species. The basic idea of "kinetic effect" is that the reaction pathways of reactive species are blocked kinetically by introducing sterically hindered substituents around the reaction centers to increase the energy barriers and lifetimes of the reactive species (Fig. 1.2). Therefore, kinetic stabilization is a very powerful method for elucidating the chemical properties of highly reactive molecules. For example, in 1981, Yoshifuji and coworkers succeeded in isolating the labile P=P double-bonded species using a sterically hindered substituent such as 1,3,5-tri-*tert*-butylphenyl on phosphorus atoms (Fig. 1.3) *(5)*. A huge number of P=P double-bonding chemistry has emerged since this pioneering study *(6–8)*.

Recently, molecules with π bonds between heavy atoms were isolated using the kinetic stabilization effect (Fig. 1.4). Sekiguchi and coworkers succeeded in isolating the Si–Si triple bonding using "Dis_2iPrSi" *(9,10)*. The Sb–Sb double bonding was synthesized by Tokitoh and coworkers using "Tbt" substituents *(11,12)*. The "Eind" groups developed by Tamao were utilized to isolate and characterize Ge=O compounds *(13,14)*.

In this chapter, the recent studies on generating π single-bonded species in which the carbon–carbon is connected by only a π single bond, and the substituent effect on the reactivity, are summarized as a case study of the kinetic stabilization effects. The details of the features of the π single bond and its molecular design are described in Section 2. In Section 3, the kinetic

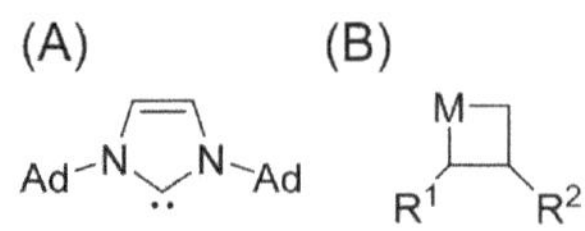

Figure 1.1 Examples of high-energy species. (A) stable singlet carbene, (B) metalacyclobutane intermediate for olefin metathesis reactions (M = Ru, Mo ...).

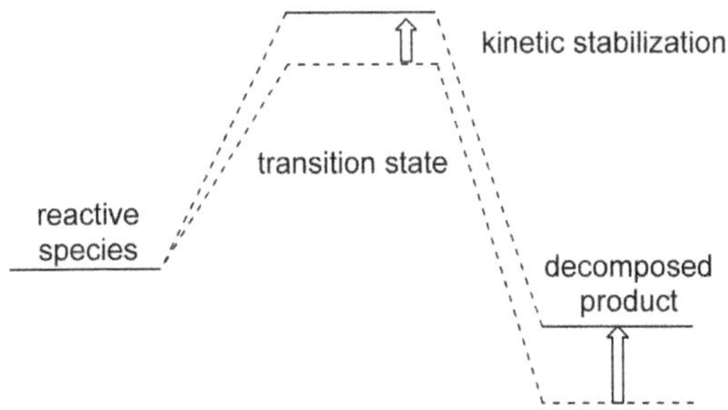

Figure 1.2 Concept of kinetic stabilization.

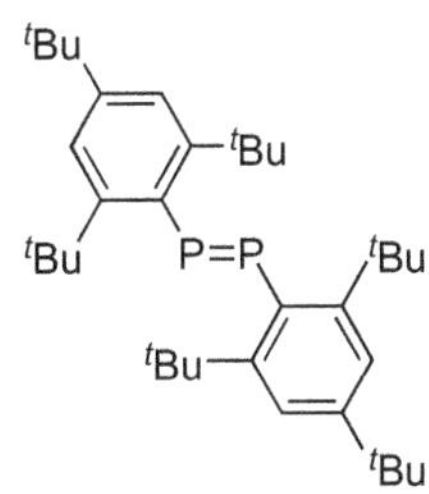

Figure 1.3 First example of kinetically stabilized molecule.

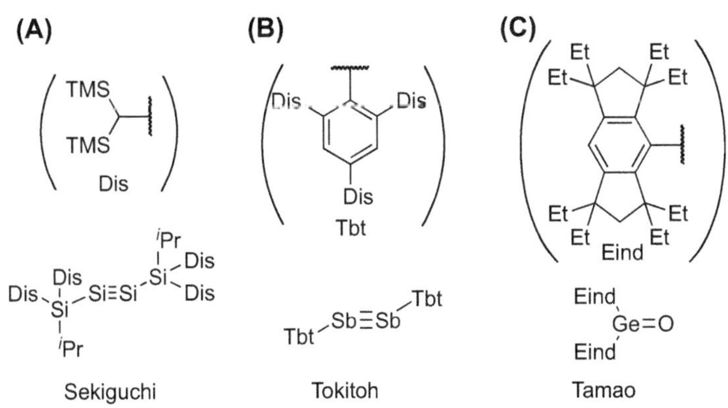

Figure 1.4 Recent examples of kinetic stabilization: disilyne (A), distibene (B), and germanone (C).

stabilization of π single-bonded compounds with bulky substituents is summarized. In Section 4, a new molecular-strain effect on the kinetic stabilization is discussed using the reactivity of the π-bonded species as a mechanistic probe.

2. π SINGLE-BONDED SPECIES

2.1 π Single Bond

In organic chemistry, carbon–carbon bonds are classified into two types of chemical bonds, σ bond and π bond. The σ bonds are much stronger than π bonds, suggesting that the σ-bonding systems play crucial rules in constructing molecules. π-bonding systems are important in characterizing chemical features such as reactivity, light absorption, conductivity, etc., because π-bonding systems are included in the frontier molecular orbital regions. The bonding (HOMO) and antibonding (LUMO) energy gap is smaller than that of σ bonds (Fig. 1.5). Therefore, π bonds are the origins of molecular properties and functions.

If carbon–carbon atoms are connected by only a π bond (π single bond, C–π–C) without σ bond, a new type of π-bonding system is constructed, producing new molecular structures with planar four-coordinate carbons and new molecular functions, owing to the extremely small HOMO–LUMO gap *(15–20)*.

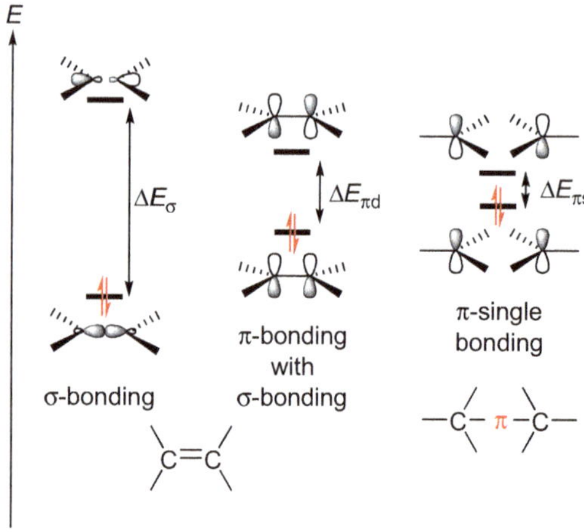

Figure 1.5 Molecular structures of ethylene and π single-bonded species, and their bonding and antibonding energy gaps (Δ*E*) in σ-bonding, π-bonding with σ-bonding, and π-single bonding: $\Delta E_{\sigma} > \Delta E_{\pi d} > \Delta E_{\pi s}$.

2.2 Construction of π Single Bonds

How can we construct π single bonds (C−π−C)? When the two p-orbitals of two methyl radicals (CH_3) are geometrically fixed in a parallel fashion (Fig. 1.5), π single bonding will be possible. Because the σ bonding is much stronger than the π bonding, the σ-bond formation should be retarded using appropriate molecular designs. The bond dissociation energy of a C−C σ bond of the ethane molecule CH_3-CH_3 is ~85 kcal/mol *(21)*. First, we discuss the molecular-strain effect on the retardation of the σ-bond formation.

Berson and coworkers first reported the matrix isolation of trimethylenemethane derivatives using a molecular strain originating from bicyclic structures (Fig. 1.6A) *(22)*. This molecular design can be applied for the kinetic stabilization of the π single-bonding case. The strain energies of cyclopropane, bicyclo[1.1.0]butane, and bicyclo[2.1.0]pentane are ~27, 64, and 54 kcal/mol, respectively *(23,24)*. The molecular strain decreases the energy difference between the σ-bonded and planar singlet diradical forms (Fig. 1.6B). The singlet states of the opened structures would possess the π single-bonded character between C1−C3 carbons.

The singlet states of cyclopentane-1,3-diyls and cyclobutane-1,3-diyls appear to satisfy the conditions for making π single bonds. However, Closs and Dougherty clarified that the parent 1,3-diyls had triplet ground states *(25−27)*. The singlet ground state is necessary to form π bonds between the C1 and C3 carbons. The bonding orbital ψ_S should be lower in energy than the antibonding orbital ψ_A (Fig. 1.7). How can we switch the ground-state spin multiplicity from triplet to singlet? The triplet ground state of the parent 1,3-diyls can be explained by the balance between the through space (TS) interaction and through bond (TB) interaction (Fig. 1.8) *(28−34)*. The TS interaction between the two p-orbitals in the C1 and C3 carbons produces two molecular orbitals, which are bonding ψ_S and

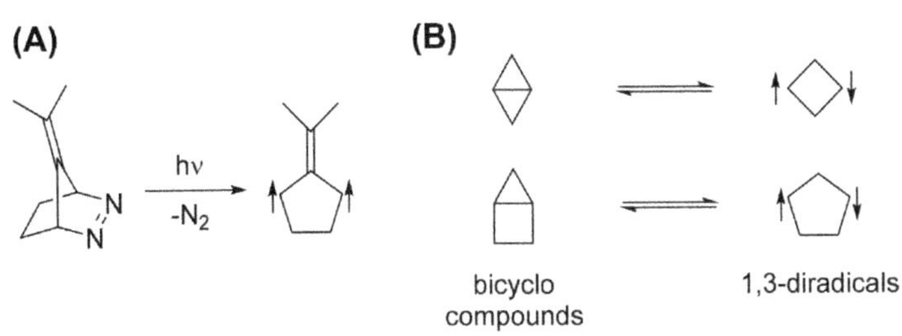

Figure 1.6 (A) Generation of trimethylenemethane with bicyclic molecular strain; (B) molecular strain induced by bicyclic structure and generation of 1,3-diradicals.

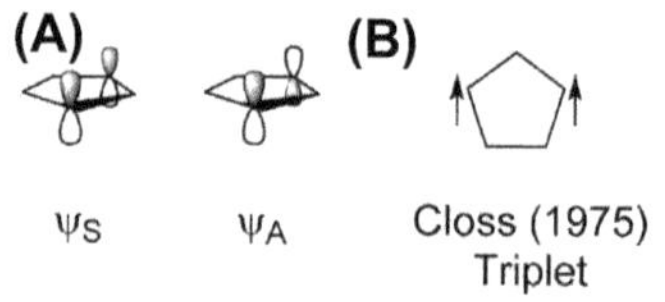

Figure 1.7 (A) Nonbonded molecular orbitals ψ_S and ψ_A; (B) parent cyclopentane-1,3-diyls, which has triplet ground state.

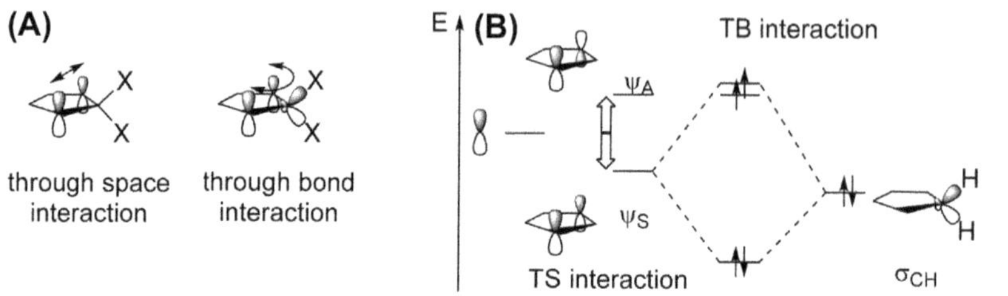

Figure 1.8 (A) Concept of through space (TS) interaction and through bond (TB) interaction in cyclopentane-1,3-diyls; (B) TB interaction makes the electronic ground state of parent cyclopentane-1,3-diyls triplet.

antibonding ψ_A, suggesting that the ground state of 1,3-diyls is singlet. The TB interaction between the pseudo p orbital of the C2 carbon and ψ_S destabilizes the symmetric ψ_S to locate ψ_S close to ψ_A in energy. Therefore, the triplet is the ground state. To make a π single bond, the energy level of the ψ_S orbital should be lower than that of the ψ_A orbital.

In 1994, Borden and coworkers reported the singlet ground state of 2,2-difluorocyclopentane-1,3-diyl diradical (Fig. 1.9) *(35,36)*. The notable substituent effect of the electron-withdrawing group (EWG) was explained by the TB interaction between the low-lying pseudo p orbital CF_2 and ψ_S, which stabilized the bonding ψ_S orbital in energy and increased the energy

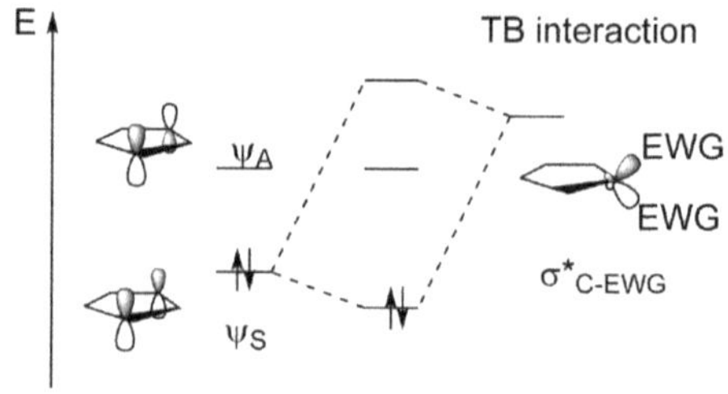

Figure 1.9 Through bond (TB) interaction of electron withdrawing group attached at C2 position generates singlet-1,3-diradicals, which has occupied ψ_S orbitals (Type-I).

gap between ψ_S and ψ_A (type-I configuration). In 1998, 2,2-dialkoxycyclopentane-1,3-diyls were also found by us to be singlet (Fig. 1.8) *(37)*. The singlet ground state was also found for the electron-donating group substituted case, that is 2,2-disilylcyclopentane-1,3-diyl diradical *(38,39)*. The most stable electronic configuration was, however, not type-I but type-II, because of the TB interaction between the high-lying pseudo p orbital $C(SiH_3)_2$ and ψ_S (Fig. 1.10). The type-I electronic configuration is suitable for constructing π single bonds (C−π−C) in the frontier orbital region. According to the natural resonance theory analysis, 2,2-dihydroxycyclopentane-1,3-diyl had 18% zwitterion character and 38% π single-bonding character (Fig. 1.11). The Mulliken bond order (BO) was calculated to be 0.58, which was between 0 and 1, at the UB3LYP/6-31G(d) level of theory, indicating that the singlet species had a weak bonding (fuzzy bonding) character ($C_1{-}\pi{-}C_3$) between C1 and C3 *(40)*.

2.3 Generation and Reactivity of Localized Singlet Diradicaloids With π Single-Bonding Character

In 1998, Adam et al. confirmed the singlet ground state of 2,2-difluorocyclopentane-1,3-diyl by generating the diradical S-**DR1** (X = F, Ar = Ph; λ_{max} at 530 nm) using the photochemical denitrogenation of **AZ1** (Table 1.1) *(41)*. The lifetime of S-**DR1** was found to be ~80 ns at room temperature in hexane. In 2000, Abe et al. found a one-order longer-lived singlet diradical S-**DR2** (X = OEt, Ar = Ph; λ_{max} at 570 nm), whose lifetime was determined to be τ ~880 ns at 293K in benzene, to give quantitatively the ring-closed σ-bonded compound **CP2** *(37)*. A small, but significant, solvent effect was observed in the ring closure reaction. The lifetime of S-**DR2** in a polar solvent such as acetonitrile was longer than

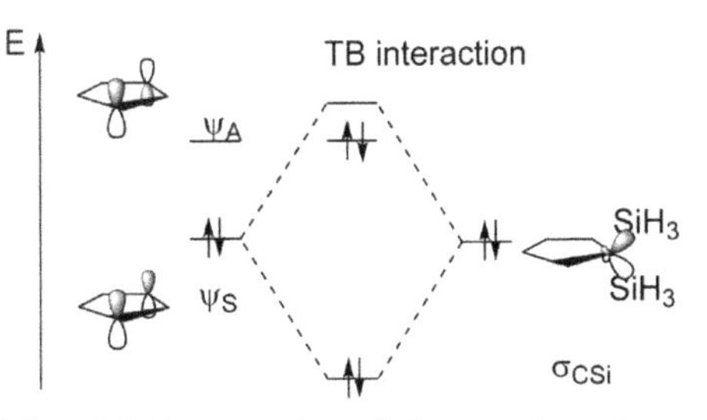

Figure 1.10 Through bond (TB) interaction of electron donating group attached at C2 position makes ψ_S orbital unstable and generates Type-II singlet-1,3-diradicals.

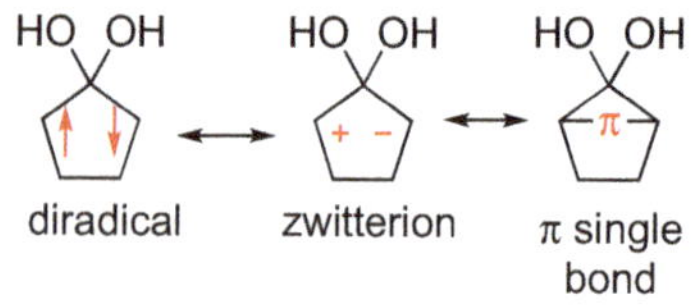

Figure 1.11 Possible mesomeric structures to express the electronic character in singlet 2,2-dihydoxycyclopentane-1,3-diyl diradicals.

that in *n*-hexane (Table 1.1), indicating that the singlet diradical had the characteristics of zwitterion (Fig. 1.11) *(40)*.

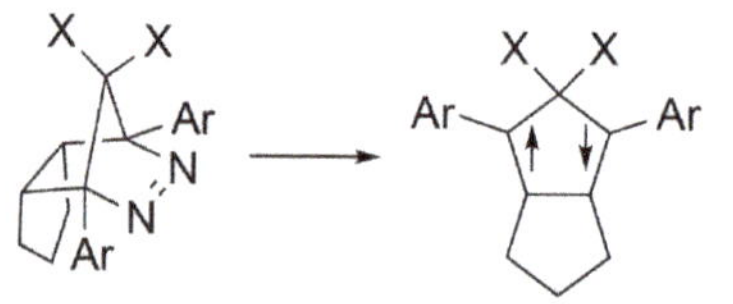

AZ1,2 S-**DR1,2**

The zwitterionic character of the singlet diradical was also proved by the substituent effect of the aryl group *(41)*. Thus, the lifetime of the electron-donating MeO-substituted diradical S-**DR5** was longer than that of S-**DR4** possessing the electron-withdrawing substituent CN group at the para position of the phenyl ring.

Table 1.1 Substituents and solvent effects on the lifetime τ_{293} of S-**DR**

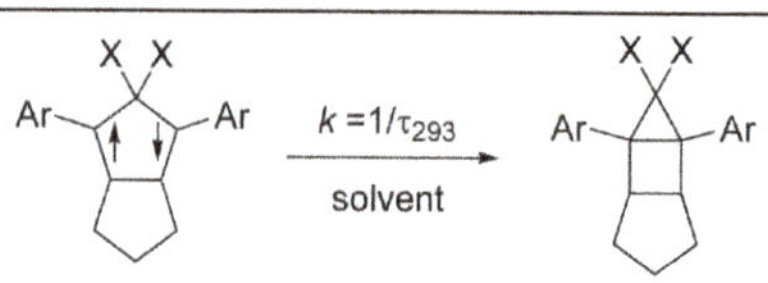

S-**DR**	X	Ar	Solvent	τ_{293} /ns	λ_{max} /nm
1	F	Ph	Hexane	80	~530
2	OEt	Ph	benzene	880	~545
2	OEt	Ph	n-hexane	520	~545
2	OEt	Ph	CH_3CN	1010	~545
3a	OMe	Ph	benzene	320	~600
4	OMe	*p*-CN-C_6H_4	benzene	625	~600
5	OMe	p-MeOC_6H_4	benzene	1050	~600

3. KINETIC STABILIZATION USING BULKY SUBSTITUENTS

3.1 Alkoxy Group Effect

As shown in the previous section, the lifetime of the 2,2-diethoxy-substituted singlet 1,3-diradical S-**DR2** (X = OEt, Ar = Ph; τ = 880 ns) was approximately three times longer than that of S-**DR3a** (X = OMe, Ar = Ph; τ = 320 ns), indicating that the size of the alkoxy group at the C2 position played an important role in determining the energy barrier for the ring-closing process. To obtain information on the alkoxy group effect, the lifetimes of S-**DR3b-d** (n, m = 3, 6, 10) were measured at 293 K in benzene *(42)*. The lifetime increased on increasing the chain length of the alkoxy group (Table 1.2). Interestingly, the absorption maximum was not affected by the alkoxy group, indicating that the kinetic stabilization of the singlet diradical was possible by using the steric hindrance of the alkyl chain to increase the energy barrier of the ring-closing reaction. The lifetimes of S-**DR3b** and S-**DR3d** were, however, nearly the same, suggesting that the steric repulsion between the phenyl ring and the alkoxy group was the key to determining the energy barrier. Indeed, no significant effect was observed on the log(A/s) value of the activation parameter, but the value of activation energy E_a was found to increase as the alkoxy chain became longer.

Table 1.2 Kinetic stabilization by increasing the alkoxy chain length in S-**DR3**

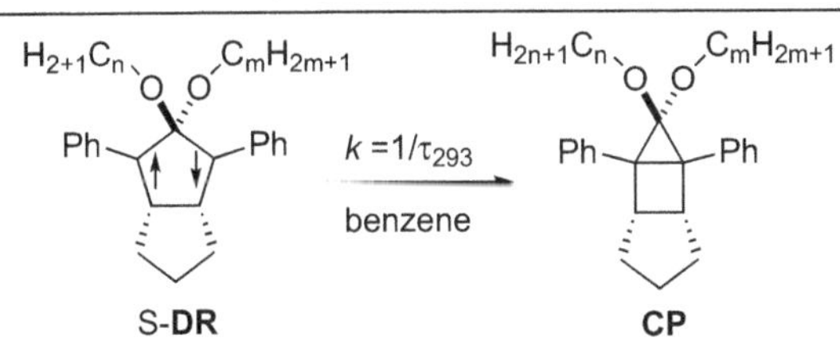

S-**DR**	n	m	τ_{293} /ns	Log A	E_a /kJ mol	λ/nm
3a	1	1	320	12.5	33.4	574
3b	3	3	1900	12.7	39.3	571
3c	6	6	2100	12.6	38.8	572
3d	10	10	2300	12.5	38.2	572
3e	3	1	936	12.8	37.8	568
3f	1	3	516	12.8	36.4	568

3.2 Kinetic Products and Thermodynamic Products in Ring Closure Reaction of S-DR3

To study the mechanism of the ring-closing reaction of the singlet diradical in more detail, asymmetrically substituted diradicals S-**DR3e** and **f** were generated and their lifetimes were compared. Surprisingly, the lifetime of S-**DR3e** with a propoxy group in the anti-configuration to the cyclopentane ring was found to be longer than that of the diastereomer S-**DR3f**, although the trans-configured **CP3e** and **f** were isolated quantitatively. Quantum chemical calculations were conducted on S-**DR3a** to clarify the notable substituent effect on the kinetic stabilization of S-**DR3e**. The transition state inv-**TS3a** was calculated to be lower in energy than ret-**TS3a** (Fig. 1.12), indicating that the steric repulsive interaction of the phenyl ring with the methylene group of the cyclopentane ring was more severe than that of the ring with the exo-alkoxy group. At room temperature (~298 K), the kinetically favored inv-**CP3a** isomerized to form the thermodynamically stable ret-**CP3a**, according to quantum chemical calculations.

The direct ^{1}H and ^{13}C NMR analyses of the photodenitrogenation of **AZ3a** were conducted to verify whether inv-**CP3a** was formed or not *(43)*.

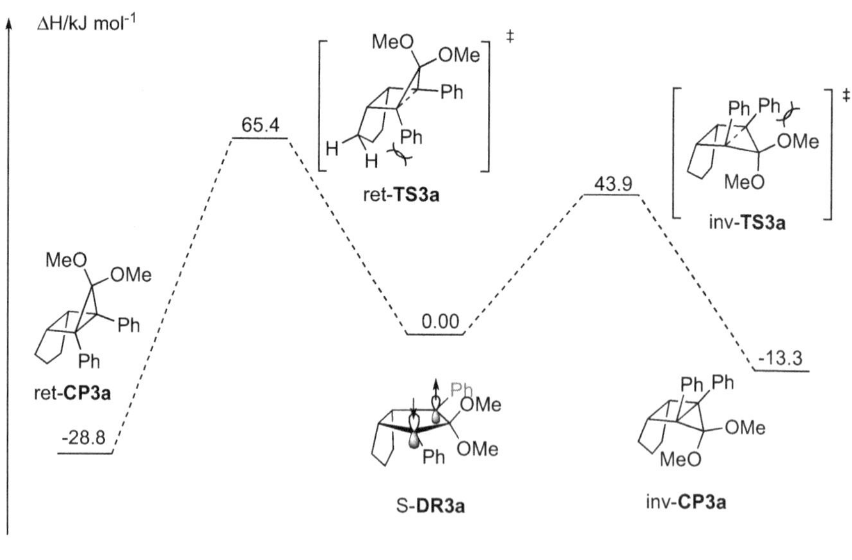

Figure 1.12 Reaction profiles of S-**DR3a** at the BS-UB3LYP/6-31G(d) level of theory/ The value ΔH is the relative thermal enthalpy calculated at 298 K and 1 atm.

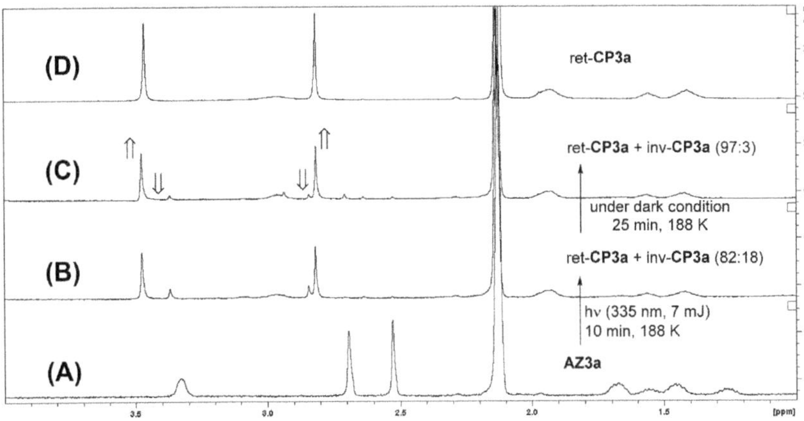

Figure 1.13 In situ ^{1}H NMR (400 MHz d 1–4 ppm) analysis of the direct photochemical denitrogenation of **AZ3a** in toluene-d_8 under 355 nm laser irradiation at 188 K. (A) ^{1}H NMR spectrum before irradiation (**AZ3a**). (B) ^{1}H NMR spectrum after 10 min of irradiation at 188K (ret-**CP3a**/inv-**CP3a** = 82/18). (C) ^{1}H NMR spectrum after 25 min in the dark at 188K, which indicates that inv-**CP3a** isomerize to ret-**CP3a**. (D) ^{1}H NMR spectrum of ret-**CP3a**, which was prepared by photoreaction at room temperature. The chemical shift of the toluene methyl group (d = 2.18 ppm) was used as reference.

When **AZ3a** was irradiated with a 355 nm laser for 10 min at 188K in toluene-d_8, weak singlet signals at δ 3.37 and 2.84 ppm, which corresponded to the OMe protons, were observed together with the MeO signals of ret-**CP3a** (Fig. 1.13). The weak signals were converted to those of signals of ret-**CP3a** under dark conditions, indicating that the weak signals isomerized to ret-**CP3a**. These newly observed signals were assigned to inv-**CP3a**. Interestingly, inv-**CP3a** was the major product of the triplet-sensitized denitrogenation reaction of **AZ3a** at 199K. Almost no signals of 3.47 and 2.82 ppm corresponding to ret-**CP3a** were observed, although only 3.37 and 2.84 ppm signals of inv-**CP3a** were observed clearly from S-**DR3a**. From the NOE measurement of that sample, a correlation was found between 3.37 and 1.25 ppm, and it was concluded that the product at low temperature was inv-**CP3a**.

The temperature dependency of the thermal isomerization process from inv-**CP3a** to ret-**CP3a** was analyzed using ^{1}H NMR spectroscopic analysis at a low temperature under dark conditions. The decay rate of inv-**CP3a** was consistent with the growth rate of ret-**CP3a**. The activation parameters were calculated by determining the temperature dependency of the decay rate of inv-**CP3a** and the growth rate of ret-**CP3a** (Table 1.3).

Table 1.3 Activation parameters for isomerization from inv-**CP3a** to ret-**CP3a**

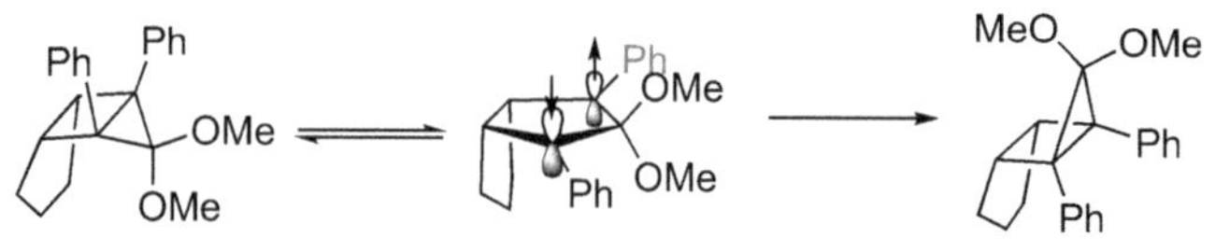

inv-**CP3a** S-**DR3a** ret-**CP3a**

	E_a /kJ mol	Log (A/s)
decay of inv-**CP3a**	60.4 ± 1.0	12.0 ± 0.1
formation of ret-**CP3a**	60.5 ± 1.5	12.0 ± 0.5

3.3 Kinetic Stabilization by Bulky Aryl Group

As clarified above, the direct observation of localized singlet diradicaloids using conventional spectroscopy was conducted using the kinetic stabilization originating from the steric repulsion of the alkoxy group. In this section, the aryl group effect on kinetic stabilization is summarized *(44)*. The lifetime of S-**DR6a-d** and the activation parameters of the ring closure reaction were determined by the LFP method (Table 1.4). The lifetime of S-**DR6a**, which had 3,5-dimethoxybenzene attached to C1 and C3, was nearly twice as long as that of the singlet diradical S-**DR3a** (X = OMe,

Table 1.4 Kinetic stabilization by introducing the bulky aryl group in S-**DR5**

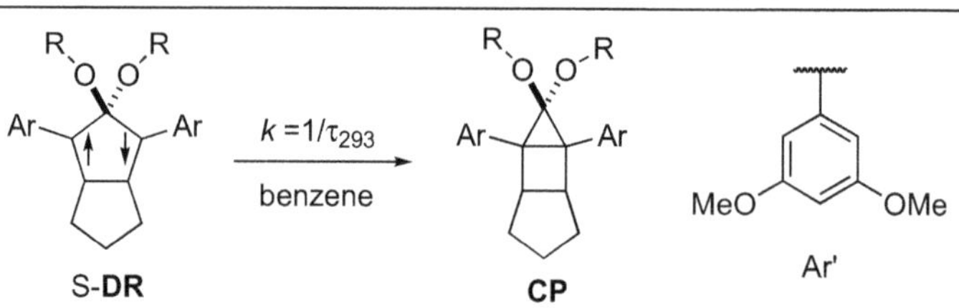

S-**DR** **CP**

S-**DR**	R	Ar	τ_{293} /ns	$\Delta S^{\ddagger}$/kJ mol/T	$\Delta H^{\ddagger}$/kJ mol	$\Delta G^{\ddagger}_{293}$/kJ mol	λ_{max}/nm
3a	Me	Ph	320	-8.1 ± 0.7	32.7 ± 0.2	35.1 ± 0.7	573
6a	Me	Ar'	580	-12.9 ± 1.0	33.0 ± 0.2	36.7 ± 0.4	583
6b	CH_2Ph	Ar'	4001	-27.8 ± 1.3	33.3 ± 0.4	40.9 ± 0.8	592
6c	CH_2Ar'	Ph	1933	-10.1 ± 1.1	36.6 ± 0.6	39.6 ± 0.6	584
6d	CH_2Ar'	Ar'	5394	-19.4 ± 1.0	36.5 ± 0.3	42.2 ± 0.7	593

Ar = Ph). However, it was shorter than that of S-**DR3d**, which had a long alkoxy group. Therefore, to increase the lifetime further, a benzylalkoxy chain was introduced into the alkoxy chain. The lifetime of S-**DR6b** was 4001 ns and ~13 times longer than that of S-**DR3a**. On the other hand, S-**DR6c**, which had a 3,5-dimethoxybenzylalkoxy group in the C2 position and a phenyl ring attached to C1 and C3, had a lifetime of 1933 ns. Therefore, introducing bulky substituents at the C1 and C3 positions is more effective than the substitution at the C2 position (compare S-**DR6b** vs. S-**DR6c**). By possessing the features of S-**DR6b** and **c**, the lifetime of the singlet diradical could be lengthened to 5394 ns.

4. KINETIC STABILIZATION BY "STRETCH EFFECT"

In Section 3, we described how the bulky substituents of the alkoxy and aryl groups kinetically stabilized S-**DR6d.** A lifetime of 5.4 μs was obtained, which was 17 times longer than that of the standard S-**DR3a**. In polymer chemistry, strained bonds were broken by ultrasound irradiation to generate diradicals and induce the rearrangement reaction *(45–48)*. If carbon–carbon bonds were stretched by molecular designs, the homolysis accelerated to produce diradical intermediates. In other words, the lifetime of singlet diradicaloids increased owing to the "stretch effect." In 2012, we proposed a concept of "stretch effect" originating from two types of molecular strains, by which the energy difference between the singlet diradical and its ring-closed compound became small to increase the lifetime of a singlet diradical (Fig 1.14) *(49)*.

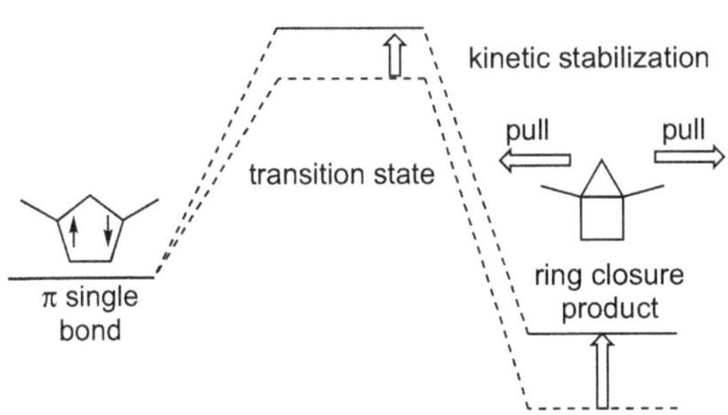

Figure 1.14 Concept of kinetic stabilization by using "stretch effect."

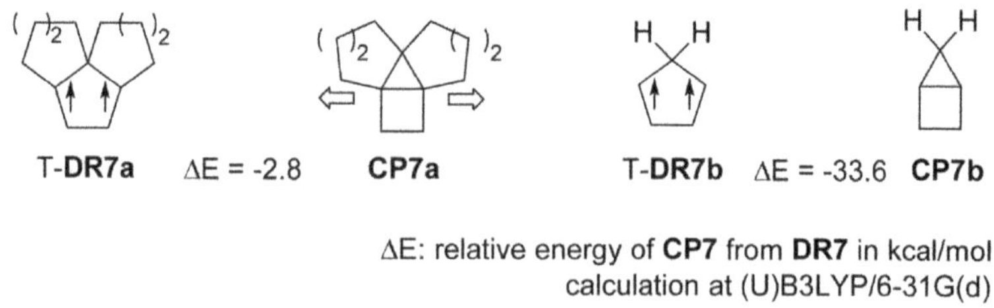

Figure 1.15 Introduction of spirobonding into T-**DR7** destabilizes ring-closure product **CP7**.

4.1 Type 1: Spirobonding

The bond dissociation energy of CH_3-CH_3 is ~85 kcal/mol. The molecular strain of bicyclic[2.1.0]pentane is found to be approximately 54 kcal/mol. Therefore, it is reasonable that the energy difference between **DR** and **CP** is calculated to be ~30 kcal/mol (Fig. 1.15). To stretch the central C1–C3 carbons, spirobonding was introduced at the C2 carbon, which was supposed to increase the molecular strain of the ring-closed structure because the trans-fused bicyclic structure was highly strained. Indeed, the strain energy of **CP7a** was calculated to be 81.5 kcal/mol. The energy difference between T-**DR7a** and **CP7a** was reduced greatly, to −2.8 kcal/mol.

Interestingly, the singlet diradical S-**DR7c** was found to be more stable, by 1.6 kcal/mol in energy, than the corresponding ring-closing compound **CP7c** (Fig. 1.16). The BO of S-**DR7c** was calculated to be 0.33 because the electron occupation numbers in ψ_S and ψ_A were calculated to be 1.33 and 0.67, respectively, at (6/6)CASSCF/ANO-RCC/VTZP.

4.2 Type 2: Macrocyclic System

Another molecular design to stretch the central bond in the bicyclic structure is shown in Fig 1.17. The macrocyclic structures **DR8a** and **b**

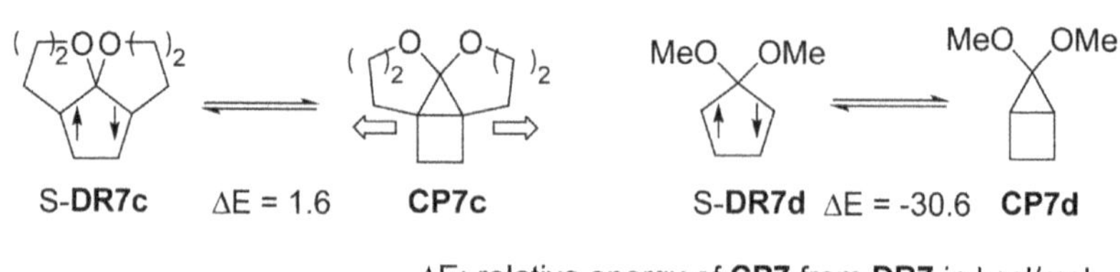

Figure 1.16 Theoretical calculations clarify that S-**DR7c** is kinetically stabilized by introducing spirobonding.

Figure 1.17 Introduction of macrocyclic system to cyclopentane-1,3-diradicals.

with planar π systems such as benzene and naphthalene are the key to kinetically stabilize the singlet diradicals.

The obtained information about the macrocyclic effect on the kinetic stabilization of diradical structures, the model compounds T-**DR8c** and T-**DR8d**, were used to calculate the energy difference with the corresponding ring-closing compounds (Table 1.5). The energy difference

Table 1.5 Introduction of rigid π system in macrocyclic system destabilizes the **CP8**.

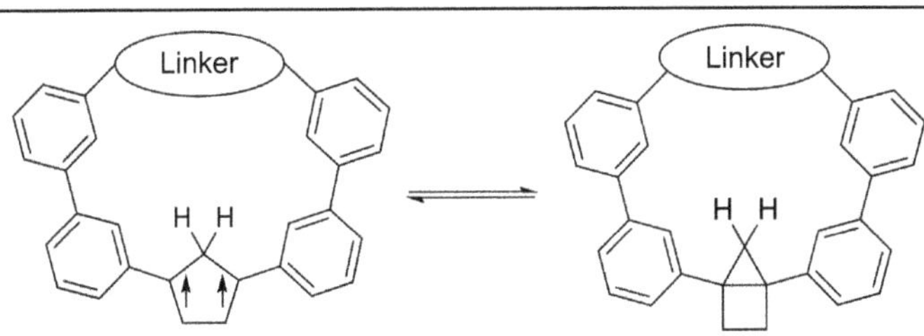

T-**DR8** **CP8**

T-**DR**	Linker	ΔE	ΔE_{ST}
8c	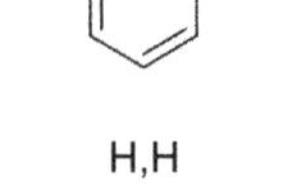	–5.4	–0.18
8d	H,H	–7.3	0.59
8e	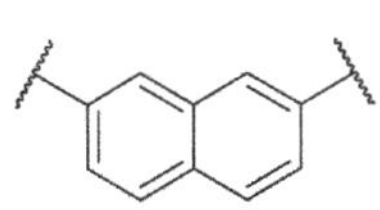	–0.8	0.55

ΔE: relative energy of **CP8** from T-**DR7** in kcal/mol.

ΔE_{ST}: relative energy of S-**DR8** from T-**DR8** in kcal/mol.
Calculation at UB3LYP/6-31G(d)

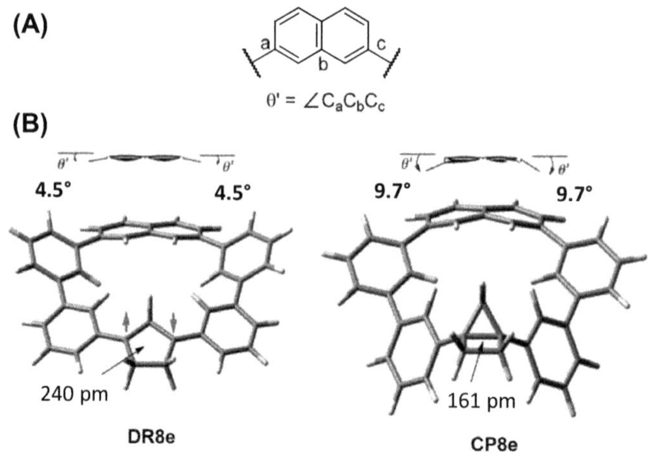

Figure 1.18 Optimized structures of (A) singlet state diradical **DR8e** and (B) ring-closed compound **CP8e**.

between T-**DR8c** and **CP8c**, in which the radical sites are connected at the meta positions of the benzene ring, was calculated to be −5.3 kcal/mol, which was smaller than that for T-**DR8d**. That is, the macrocyclic structure was suggested to be sufficient for the kinetic stabilization of the radical form. When the 2,7-naphtyl moiety was introduced in the macrocyclic system, a notable decrease of −0.80 kcal/mol was observed in the energy difference between T-**DR8e** and **CP8e**.

To clarify the molecular stretch effect in **DR8e**, the angle θ' in the diradical state was compared with the angle in the corresponding ring-closing compound **CP8e** (Fig. 1.18). In the diradical state, the angle θ' was found to be 4.5 degrees, whereas, the angle was found to increase to

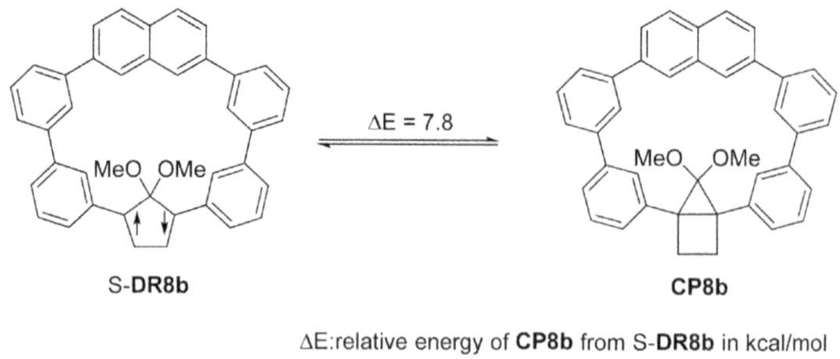

Figure 1.19 S-**DR8b** is kinetically stabilized by introducing macrocyclic π system.

9.7 degrees by the ring closure in **CP8e**. The deviation from the planarity and size of the π system could be the main reason for the small energy difference.

When the singlet ground state diradical moiety was introduced in the macrocyclic system, S-**DR8b** was calculated to be 7.8 kcal/mol stable in energy than **CP8b** (Fig. 1.19). The singlet diradical S-**DR8b** can be a target for new experimental studies.

5. SUMMARY

Localized singlet diradicals are key intermediates for homolytic bond-forming and bond-breaking processes. Direct detection of the reactive intermediates is necessary to understand the chemical reactions. In this chapter, the molecular design of kinetic stabilization in localized singlet diradicals was discussed in detail. First, a bicyclic structure was proposed for stabilizing the localized 1,3-diradicals. This proposal led to the generation of S-**DR1** and **2**. To increase the lifetime of S-**DR**, the alkoxy chain of S-**DR3** was lengthened, and bulking of the aryl group in S-**DR6** was performed. As a result, we obtained S-**DR6d**, which had a lifetime of 5.4 μs. These results showed us that it was very important to identify the reaction pathway and to introduce bulky substituents that raised the activation barrier. Another approach, the "stretch effect," was discussed using quantum calculations, to increase the activation energy for the ring-closure reaction of S-**DR**. It was confirmed that stretching the radical carbons and lengthening the distances made the radical—radical coupling severe.

Why do we need to make singlet diradicals stable? The answer would be to understand further the bond-homolysis process and to make π single-bond species (C—π—C). π Single bonds have very small HOMO—LUMO energy gaps. If we could obtain such new molecules, it could replace polycyclic aromatic hydrocarbons and could be applied to semiconducting materials such as OELs, OFETs, organic solar cells, etc. *(50—52)*. Therefore, we are now trying to isolate singlet diradicals.

REFERENCES

1. Arduengo, A. J.; Harlow, R. L.; Kline, M. A Stable Crystalline Carbene. *J. Am. Chem. Soc.* **1991,** *113* (1), 361—363.
2. Fortman, G. C.; Nolan, S. P. N-Heterocyclic Carbene (NHC) Ligands and Palladium in Homogeneous Cross-Coupling Catalysis: A Perfect Union. *Chem. Soc. Rev.* **2011,** *40* (10), 5151.

3. Chauvin, Y. Olefin Metathesis: The Early Days (Nobel Lecture). *Angew. Chem. Int. Ed.* **2006,** *45* (23), 3740–3747.
4. Moss, R. A., Platz, M., Jones, M., Eds. *Reactive Intermediate Chemistry;* Wiley-Interscience: Hoboken, NJ, 2004.
5. Yoshifuji, M.; Shima, I.; Inamoto, N.; Hirotsu, K.; Higuchi, T. Synthesis and Structure of Bis (2, 4, 6-Tri-Tert-Butylphenyl) Diphosphene: Isolation of a True Phosphobenzene. *J. Am. Chem. Soc.* **1981,** *103* (15), 4587–4589.
6. Tsurusaki, A.; Nagahora, N.; Sasamori, T.; Matsuda, K.; Kanemitsu, Y.; Watanabe, Y.; Hosoi, Y.; Furukawa, Y.; Tokitoh, N. Synthesis, Structures, and Reactivity of Kinetically Stabilized Anthryldiphosphene Derivatives. *Bull. Chem. Soc. Jpn.* **2010,** *83* (5), 456–478.
7. Shah, S.; Burdette, S. C.; Swavey, S.; Urbach, F. L.; Protasiewicz, J. D. Alkali Metal Induced Rupture of a Phosphorus–Phosphorus Double Bond. Electrochemical and EPR Investigations of New Sterically Protected Diphosphenes and Radical Anions $[ArPPAr]^{\bullet-}$. *Organometallics* **1997,** *16* (15), 3395–3400.
8. Dutan, C.; Shah, S.; Smith, R. C.; Choua, S.; Berclaz, T.; Geoffroy, M.; Protasiewicz, J. D. Sterically Encumbered Diphosphaalkenes and a Bis(Diphosphene) as Potential Multiredox-Active Molecular Switches: EPR and DFT Investigations. *Inorg. Chem.* **2003,** *42* (20), 6241–6251.
9. Sekiguchi, A. A. Stable Compound Containing a Silicon-Silicon Triple Bond. *Science* **2004,** *305* (5691), 1755–1757.
10. Sekiguchi, A.; Ichinohe, M.; Kinjo, R. The Chemistry of Disilyne with a Genuine Si–Si Triple Bond: Synthesis, Structure, and Reactivity. *Bull. Chem. Soc. Jpn.* **2006,** *79* (6), 825–832.
11. Tokitoh, N.; Arai, Y.; Sasamori, T.; Okazaki, R.; Nagase, S.; Uekusa, H.; Ohashi, Y. A Unique Crystalline-State Reaction of an Overcrowded Distibene with Molecular Oxygen: The First Example of a Single Crystal to a Single Crystal Reaction with an External Reagent. *J. Am. Chem. Soc.* **1998,** *120* (2), 433–434.
12. Sasamori, T.; Tokitoh, N. A New Family of Multiple-Bond Compounds between Heavier Group 14 Elements. *Bull. Chem. Soc. Jpn.* **2013,** *86* (9), 1005–1021.
13. Matsuo, T.; Tamao, K. Fused-Ring Bulky "Rind" Groups Producing New Possibilities in Elemento-Organic Chemistry. *Bull. Chem. Soc. Jpn.* **2015,** *88* (9), 1201–1220.
14. Li, L.; Fukawa, T.; Matsuo, T.; Hashizume, D.; Fueno, H.; Tanaka, K.; Tamao, K. A Stable Germanone as the First Isolated Heavy Ketone with a Terminal Oxygen Atom. *Nat. Chem.* **2012,** *4* (5), 361–365.
15. Hoffmann, R.; Alder, R. W.; Wilcox, C. F. Planar Tetracoordinate Carbon. *J. Am. Chem. Soc.* **1970,** *92* (16), 4992–4993.
16. Cooper, O. J.; Wooles, A. J.; McMaster, J.; Lewis, W.; Blake, A. J.; Liddle, S. T. A Monomeric Dilithio Methandiide with a Distorted Trans-Planar Four-Coordinate Carbon. *Angew. Chem. Int. Ed.* **2010,** *49* (32), 5570–5573.
17. Collins, J. B.; Dill, J. D.; Jemmis, E. D.; Apeloig, Y.; Schleyer, P. v. R.; Seeger, R.; Pople, J. A. Stabilization of Planar Tetracoordinate Carbon. *J. Am. Chem. Soc.* **1976,** *98* (18), 5419–5427.
18. Deva Priyakumar, U.; Narahari Sastry, G. A System with Three Contiguous Planar Tetracoordinate Carbons Is Viable: A Computational Study on a C6H62= Isomer. *Tetrahedron Lett.* **2004,** *45* (7), 1515–1517.
19. Priyakumar, U. D.; Reddy, A. S.; Sastry, G. N. The Design of Molecules Containing Planar Tetracoordinate Carbon. *Tetrahedron Lett.* **2004,** *45* (12), 2495–2498.
20. Sateesh, B.; Srinivas Reddy, A.; Narahari Sastry, G. Towards Design of the Smallest Planar Tetracoordinate Carbon and Boron Systems. *J. Comput. Chem.* **2007,** *28* (1), 335–343.

21. Pedley, J. B.; Naylor, R. D.; Kirby, S. P.; Pedley, J. B. *Thermochemical Data of Organic Compounds,* 2nd ed.; Chapman and Hall: London, New York, 1986.
22. Berson, J. A.; Bushby, R. J.; McBride, J. M.; Tremelling, M. 2-Isopropylidenecyclopentane-1, 3-Diyl. Preparation, Properties, and Reactions of a Distorted Trimethylenemethane. Direct Evidence for a Triplet Reaction. *J. Am. Chem. Soc.* **1971,** *93* (6), 1544–1546.
23. Wiberg, K. B.; Bartley, W. J.; Lossing, F. P. On the Strain Energy in Cyclopropene and the Heat of Formation of the $C_3H_3^+$ Ion. *J. Am. Chem. Soc.* **1962,** *84* (20), 3980–3981.
24. Turner, R. B.; Goebel, P.; Mallon, B. J.; Doering, W. von E.; Coburn, J. F.; Pomerantz, M. Heats of Hydrogenation. VIII. Compounds with Three-and Four-Membered Rings. *J. Am. Chem. Soc.* **1968,** *90* (16), 4315–4322.
25. Buchwalter, S. L.; Closs, G. L. Electron Spin Resonance Study of Matrix Isolated 1,3-Cyclopentadiyl, a Localized 1,3-Carbon Biradical. *J. Am. Chem. Soc.* **1975,** *97* (13), 3857–3858.
26. Buchwalter, S. L.; Closs, G. L. Electron Spin Resonance and CIDNP Studies on 1,3-Cyclopentadiyls. A Localized 1,3 Carbon Biradical System with a Triplet Ground State. Tunneling in Carbon-Carbon Bond Formation. *J. Am. Chem. Soc.* **1979,** *101* (16), 4688–4694.
27. Dougherty, D. A. Spin Control in Organic Molecules. *Acc. Chem. Res.* **1991,** *24* (3), 88–94.
28. Hoffmann, R. Trimethylene and the Addition of Methylene to Ethylene. *J. Am. Chem. Soc.* **1968,** *90* (6), 1475–1485.
29. Hoffmann, R. Interaction of Orbitals Through Space and Through Bonds. *Acc. Chem. Res.* **1971,** *4* (1), 1–9.
30. Goldberg, A. H.; Dougherty, D. A. Effects of Through-bond and Through-Space Interactions on Singlet-Triplet Energy Gaps in Localized Biradicals. *J. Am. Chem. Soc.* **1983,** *105* (2), 284–290.
31. Abe, M.; Ishihara, C.; Nojima, M. DFT Prediction of Ground-State Spin Multiplicity of Cyclobutane-1,3-Diyls: Notable Effects of Two Sets of Through-Bond Interactions. *J. Org. Chem.* **2003,** *68* (4), 1618–1621.
32. Abe, M.; Ye, J.; Mishima, M. The Chemistry of Localized Singlet 1,3-Diradicals (Biradicals): From Putative Intermediates to Persistent Species and Unusual Molecules with a π-Single Bonded Character. *Chem. Soc. Rev.* **2012,** *41* (10), 3808.
33. Abe, M. Diradicals. *Chem. Rev.* **2013,** *113* (9), 7011–7088.
34. Abe, M.; Akisaka, R. Is π-Single Bonding (C–π–C) Possible? A Challenge in Organic Chemistry. *Chem. Lett.* **2017,** *46* (11), 1586–1592.
35. Getty, S. J.; Hrovat, D. A.; Borden, W. T. Ab Initio Calculations on the Stereomutation of 1,1-Difluorocyclopropane. Prediction of a Substantial Preference for Coupled Disrotation of the Methylene Groups. *J. Am. Chem. Soc.* **1994,** *116* (4), 1521–1527.
36. Xu, J. D.; Hrovat, D. A.; Borden, W. T. Ab Initio Calculations of the Potential Surfaces for the Lowest Singlet and Triplet States of 2,2-Difluorocyclopentane-1,3-Diyl. The Singlet Diradical Lies Below the Triplet. *J. Am. Chem. Soc.* **1994,** *116* (12), 5425–5427.
37. Abe, M.; Adam, W.; Heidenfelder, T.; Nau, W. M.; Zhang, X. Intramolecular and Intermolecular Reactivity of Localized Singlet Diradicals: The Exceedingly Long-Lived 2,2-Diethoxy-1,3-Diphenylcyclopentane-1,3-Diyl. *J. Am. Chem. Soc.* **2000,** *122* (9), 2019–2026.
38. Johnson, W. T. G.; Hrovat, D. A.; Skancke, A.; Borden, W. T. Ab Initio Calculations Find 2,2-Disilylcyclopentane-1,3-Diyl Is a Singlet Diradical with a High Barrier to Ring Closure. *Theor. Chem. Acc.* **1999,** *102* (1–6), 207–225.
39. Abe, M.; Kawanami, S.; Ishihara, C.; Nojima, M. 2-Silyl Group Effect on the Reactivity of Cyclopentane-1,3-Diyls. Intramolecular Ring-Closure versus Silyl Migration. *J. Org. Chem.* **2004,** *69* (17), 5622–5626.

40. Abe, M.; Adam, W.; Hara, M.; Hattori, M.; Majima, T.; Nojima, M.; Tachibana, K.; Tojo, S. On the Electronic Character of Localized Singlet 2,2-Dimethoxycyclopentane-1,3-Diyl Diradicals: Substituent Effects on the Lifetime. *J. Am. Chem. Soc.* **2002,** *124* (23), 6540–6541.
41. Adam, W.; Borden, W. T.; Burda, C.; Foster, H.; Heidenfelder, T.; Heubes, M.; Hrovat, D. A.; Kita, F.; Lewis, S. B.; Scheutzow, D.; Wirz, J. Transient Spectroscopy of a Derivative of 2,2-Difluoro-1,3-Diphenylcyclopentane-1,3-DiylA Persistent Localized Singlet 1,3-Diradical. *J. Am. Chem. Soc.* **1998,** *120* (3), 593–594.
42. Nakagaki, T.; Sakai, T.; Mizuta, T.; Fujiwara, Y.; Abe, M. Kinetic Stabilization and Reactivity of π Single-Bonded Species: Effect of the Alkoxy Group on the Lifetime of Singlet 2,2-Dialkoxy-1,3-Diphenyloctahydropentalene-1,3-Diyls. *Chem. Eur. J.* **2013,** *19* (31), 10395–10404.
43. Abe, M.; Tada, S.; Mizuno, T.; Yamasaki, K. Impact of Diradical Spin State (Singlet vs Triplet) and Structure (Puckered vs Planar) on the Photodenitrogenation Stereoselectivity of 2,3-Diazabicyclo[2.2.1]Heptanes. *J. Phys. Chem. B* **2016,** *120* (29), 7217–7226.
44. Ye, J.; Fujiwara, Y.; Abe, M. Substituent Effect on the Energy Barrier for σ-Bond Formation from π-Single-Bonded Species, Singlet 2,2-Dialkoxycyclopentane-1,3-Diyls. *Beilstein J. Org. Chem.* **2013,** *9,* 925–933.
45. Beyer, M. K.; Clausen-Schaumann, H. Mechanochemistry: The Mechanical Activation of Covalent Bonds. *Chem. Rev.* **2005,** *105* (8), 2921–2948.
46. Hickenboth, C. R.; Moore, J. S.; White, S. R.; Sottos, N. R.; Baudry, J.; Wilson, S. R. Biasing Reaction Pathways with Mechanical Force. *Nature* **2007,** *446* (7134), 423.
47. Lenhardt, J. M.; Ong, M. T.; Choe, R.; Evenhuis, C. R.; Martinez, T. J.; Craig, S. L. Trapping a Diradical Transition State by Mechanochemical Polymer Extension. *Science* **2010,** *329* (5995), 1057–1060.
48. Lenhardt, J. M.; Black, A. L.; Craig, S. L. Gem-dichlorocyclopropanes as Abundant and Efficient Mechanophores in Polybutadiene Copolymers under Mechanical Stress. *J. Am. Chem. Soc.* **2009,** *131* (31), 10818–10819.
49. (a) Abe, M.; Furunaga, H.; Ma, D.; Gagliardi, L.; Bodwell, G. J. Stretch Effects Induced by Molecular Strain on Weakening σ-Bonds: Molecular Design of Long-Lived Diradicals (Biradicals). *J. Org. Chem.* **2012,** 77 (17), 7612–7619.
(b) Harada, Y.; Wang, Z.; Hatano, S.; Abe, M. Extremely Long-lived Localized Singlet Diradicals in a Macrocyclic Structure: A Case Study of the Stretch Effect. *Chem. Eur. J.* **2018**, in press.
50. Anthony, J. E. The Larger Acenes: Versatile Organic Semiconductors. *Angew. Chem. Int. Ed.* **2008,** *47* (3), 452–483.
51. Yamashita, Y. Organic Semiconductors for Organic Field-Effect Transistors. *Sci. Technol. Adv. Mater.* **2009,** *10* (2), 024313.
52. Kubozono, Y.; He, X.; Hamao, S.; Teranishi, K.; Goto, H.; Eguchi, R.; Kambe, T.; Gohda, S.; Nishihara, Y. Transistor Application of Phenacene Molecules and Their Characteristics. *Eur. J. Inorg. Chem.* **2014,** *2014* (24), 3806–3819.

CHAPTER 2

Dynamic Kinetic Resolution for Asymmetric Synthesis: Synergy of Hydrolase and Metal Catalysts

Shuji Akai[1], Gamal A.I. Moustafa[1,2]
[1]Graduate School of Pharmaceutical Sciences, Osaka University, Osaka, Japan; [2]Department of Medicinal Chemistry, Faculty of Pharmacy, Minia University, Minia, Egypt

Contents

1. INTRODUCTION

Optically pure organic compounds are used in various products, such as pharmaceuticals *(1)*, agrochemicals, perfumes, and polymeric materials *(2)*. With the ever-increasing demand for optically active compounds, there is a growing need for the development of environmentally benign asymmetric synthetic methods. During the latest few decades, transition-metal catalyses and organomolecule catalyses have significantly enriched the field of asymmetric synthesis; however, there are still few instances in which these artificial catalysts surpass natural enzymes in terms of enantioselectivity, chemoselectivity, and catalyst turnover at room temperature and normal pressure *(3)*. Furthermore, enzymes, which can be mass-produced by fermentation and biodegraded to be returned to nature, are less environmental burdens and are not depleted. On the other hand, the applicability of enzymatic reactions is rather still encumbered by some limitations. Undoubtedly, if we could expand the use of natural enzymes by various

Kinetic Control in Synthesis and Self-Assembly
ISBN 978-0-12-812126-9
https://doi.org/10.1016/B978-0-12-812126-9.00002-X

means, unparalleled asymmetric transformations would be realized to meet the abovementioned requests.

Among various natural enzymes, hydrolases such as lipases and esterases, which natively hydrolyze the ester bonds of lipids, have had by far the largest number of applications in the field of synthetic organic chemistry *(3)*. This is because hydrolases are stable, do not require any coenzymes, and show high catalytic activity, chemoselectivity, and enantioselectivity for a broad range of unnatural substrates. Furthermore, when used in organic solvents, hydrolases, in particular lipases, also exhibit high or even higher catalytic activities for ester bond formation from carboxylic acids and alcohols and also for transesterification of carboxylates with alcohols. These esterification reactions in organic media have some advantages over the natural hydrolysis in aqueous media because of high solubility and stability of organic substrates, particularly moisture-sensitive molecules, in organic solvents, facile operation, and uncomplicated purification of products (Fig. 2.1).

Typical examples of the lipase-catalyzed esterification and transesterification include the kinetic resolution (KR) of racemic alcohols and enantioselective desymmetrization of prochiral diols. The former reactions are the most widely employed to separate the enantiomers of racemic materials, each in up to 50% yields. The latter method has a particular benefit because it can convert prochiral molecules into homochiral forms in up to quantitative yields; however, it has a limitation in the substrate structures (Fig. 2.2).

In order to overcome the inherent issue of the maximum 50% yields of the products, dynamic kinetic resolution (DKR) has been devised to produce optically pure products in up to quantitative yields *(4,5)*. This chapter mainly introduces basic concepts of hydrolase-catalyzed DKR of racemic alcohols and its synthetic applications following an introduction about the general aspects of hydrolase-catalyzed KR.

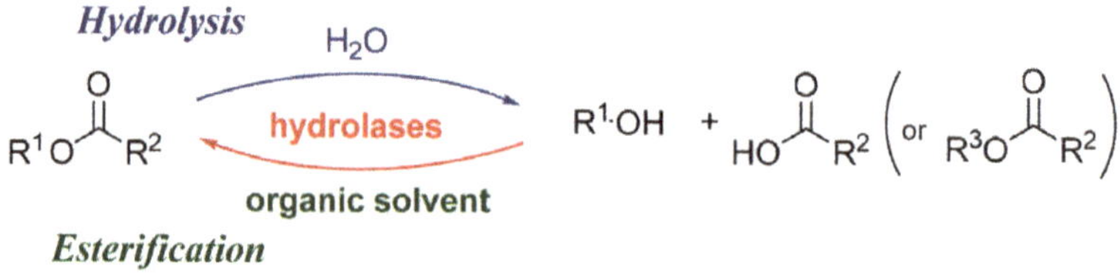

Figure 2.1 Hydrolases-catalyzed two-way reactions.

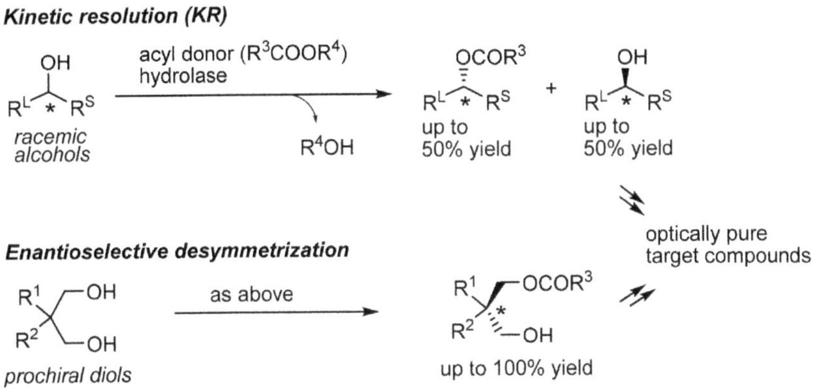

Figure 2.2 Typical uses of lipases for asymmetric syntheses.

2. GENERAL ASPECTS OF HYDROLASE-CATALYZED KINETIC RESOLUTION AND DYNAMIC KINETIC RESOLUTIONS

Enzymatic KR of racemic secondary alcohols has been performed with an acyl donor and a hydrolase in an organic solvent. A range of hydrolases, such as lipases, esterases, and proteases, are commercially available as solids immobilized on a support, and by choosing suitable hydrolases for each substrate, optically active compounds with >99% enantiomeric excess (ee) are obtainable for both enantiomers in up to 50% yield (Fig. 2.3A) *(1)*. The success of KR is dependent on the substantial difference in the reaction rate of the two enantiomers ($k_A \gg k_B$), and most of the lipases preferentially catalyze the esterification of *R*-enantiomers which match the shape of the reactive site of enzymes.

The structure of the acyl donor is another key factor to obtain high yields of the products because the lipase-catalyzed reaction is basically reversible and the generated R—OH may react with the products to cause the reverse transesterification. Therefore, vinyl esters such as vinyl and isopropenyl esters have often been used because of their high reactivity and the formation of enols as leaving groups, which immediately and irreversibly change into carbonyl compounds to completely suppress the reverse reaction. However, the generated acetaldehyde and acetone may react with lipases and/or suppress the reactions. In such cases, 1-ethoxyvinyl esters are better alternative acyl donors which are more reactive and produce harmless ethyl acetate as a side product *(6)*.

Extensive efforts have been devoted to enhancing the efficiency and practical applicability and minimizing environmental burden of KR by

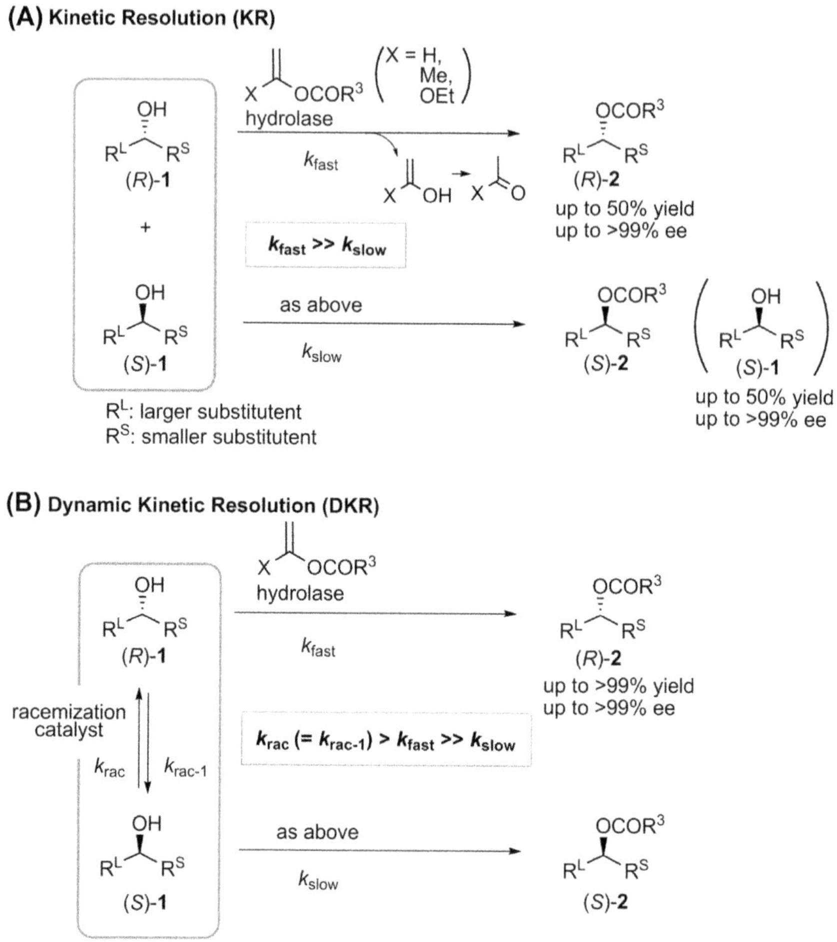

Figure 2.3 Concept of kinetic resolution (A) and dynamic kinetic resolution (B) of racemic secondary alcohols **1**.

developing new solvent systems such as supercritical fluids *(7)* and ionic liquids *(8)*, improving catalytic activity and enantioselectivity by additives *(9)* and sol–gel immobilization *(10)*.

Although the KR is a fundamental protocol to completely separate two enantiomers of racemates, the yield of each enantiomer can never exceed 50%. If the rapid interconversion of two enantiomers of the substrate could occur under the KR conditions in one flask, the racemate would be theoretically converted into a single optically pure ester in 100% yield. This method is called DKR and has considerably been developed in the last few decades *(4)*. DKR is performed via the combination of hydrolase-catalyzed KR and the in situ racemization of the remaining less reactive enantiomers using another catalyst, or without any additional catalyst in some cases. In order to achieve effective DKR for preparing optically active products with >98% ee, all of the following requirements should be fulfilled (Fig. 2.3.B): (1) the KR should be highly enantioselective (E value $= k_{fast}/k_{slow} > 100$); (2) the interconversion of two enantiomers, i.e., the racemization (k_{rac}), should be faster than the fast-reacting enantiomer's transformation ($k_{rac} > k_{fast}$); (3) the racemization should occur at around room temperature at which hydrolases show the best catalytic performance and should not be affected by acyl donors or by-products generated from acyl donors; (4) nonenzymatic acylation should not proceed because it produces racemic ester resulting in a drop of the optical purity of the product; (5) the optically active products should not racemize or decompose under the reaction conditions; and (6) hydrolases and racemization catalysts should be compatible with each other under the same reaction conditions.

In KR, the optical purities of products and remaining substrates continuously change according to the conversion, and therefore, the enantioselectivity should be discussed based on the E value designated by the following formula *(11)*:

$$E = \frac{\ln(1-c)(1-ee_S)}{\ln(1-c)(1+ee_S)} = \frac{\ln(1-c(1-ee_P))}{\ln(1-c(1+ee_P))}$$

c=conversion, ee_S=optical purity of recovered substrate, ee_P=optical purity of product.

For instance, in the case of KR with the E value of 100, the optical purity (98% ee) of the product at 1% conversion gradually decreases to 93% ee at 50% conversion and then rapidly drops toward 0% ee (Fig. 2.4). With sufficiently rapid racemization, the substrate can be maintained as a racemate (0% ee) at any conversion, which results in the complete retention of the highest optical purity (98% ee) of the product throughout the DKR process. Similarly, in the case of DKR with the E value of 20, the optical purity of the product can maintain 90% ee but never be improved, and the product with the same optical purity is obtained after completion of DKR. These facts mean that high enantioselectivity (E = >100) and rapid racemization ($k_{rac} > k_{fast}$) are inevitable to produce the products with >98% ee.

The pioneering work on the lipase-catalyzed DKR of secondary alcohols was reported in 1992 by Oda et al. The authors appropriately

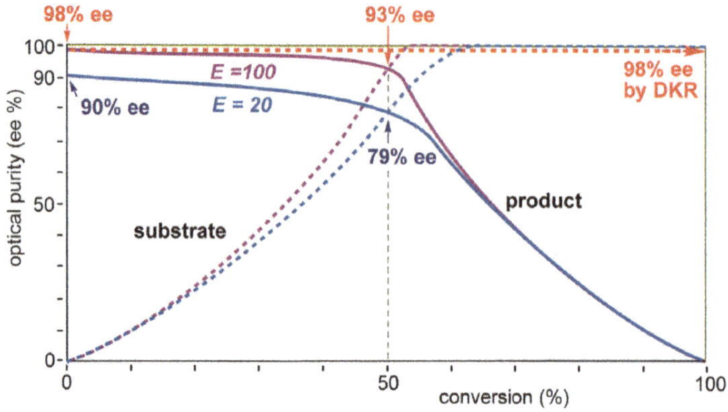

Figure 2.4 Changes of optical purity of the product and recovered substrate along with the conversion of kinetic resolution. *DKR*, dynamic kinetic resolution.

employed the inherent nature of the cyanohydrins, which readily racemize through reversible formation of the cyanohydrins from aldehydes and acetone cyanohydrin by the aid of a basic resin. Thus, the lipase-catalyzed KR of the cyanohydrin **3**, in situ generated from the aldehyde, was performed in the presence of the resin to produce (*S*)-acetate **4** (Fig. 2.5) *(12)*.

More widely applicable DKR protocols have been developed by embedding powerful racemization processes enabled by specific metal catalysts; however, the compatibility between hydrolases and racemization catalysts often becomes a critical issue because these catalysts have quite

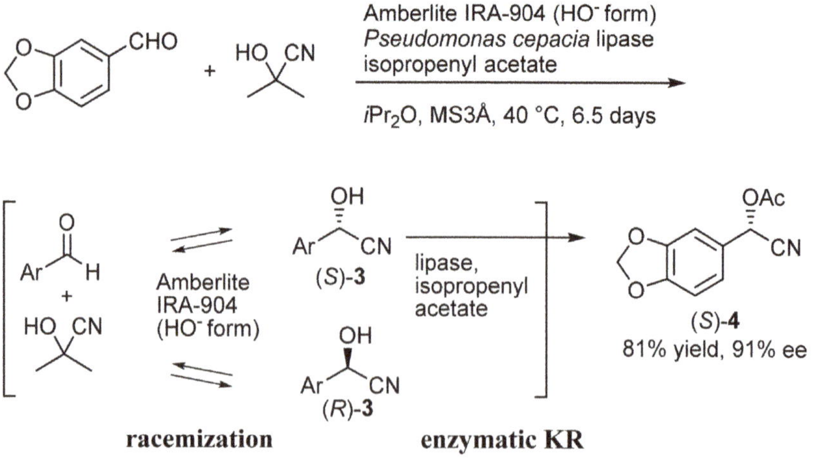

Figure 2.5 Lipase-catalyzed dynamic kinetic resolution of in situ generated cyanohydrin **3**. *KR*, kinetic resolution

distinct natures and best operate under very different conditions *(13)*. For instance, the racemization catalyst may interfere with the lipase-catalyzed resolution and/or the lipases with their considerable water content, and the added surfactants may nullify the racemization process, or both processes may be inhibited by the by-products of the KR and/or racemization. Therefore, intensive efforts have been devoted to addressing these mutual inactivation issues by the creation of highly active racemization catalysts with sufficient compatibility with hydrolases.

3. HYDROLASE/METAL COMBO-CATALYZED DYNAMIC KINETIC RESOLUTION

The seminal hydrolase/metal combo-catalyzed DKR was published in 1996 by Williams et al. through the enzymatic hydrolytic KR of pseudosymmetric allyl acetate (±)-**5** coupled with the Pd(II)-catalyzed racemization through the 1,3-transposition of the acetoxy group, although it took 19 days to produce 81% yield of (*S*)-**6** (96% ee) (Fig. 2.6) *(14)*.

At the same time, a variant of DKR was invented by the same group using a redox reaction for the racemization *(15)*. A typical example of this DKR is shown in Fig. 2.7A, in which 1-phenylethanol ((±)-**1a**) was converted into 1-phenylethyl acetate ((*R*)-**2a**) (98% ee) with 60% conversion by a combination $Rh_2(OAc)_4$ as a redox catalyst and *Pseudomonas fluorescens* lipase; however, the reaction needed a stoichiometric amount of acetophenone to promote the racemization and took 3 days for 60% conversion. Bäckvall made significant improvements of this DKR protocol using a Ru complex **7a**, which enabled effective racemization without using the corresponding ketone and produced optically active esters **2** (up to 88% yield, up to 99% ee) from racemic alcohols **1** *(16)*. This breakthrough has drawn remarkable attention from many research groups and more than a dozen of Ru complexes possessing different ligands have been devised to perform the lipase-catalyzed DKR (some typical examples of the

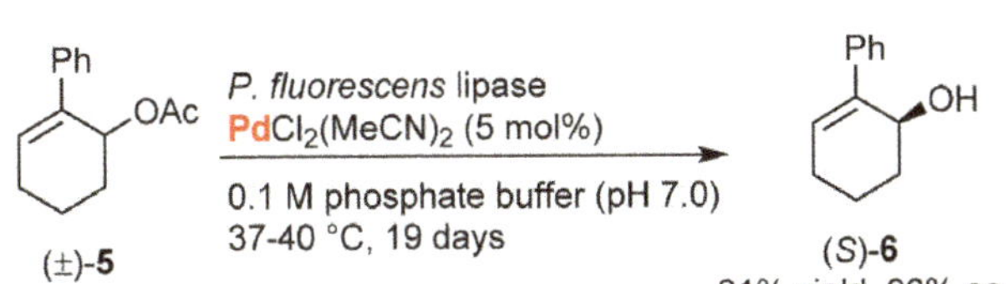

Figure 2.6 Lipase/palladium combo-catalyzed dynamic kinetic resolution of allyl acetate (±)-**5**.

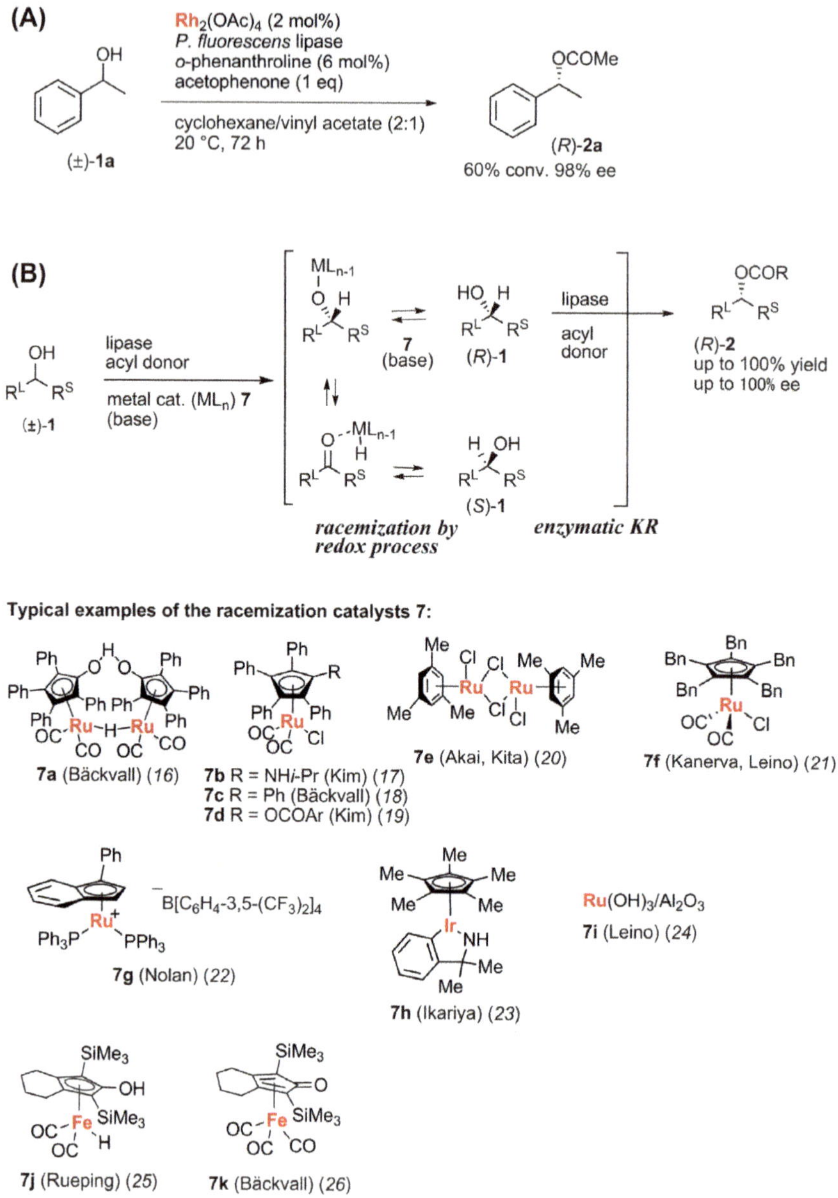

Figure 2.7 (A) Lipase-catalyzed dynamic kinetic resolution (DKR) of 1-phenylethanol (±)-**1a** using $Rh_2(OAc)_4$ as a racemization catalyst. (B) Lipase-catalyzed DKR of sec-alcohols (±)-1 using a variety of redox catalysts **7**. R^L: larger substituent, R^S: smaller substituent; *KR*, kinetic resolution.

Ru catalysts **7b**–**7g** for DKR are shown in Fig. 2.7B *(4,17–22))*. Recently, Ir **7h** *(23)*, Ru **7i** *(24)*, and Fe **7j**, **7k** *(25,26)* complexes were reported to be applicable to DKR in which the racemization proceeded through a similar redox process. The synthetic advantages of the metal–lipase combo-catalyzed DKR protocols have been demonstrated in the DKR of diverse secondary alcohols possessing other functional groups, such as allylic alcohols, homoallylic alcohols, propargyl alcohols, chloro alcohols, amino alcohols, *N*-heterocyclic amino alcohols, azido alcohols, α-hydroxy ketones, hydroxyl esters, amides, and hydroxyl nitriles *(4)*.

Most natural lipases preferentially catalyze the esterification of (*R*)-secondary alcohols (*R*)-**1**, where the larger substituents R^L have in general higher priority than the smaller substituents R^S, thus the lipase-catalyzed DKR affords (*R*)-esters (*R*)-**2** in up to quantitative yields. This means that the production of (*S*)-**2** by lipase-catalyzed DKR has been very much limited, and only some proteases, such as subtilisin *(27)*, *Candida antarctica* lipase A *(28)*, or a mutated variant of *C. antarctica* lipase B *(29)* are known to be applicable to *S*-selective DKR to afford (*S*)-esters (Fig. 2.8). In addition, the same combination of enzymes and Ru complexes has been effectively applied for the DKR of racemic primary amines *(4,30)*.

An alternative and very different DKR system was reported by Jacobs et al. using $O{=}VSO_4 \cdot 5H_2O$ as a racemization catalyst. The racemization of 1-phenethyl alcohol ((±)-**1a**) proceeded at 80°C in *n*-octane through the formation of a benzyl cation intermediate and was combined with lipase-catalyzed KR to afford the optically pure ester (*R*)-**2b** in 93% yield (Fig. 2.9A). This DKR was performed by using a special reactor with a

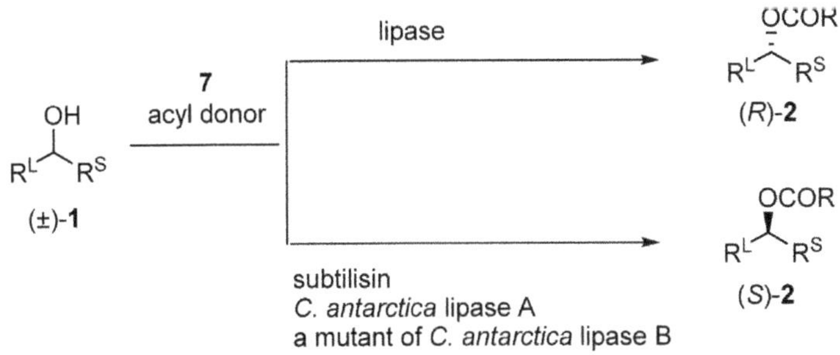

Figure 2.8 Enantiodivergent dynamic kinetic resolution of racemic secondary alcohols (±)-**1** to produce both (*R*)- and (*S*)-esters **2** by choosing lipases or subtilisin. R^L: larger substituent, R^S: smaller substituent.

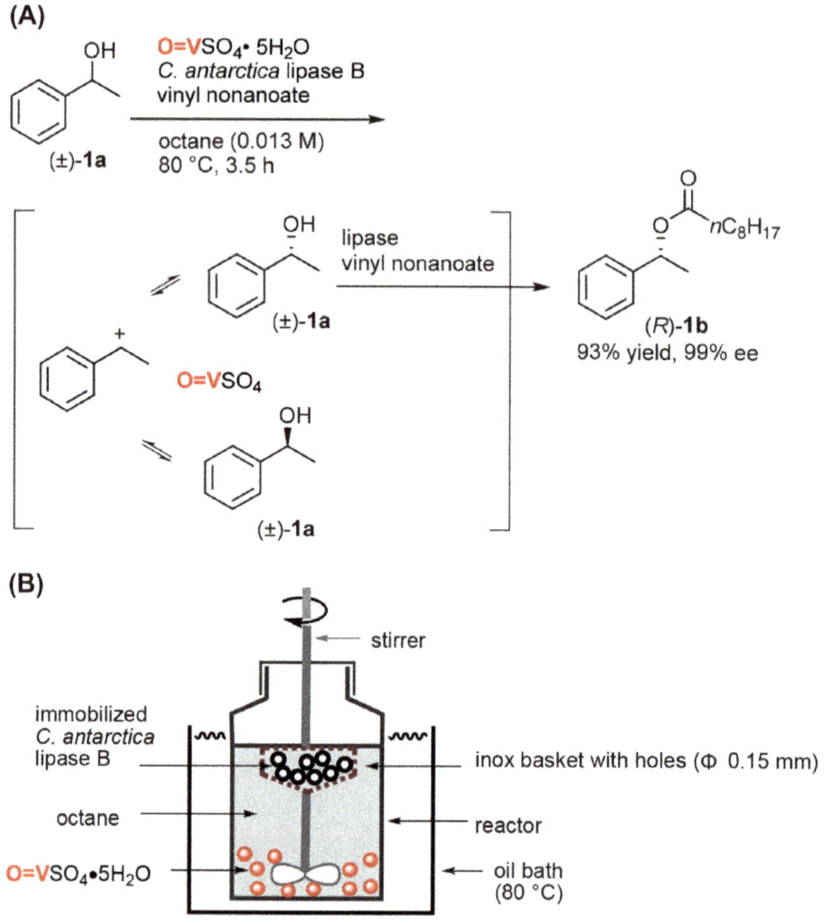

Figure 2.9 (A) Dynamic kinetic resolution (DKR) of 1-phenylethanol (±)-**3a** using the combination of O=VSO_4 and *Candida antarctica* lipase B. (B) A reactor for DKR that separates O=VSO_4 from lipase.

porous inox basket, in which commercial immobilized *C. antarctica* lipase B was put. Insoluble solid O=$VSO_4 \cdot 5H_2O$ was placed at the bottom of the reactor, and thereby, these two solid catalysts were separated to suppress the mutual deactivation (Fig. 2.9B) *(31)*.

Meanwhile, Akai et al. employed a homogeneous catalyst O=V($OSiPh_3$)$_3$ **10A** *(32)* or a polymer-bound vanadyl phosphate **10B** *(33)* for the racemization of the lipase-catalyzed DKR of racemic allyl alcohols **8**. The oxovanadium compounds (**10A** and **10B**) catalyze the racemization of

8 by forming allyl cation intermediates **11** and generate a mixture of racemates of two regioisomers (**8** and **9**), and a lipase conducts an enantio- and chemoselective esterification of the less hindered alcohol **8** to afford the esters (*R*)-**12**. Therefore, this DKR has a synthetic advantage of the use of regioisomers **9** as equivalent substrates of **8** to produce the same products (*R*)-**12** (Fig. 2.10). However, the racemization activities of **10A** and **10B**

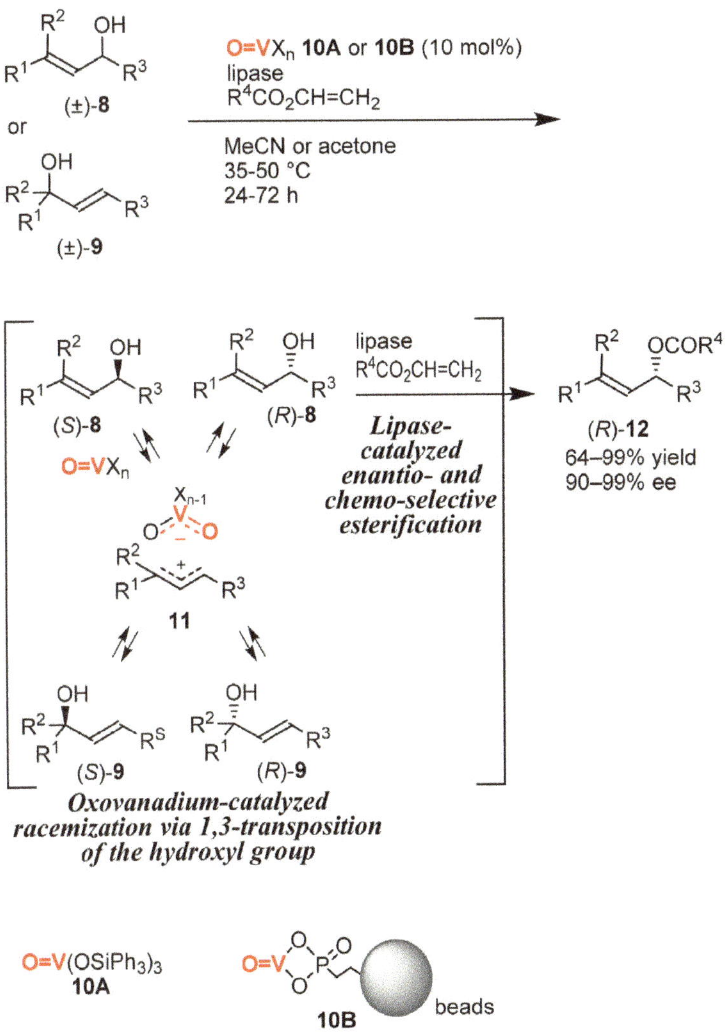

Figure 2.10 Dynamic kinetic resolution of racemic allyl alcohols (**8** and **9**) using the combination of oxovanadium compounds (**10A** or **10B**) and lipases.

were moderate, and the combined use of these vanadium compounds with lipases for extended periods caused mutual deactivation to produce moderate yields (60%–80%) of the products **12**.

This problem has been completely overcome by Akai *et al.* by inventing an originally designed new oxovanadium catalyst V-MPS3, in which the oxovanadium species was covalently bound to the inner surface of mesoporous silica (MPS) with a pore size of about 3 nm (Fig. 2.11A) *(34)*. Its small pore size attained separation of the racemization and KR sites and dramatically improved the compatibility between the oxovanadium moiety and lipases (Fig. 2.11B). V-MPS3 exhibited much higher racemization activity than the previous catalysts (**10A** and **10B**), and the combination of V-MPS3 (1 mol% of the vanadium amount) and lipases demonstrated the production of optically pure esters (*R*)-**12** in 90%–quantitative yields, whose results were superior to those obtained using **10A** or **10B**. In addition, a mixture of powdery V-MPS3 and immobilized *C. antarctica* lipase B was reused for up to six times without any loss of catalytic activities.

Akai et al. also prepared a similar catalyst, V-MPS4, using MPS with a pore size of about 4 nm to find out that V-MPS4 exhibited the higher racemization activity for larger alcohols, such as a molecule with maximum length of 1.4 nm, than V-MPS3. The combination of V-MPS4 and *C. antarctica* lipase B produced higher yields of esters **12** than that of V-MPS3 and the lipase *(35)*. An appropriate combination of commercially available immobilized lipases (such as *C. antarctica* lipase B and *Burkholderia cepacia* lipase) with V-MPS3 or V-MPS4 led to the effective DKR of a wide range of allylic alcohols (**8** and **9**), producing the corresponding optically active allylic esters **12** in excellent chemical and optical yields (some examples are given in Fig. 2.11C) *(35,36)*.

Hou et al. designed and prepared a series of Ru(II)-supported MPSs, which have hydrophilic or hydrophobic environments and different pore sizes. Ru(II) moieties were immobilized on the inner surface of MPS with a linked ligand having an *N*-heterocyclic carbene. Among a series of Ru(II)-supported MPSs, Ru(II)-supported MPS SBA-15, Ru/SBA-15 (pore diameter 1.5 nm) exhibited the highest racemization activity for DKR of (±)-**1a** in the presence of Novozyme 435 (same as *C. antarctica* lipase B) to afford the ester (*R*)-**2a** (up to 96% yield and 99% ee) (Fig. 2.12A and B). These combined catalysts were reused for at least seven times *(37)*.

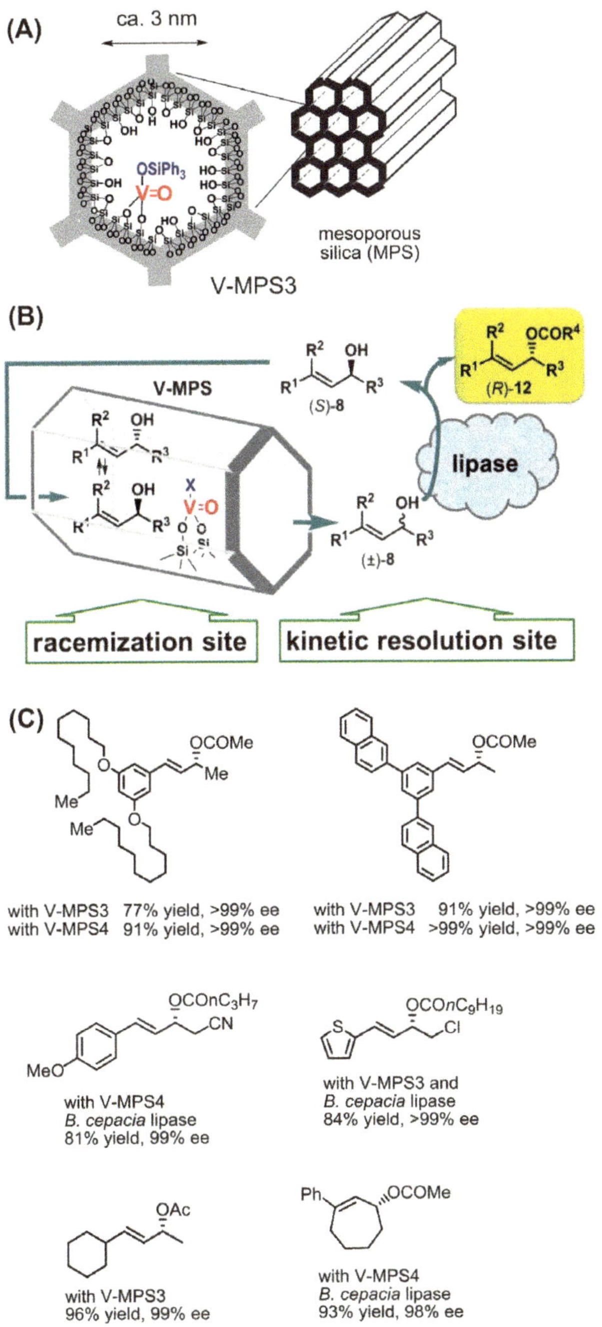

Figure 2.11 (A) Structure of V-MPS3. (B) The concept of reaction site separation for dynamic kinetic resolution (DKR) using the narrow pores of MPS3. (C) Typical examples of DKR products **12** obtained from allylic alcohols using V-MPS and *Candida antarctica* lipase B, except where otherwise noted.

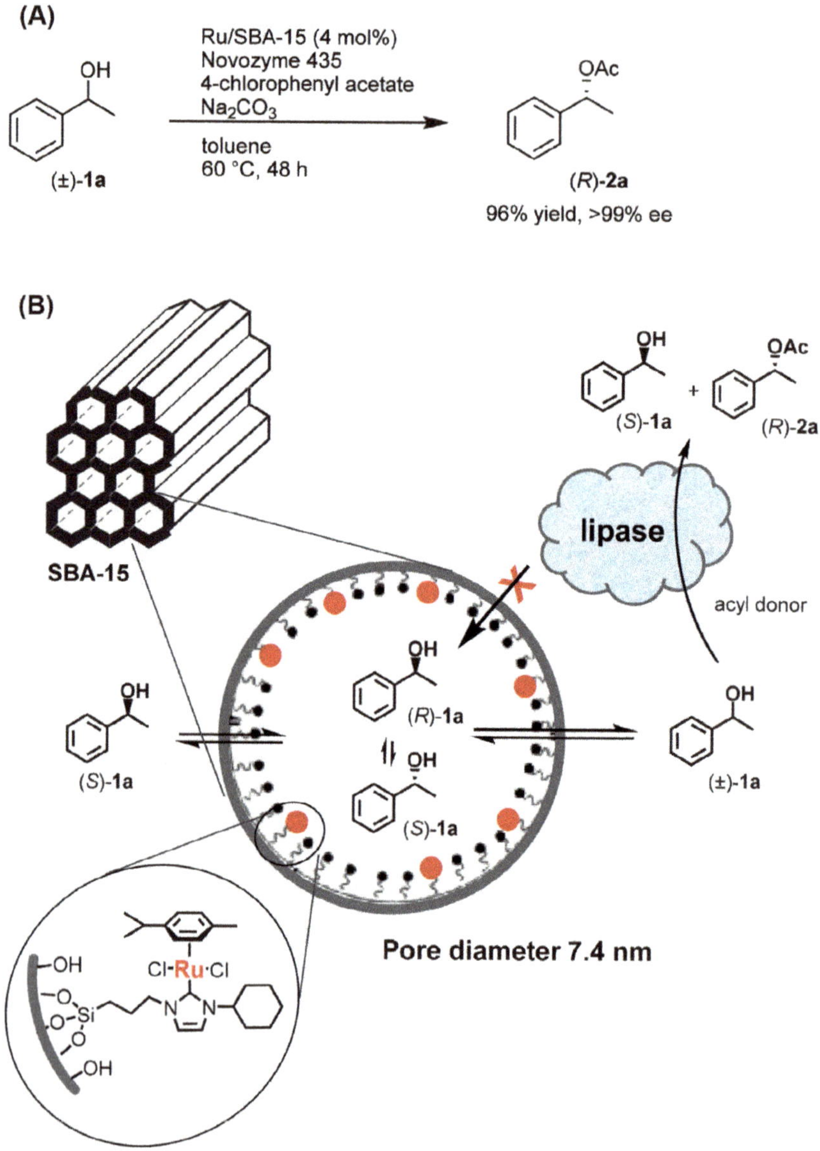

Figure 2.12 (A) Ru/SBA-15 and lipase combo-catalyzed dynamic kinetic resolution (DKR) of (±)-**1a**. (B) Structure of Ru/SBA-15 and schematic explanation of the DKR process.

4. SYNTHETIC APPLICATIONS OF DYNAMIC KINETIC RESOLUTION

Some synthetic applications of the abovementioned DKR developed by Akai et al. have been shown by the asymmetric total synthesis of naturally occurring compounds.

The first example is the asymmetric synthesis of the synthetic key intermediate (*R*)-**13** of an antibacterial agent (+)-tanikolide, whose total synthesis was originally reported by Kita, Fujioka et al. *(38)*. They first applied the Corey–Bakshi–Shibata (CBS) asymmetric reduction to the enone **15**; however, the optical purity of the product (*R*)-**13** was 77% ee. Therefore, they took a detour to conduct similar CBS reduction of **16** with a bulky iodo group at the α-position to obtain a better optical purity (98% ee) of (*R*)-**13** after reductive removal of the iodo group. This overall transformation produced (*R*)-**13** (98% ee) in a total yield of 31% from β-ethoxycyclohexenone (**14**) (Fig. 2.13A). On the other hand, Akai *et al.* applied their DKR method to tertiary alcohol (±)-**19** derived from cyclohexenone (**18**), and (*R*)-**13** (>99% ee) was prepared in 85% yield after hydrolysis of the ester (*R*)-**20** (Fig. 2.13B) *(34)*. These results have disclosed some advantages of the Akai's DKR method: (1) the asymmetric reduction protocol inherently requires the use of the enone **14** with a higher oxidation state compared with **18**. On the other hand, the DKR causes no change of the oxidation state between substrates and products and allows

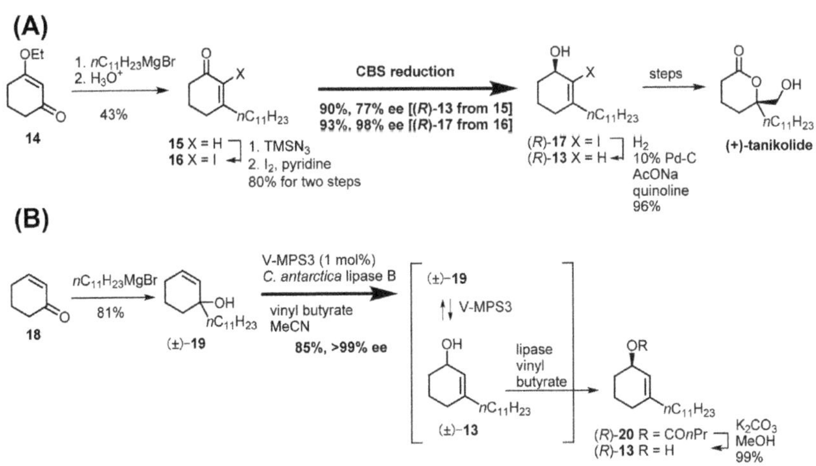

Figure 2.13 (A) Asymmetric total synthesis of (+)-tanikolide by Kita–Fujioka. (B) Akai's synthesis of the key intermediate (*R*)-**13**. *CBS*, Corey–Bakshi–Shibata

the use of inexpensive **18** as the substrate. (2) The oxovanadium-catalyzed racemization proceeds along with the 1,3-transposition of the hydroxyl group of allylic alcohols, which enables the use of tertiary alcohol (±)-**19**, readily available from **18**, as a substrate for DKR. This is one of the most distinctive advantages of Akai's DKR because similar migratory DKR is not attainable by the Ru/lipase-integrated DKR. Thus, Akai's DKR method can shorten the entire process, improve the total yield (the yield of (*R*)-**13** was doubled compared with the Kita–Fujioka method), and reduce wastes.

As have been presented in Fig. 2.13 and others, most of the acyl groups of optically active esters obtained by KR or DKR of racemic alcohols are removed in subsequent transformation processes. On the other hand, Akai et al. have demonstrated effective utilization of the introduced acyl group to construct carbon framework. An example is illustrated as follows.

The DKR of the tertiary alcohol (±)-**24**, obtained by the reaction of **18** and an organolithium compound **21**, was carried out to give the ester (*R*)-**25**, the acyl moiety of which was then applied to the Ireland–Claisen

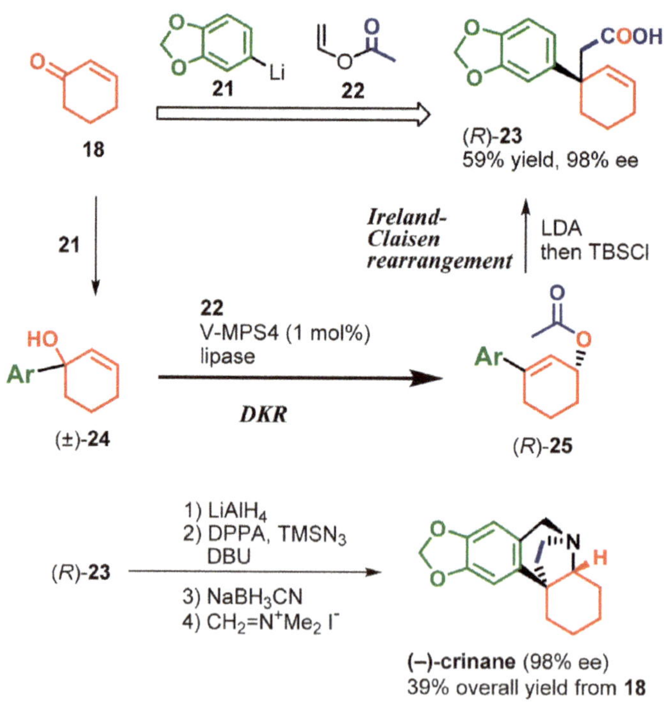

Figure 2.14 Enantioselective assembly of enone **18**, an organolithium reagent **21**, and vinyl ester **22** to prepare a compound **23** having an all-carbon quaternary stereogenic center and its application to asymmetric total synthesis of (−)-crinane.

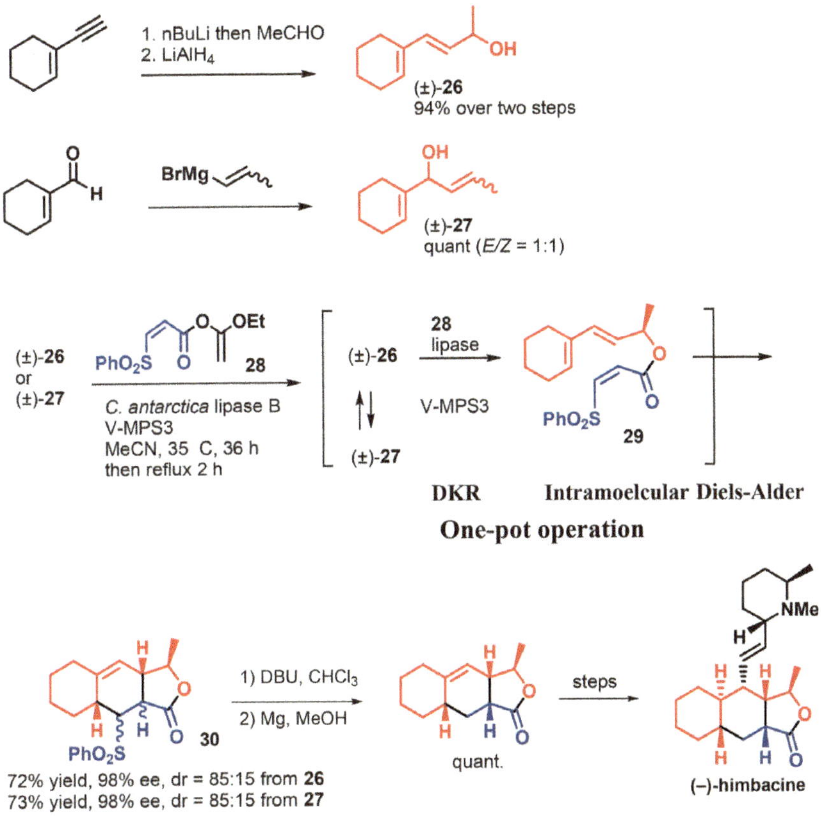

Figure 2.15 Total synthesis of (-)-himbacine. *DKR*, dynamic kinetic resolution.

rearrangement to give an optically active carboxylic acid (*R*)-**23** having an all-carbon quaternary stereogenic carbon center. The overall transformation allowed the asymmetric construction of all-carbon quaternary carbons by enantioselectively assembling two achiral components (**21** and **22**) onto the carbonyl carbon of a prochiral raw material **18**. By applying this method, (-)-crinane (98% ee) was synthesized from **18** in a total of seven steps and 39% yield (Fig. 2.14) *(36)*.

Another example of the effective use of the installed acyl moiety is shown in the total synthesis of (-)-himbacine. Thus, a racemic alcohol **26** containing a diene moiety and an ester **28** having a dienophile moiety were reacted with *C. antarctica* lipase B and V-MPS3 to cause DKR to form an optically active ester (*R*)-**29**. Without isolation of **29**, the reaction mixture was directly refluxed to undergo an intramolecular Diels–Alder reaction to obtain a tricyclic compound **30** (98% ee) in 72% yield. From the racemic alcohol **27** (*E*/*Z* mixture) obtained from a commercially available aldehyde and a Grignard reagent in one step, **30** (98% ee) was obtained in the same yield by the same reaction. The latter results have proved that VMPS3 catalyzed the 1,3-migration of the hydroxyl group along with *Z*-to-*E* isomerization and racemization to generate **26**, which was then subjected to one-pot DKR/intramolecular Diels–Alder reaction. Subsequently, total synthesis of (-)-himbacine was accomplished by further elaboration of **30** (Fig. 2.15) *(39)*. Thus, the appropriate design of the substrates and the acylating agents enables effective use of the acyl groups introduced by DKR in the construction of complicated molecular skeletons (similar utilizations of the installed acyl groups for constructing a molecular architecture has also been presented by Akai et al. *(20,40)*). In addition, the regioisomeric allylic alcohols (**26** and **27**) can be used as equivalent substrates; this is also another advantage of this DKR that enables more flexible choice of substrates when designing a synthetic route.

5. SUMMARY

Simultaneous use of different kinds of catalysts, such as lipases (biocatalysts) and metal catalysts, in one flask has made an enormous progress in the field of asymmetric synthesis to convert readily available racemic alcohols into enentio-enriched substances almost quantitatively. In some cases, the poor compatibility of these two catalysts is a serious problem, for which the use of nanoscale pores of MPS to separate their reaction sites has proved to be a practical solution to this compatibility issue.

Enzymes, which can be mass-produced by fermentation and biodegraded back to nature, are very attractive regenerable catalysts that do not deplete. In addition, with the remarkable advances of genetic engineering and protein engineering in the latest decade, it has become possible to rationally design and produce enzymes according to purpose. In this way, the development of a methodology for synthesizing optically active compounds more economically, safely, and efficiently by taking advantage of the excellent catalytic function of enzymes is important in promoting the manufacture of optically active medicines, agricultural chemicals, and functional materials. Further improvement and extension of the enzyme-catalyzed DKR methods by the collaboration with transition-metal catalysts or organometallic catalysts is more desired than ever.

REFERENCES

1. For reviews:
(a) Brabcova, J.; Filice, M.; Gutarra, M.; Palomo, J. M. Synthesis of Enantiopure Drugs and Drug Intermediates by Immobilized Lipase-Catalysis. *Curr. Bioact. Compd.* **2013,** *9,* 113–136.
(b) Jose, C.; Toledo, M. V.; Briand, L. E. Enzymatic Kinetic Resolution of Racemic Ibuprofen: Past, Present and Future. *Crit. Rev. Biotechnol.* **2016,** *36,* 891–903. For a recent paper.
(c) Kumar, R.; Banoth, L.; Banerjee, U. C.; Kaur, J. Enantiomeric Separation of Pharmaceutically Important Drug Intermediates Using a Metagenomic Lipase and Optimization of Its Large Scale Production. *Int. J. Biol. Macromol.* **2017,** *95,* 995–1003.
2. For review: Yang, Y.; Zhang, J.; Wu, Di; Xing, Z.; Zhou, Y.; Shi, W.; Li, Q. Chemoenzymatic Synthesis of Polymeric Materials Using Lipases as Catalysts: A Review *Biotechnol. Adv.* **2014,** *32,* 642–651.
3. Faber, K. *Biotransformations in Organic Chemistry; A Textbook,* 7th ed.; Springer: New York, 2017.
4. Some recent reviews:
(a) Akai, S. Dynamic Kinetic Resolution of Racemic Allylic Alcohols via Hydrolase-Metal Combo catalysis: An Effective Method for the Synthesis of Optically Active Compounds. *Chem. Lett.* **2014,** *43,* 746–754.
(b) Takizawa, S.; Gröger, H.; Sasai, H. Vanadium in Asymmetric Synthesis: Emerging Concepts in Catalyst Design and Applications. *Chem. Eur. J.* **2015,** *21,* 8992–8997.
(c) Verho, O.; Bäckvall, J.-E. Chemoenzymatic Dynamic Kinetic Resolution: A Powerful Tool for the Preparation of Enantiomerically Pure Alcohols and Amines. *J. Am. Chem. Soc.* **2015,** *137,* 3996–4009.
(d) Långvik, O.; Saloranta, T.; Murzin, D. Y.; Leino, R. Heterogeneous Chemoenzymatic Catalyst Combinations for One-Pot Dynamic Kinetic Resolution Applications. *ChemCatChem* **2015,** *7,* 4004–4015.
(e) de Miranda Amanda, S.; Miranda Leandro, S. M.; de Souza Rodrigo, O. M. A. Lipases: Valuable Catalysts for Dynamic Kinetic Resolutions. *Biotechnol. Adv.* **2015,** *33,* 372–393.
(f) Akai, S. In *Future Directions in Biocatalysis;* Matsuda, T., Ed., 2nd ed.; Elsevier: Amsterdam, 2017; p 337.
(g) Seddigi, Z. S.; Malik, M. S.; Ahmed, S. A.; Babalghith, A. O.; Kamal, A. Lipases in Asymmetric Transformations: Recent Advances in Classical Kinetic Resolution and Lipase-Metal Combinations for Dynamic Processes. *Coord. Chem. Rev.* **2017,** *348,* 54–70.
5. For recent reviews on general DKR methodologies, see:
(a) Xie, J.-H.; Zhou, Q.-L. Catalytic Asymmetric Hydrogenation of α-Substituted Ketones and Aldehydes via Dynamic Kinetic Resolution: Efficient Approach to Chiral Alcohols. *Aldrichim Acta* **2015,** *48,* 33–40.
(b) Echeverria, P.-G.; Ayad, T.; Phansavath, P.; Ratovelomanana-Vidal, V. Recent Developments in Asymmetric Hydrogenation and Transfer Hydrogenation of Ketones and Imines through Dynamic Kinetic Resolution. *Synthesis* **2016,** *48,* 2523–2539.
(c) Wencel-Delord, J.; Colobert, F. Stereoselective Metal-Catalyzed C-C Bond Coupling Reactions by Stereoconvergence, Dynamic Kinetic Asymmetric Transformation, or Dynamic Kinetic Resolution. *Synthesis* **2016,** *48,* 2981–2996.
(d) Li, P.; Hu, X.; Dong, X.-Q.; Zhang, X. Recent Advances in Dynamic Kinetic Resolution by Chiral Bifunctional (Thio)urea- and Squaramide-Based Organocatalysts. *Molecules* **2016,** *21,* 1327/1–1327/14.
(e) Pellissier, H. Recent Developments in Organocatalytic Dynamic Kinetic Resolution. *Tetrahedron* **2016,** *72,* 3133–3150.

(f) Bhat, V.; Welin, E. R.; Guo, X.; Stoltz, B. M. Advances in Stereoconvergent Catalysis from 2005 to 2015: Transition-Metal-Mediated Stereoablative Reactions, Dynamic Kinetic Resolutions, and Dynamic Kinetic Asymmetric Transformations. *Chem. Rev.* **2017,** *117,* 4528–4561.
6. Kita, Y.; Akai, S. 1-Alkoxyvinyl Esters: Renaissance of Half-Century-Old Acyl Donors with Potential Applicability. *Chem. Rec.* **2004,** *4,* 363–372.
7. (a) Szekely, E.; Utczas, M.; Simandi, B. Kinetic Enzymatic Resolution in $scCO_2$ - Design of Continuous Reactor Based on Batch Experiments. *J. Supercrit. Fluids* **2013,** *79,* 127–132.
(b) Hoang, H. N.; Nagashima, Y.; Mori, S.; Kagechika, H.; Matsuda, T. CO_2-Expanded Bio-based Liquids as Novel Solvents for Enantioselective Biocatalysis. *Tetrahedron* **2017,** *73,* 2984–2989.
8. (a) Shimomura, K.; Harami, H.; Matsubara, Y.; Nokami, T.; Katada, N.; Itoh, T. Lipase-Mediated Dynamic Kinetic Resolution (DKR) of Secondary Alcohols in the Presence of Zeolite Using an Ionic Liquid Solvent System. *Catal. Today* **2015,** *255,* 41–48.
(b) Sandig, B.; Buchmeiser, M. R. Highly Productive and Enantioselective Enzyme Catalysis under Continuous Supported Liquid-Liquid Conditions Using a Hybrid Monolithic Bioreactor. *ChemSusChem* **2016,** *9,* 2917–2921.
(c) Belafriekh, A.; Secundo, F.; Serra, S.; Djeghaba, Z. Enantioselective Enzymatic Resolution of Racemic Alcohols by Lipases in Green Organic Solvents. *Tetrahedron Asymmetry* **2017,** *28,* 473–478.
9. (a) Kadotani, S.; Inagaki, R.; Nishihara, T.; Nokami, T.; Itoh, T. Enhanced Activity of a Lipase by the Coating with a Quaternary Ammonium Alkyl-PEG Sulfate Ionic Liquid and Cooperative Activation with an Amino Acid. *ACS Sustain. Chem. Eng.* **2017,** *5,* 8541–8545.
(b) Nishihara, T.; Shiomi, A.; Kadotani, S.; Nokami, T.; Itoh, T. Remarkably Improved Stability and Enhanced Activity of a *Burkholderia epacia* Lipase by Coating with a Triazolium Alkyl-PEG Sulfate Ionic Liquid. *Green Chem.* **2017,** *19,* 5250–5256.
10. Tielmann, P.; Kierkels, H.; Zonta, A.; Ilie, A.; Reetz, M. T. Increasing the Activity and Enantioselectivity of Lipases by Sol-gel Immobilization: Further Advancements of Practical Interest. *Nanoscale* **2014,** *6,* 6220–6228.
11. Chen, C.-S.; Fujimoto, Y.; Girdaukas, G.; Sih, C. J. Quantitative-Analyses of Biochemical Kinetic Resolutions of Enantiomers. *J. Am. Chem. Soc.* **1982,** *104,* 7294–7299.
12. Inagaki, M.; Hiratake, J.; Nishioka, T.; Oda, J. One-Pot Synthesis of Optically-Active Cyanohydrin Acetates from Aldehydes Via Lipase-Catalyzed Kinetic Resolution Coupled with Insitu Formation and Racemization of Cyanohydrins. *J. Org. Chem.* **1992,** *57,* 5643–5649.
13. Martín-Matute, B.; Bäckvall, J.-E. Dynamic Kinetic Resolution Catalyzed by Enzymes and Metals. *Curr. Opin. Chem. Biol.* **2007,** *11,* 226–232.
14. (a) Allen, J. V.; Williams, J. M. J. Dynamic Kinetic Resolution with Enzyme and Palladium Combinations. *Tetrahedron Lett.* **1996,** *37,* 1859–1862. See also.
(b) Choi, Y. K.; Choi, K.; Suh, J. H.; Lee, D.; Lim, I. T.; Jung, J. Y.; Kim, M.-J. Dynamic Kinetic Resolution of Acyclic Allylic Acetates Using Lipase and Palladium. *J. Org. Chem.* **1999,** *64,* 8423–8424.
15. Dinh, P. M.; Howarth, J. A.; Hudnott, A. R.; Williams, J. M. J.; Harris, W. Catalytic Racemisation of Alcohols: Applications to Enzymatic Resolution Reactions. *Tetrahedron Lett.* **1996,** *37,* 7623–7626.
16. (a) Larsson, A. L. E.; Persson, B. A.; Bäckvall, J.-E. Enzymatic Resolution of Alcohols Coupled with Ruthenium-Catalyzed Racemization of the Substrate Alcohol. *Angew. Chem. Int. Ed.* **1997,** *36,* 1211–1212.

(b) Persson, B. A.; Larsson, A. L. E.; Le Ray, M.; Bäckvall, J.-E. Ruthenium- and Enzyme-catalyzed Dynamic Kinetic Resolution of Secondary Alcohols. *J. Am. Chem. Soc.* **1999,** *121,* 1645–1650.

17. Choi, J. H.; Kim, Y. H.; Nam, S. H.; Shin, S. T.; Kim, M.-J.; Park, J. Aminocyclopentadienyl Ruthenium Chloride: Catalytic Racemization and Dynamic Kinetic Resolution of Alcohols at Ambient Temperature. *Angew. Chem. Int. Ed.* **2002,** *41,* 2373–2376.
18. Martín-Matute, B.; Edin, M.; Bogár, K.; Bäckvall, J.-E. Highly Compatible Metal and Enzyme Catalysts for Efficient Dynamic Kinetic Resolution of Alcohols at Ambient Temperature. *Angew. Chem. Int. Ed.* **2004,** *43,* 6535–6539.
19. (a) Kim, M.-J.; Choi, Y. K.; Kim, S.; Kim, D.; Han, K.; Ko, S.-B.; Park, J. Highly Enantioselective Dynamic Kinetic Resolution of 1,2-Diarylethanols by a Lipase-Ruthenium Couple. *Org. Lett.* **2008,** *10,* 1295–1298.
   (b) Lee, J. H.; Kim, N.; Kim, M.-J.; Park, J. Substituent Effect on Catalytic Activities of [{η^5-$Ar_4C_4COC(=O)Ar$}$Ru(CO)_2Cl$] in Racemization and DKR of Secondary Alcohols. *ChemCatChem* **2011,** *3,* 354–359.
20. Akai, S.; Tanimoto, K.; Kita, Y. Lipase-Catalyzed Domino Dynamic Kinetic Resolution of Racemic 3-Vinylcyclohex-2-en-1-ols/Intramolecular Diels-Alder Reaction: One-Pot Synthesis of Optically Active Polysubstituted Decalins. *Angew. Chem. Int. Ed.* **2004,** *43,* 1407–1410.
21. (a) Mavrynsky, D.; Päiviö, M.; Lundell, K.; Sillanpää, R.; Kanerva, L. T.; Leino, R. Dicarbonylchloro(pentabenzylcyclopentadienyl)ruthenium as Racemization Catalyst in the Dynamic Kinetic Resolution of Secondary Alcohols. *Eur. J. Org Chem.* **2009,** 1317–1320.
   (b) Päivö, M.; Mavrynsky, D.; Leino, R.; Kanerva, L. T. Dynamic Kinetic Resolution of a Wide Range of Secondary Alcohols: Cooperation of Dicarbonylchlorido(pentabenzylcyclopentadienyl)ruthenium and CAL-B. *Eur. J. Org. Chem.* **2011,** 1452–1457.
22. Fernandez, J. A.; Manzini, S.; Nolan, S. P. A Cationic Ruthenium Complex for the Dynamic Kinetic Resolution of Secondary Alcohols. *Chem. Eur J.* **2014,** *20,* 13132–13135.
23. Sato, Y.; Kayaki, Y.; Ikariya, T. Efficient Dynamic Kinetic Resolution of Racemic Secondary Alcohols by a Chemoenzymatic System Using Bifunctional Iridium Complexes with C-N Chelate Amido Ligands. *Chem. Commun.* **2012,** 3635–3637.
24. Langvik, O.; Sandberg, T.; Warna, J.; Murzin, D. Y.; Leino, R. One-Pot Synthesis of (*R*)-2-Acetoxy-1-indanone from 1,2 Indanedione Combining Metal Catalyzed Hydrogenation and Chemoenzymatic Dynamic Kinetic Resolution. *Catal. Sci. Technol.* **2015,** *5,* 150–160.
25. (a) El-Sepelgy, O.; Alandini, N.; Rueping, M. Merging Iron Catalysis and Biocatalysis Iron Carbonyl Complexes as Efficient Hydrogen Autotransfer Catalysts in Dynamic Kinetic Resolutions. *Angew. Chem. Int. Ed.* **2016,** *55,* 13602–13605.
   (b) El-Sepelgy, O.; Brzozowska, A.; Rueping, M. Asymmetric Chemoenzymatic Reductive Acylation of Ketones by a Combined Iron-Catalyzed Hydrogenation-Racemization and Enzymatic Resolution Cascade. *ChemSusChem* **2017,** *10,* 1664–1668.
26. Gustafson, K. P. J.; Guethmundsson, A.; Lewis, K.; Bäckvall, J.-E. Chemoenzymatic Dynamic Kinetic Resolution of Secondary Alcohols Using an Air- and Moisture-Stable Iron Racemization Catalyst. *Chem. Eur J.* **2017,** *23,* 1048–1051.
27. (a) Kim, M. J.; Chung, Y. I.; Choi, Y. K.; Lee, H. K.; Kim, D.; Park. (*S*)-Selective Dynamic Kinetic Resolution of Secondary Alcohols by the Combination of Subtilisin and an Aminocyclopentadienylruthenium Complex as the Catalysts. *J. Am. Chem. Soc.* **2003,** *125,* 11494–11495.

(b) Choi, J. H.; Choi, Y. K.; Kim, Y. H.; Park, E. S.; Kim, E. J.; Kim, M.-J.; Park, J. Aminocyclopentadienyl Ruthenium Complexes as Racemization Catalysts for Dynamic Kinetic Resolution of Secondary Alcohols at Ambient Temperature. *J. Org. Chem.* **2004,** *69,* 1972–1977.
(c) Boren, L.; Martín-Matute, B.; Xu, Y.; Cordova, A.; Bäckvall, J.-E. (*S*)-Selective Kinetic Resolution and Chemoenzymatic Dynamic Kinetic Resolution of Secondary Alcohols. *Chem. Eur J.* **2006,** *12,* 225–232.
(d) Norinder, J.; Bogar, K.; Kanupp, L.; Bäckvall, J.-E. An Enantioselective Route to Alpha-Methyl Carboxylic Acids via Metal and Enzyme Catalysis. *Org. Lett.* **2007,** *9,* 5095–5098.
(e) Thalen, L. K.; Sumic, A.; Bogar, K.; Norinder, J.; Persson, A. K.Å.; Bäckvall, J.-E. Enantioselective Synthesis of Alpha-Methyl Carboxylic Acids from Readily Available Starting Materials via Chemoenzymatic Dynamic Kinetic Resolution. *J. Org. Chem.* **2010,** *75,* 6842–6847.

28. Kim, S.; Choi, Y. K.; Hong, J.; Park, J.; Kim, M.-J. *Candida Antarctica* Lipase a and Pseudomonas Stutzeri Lipase as a Pair of Stereocomplementary Enzymes for the Resolution of 1,2-Diarylethanols and 1,2-Diarylethanamines. *Tetrahedron Lett.* **2013,** *54,* 1185–1188.
29. Engström, K.; Vallin, M.; Syrén, P.-O.; Hult, K.; Bäckvall, J.-E. Mutated Variant of *Candida Antarctica* Lipase B in (*S*)-Selective Dynamic Kinetic Resolution of Secondary Alcohols. *Org. Biomol. Chem.* **2011,** *9,* 81–82.
30. For reviews, see:
(a) Kim, Y.; Park, J.; Kim, M.-J. Dynamic Kinetic Resolution of Amines and Amino Acids by Enzyme-Metal Cocatalysis. *ChemCatChem* **2011,** *3,* 271–277.
(b) Jiang, C.-J.; Cheng, G.-L. Racemization of Amines and Related Derivatives Leading to Their Dynamic Kinetic Resolution. *Curr. Org. Chem.* **2013,** *17,* 1225–1234.
31. (a) Wuyts, S.; De Temmerman, K.; De Vos, D. E.; Jacobs, P. A. Acid Zeolites as Alcohol Racemization Catalysts: Screening and Application in Biphasic Dynamic Kinetic Resolution. *Chem. Eur J.* **2005,** *11,* 386–397.
(b) Wuyts, S.; Wahlen, J.; Jacobs, P. A.; De Vos, D. E. Heterogeneous Vanadium Catalysts for Racemization and Chemoenzymatic Dynamic Kinetic Resolution of Benzylic Alcohols. *Green Chem.* **2007,** *9,* 1104–1108.
32. Akai, S.; Tanimoto, K.; Kanao, Y.; Egi, M.; Yamamoto, T.; Kita, Y. A Dynamic Kinetic Resolution of Allyl Alcohols by the Combined Use of Lipases and [$VO(OSiPh_3)_3$]. *Angew. Chem. Int. Ed.* **2006,** *45,* 2592–2595.
33. Akai, S.; Hanada, R.; Fujiwara, N.; Kita, Y.; Egi, M. One-Pot Synthesis of Optically Active Allyl Esters via Lipase-Vanadium Combo Catalysis. *Org. Lett.* **2010,** *12,* 4900–4903.
34. Egi, M.; Sugiyama, K.; Saneto, M.; Hanada, R.; Kato, K.; Akai, S. A Mesoporous-Silica-Immobilized Oxovanadium Cocatalyst for the Lipase-Catalyzed Dynamic Kinetic Resolution of Racemic Alcohols. *Angew. Chem. Int. Ed.* **2013,** *52,* 3654–3658.
35. Sugiyama, K.; Oki, Y.; Kawanishi, S.; Kato, K.; Ikawa, T.; Egi, M.; Akai, S. Spatial Effects of Oxovanadium-Immobilized Mesoporous Silica on Racemization of Alcohols and Application in Lipase-Catalyzed Dynamic Kinetic Resolution. *Catal. Sci. Technol.* **2016,** *6,* 5023–5030.
36. Kawanishi, S.; Sugiyama, K.; Oki, Y.; Ikawa, T.; Akai, S. Preparation of Optically Active Cycloalkenes Bearing All-Carbon Quaternary Stereogenic Centres via Lipase–Oxovanadium Combo-Catalysed Dynamic Kinetic Resolution. *Green Chem.* **2017,** *19,* 411–417.
37. Cao, H.; Zhu, X.-H.; Wang, D.; Sun, Z.; Deng, Y.; Hou, X.-F.; Zhao, D. Selectivity Enhancement in Dynamic Kinetic Resolution of Secondary Alcohols through

Adjusting the Micro-Environment of Metal Complex Confined in Nanochannels: A Promising Strategy for Tandem Reactions. *ACS Catal.* **2015,** *5,* 27–33.
38. Fujioka, H.; Matsuda, S.; Horai, M.; Fujii, E.; Morishita, M.; Nishiguchi, N.; Hata, K.; Kita, Y. Facile and Efficient Synthesis of Lactols by a Domino Reaction of 2,3-Epoxy Alcohols with a Hypervalent Iodine(III) Reagent and its Application to the Synthesis of Lactones and the Asymmetric Synthesis Oo (+)-Tanikolide. *Chem. Eur J.* **2007,** *13,* 5238–5248.
39. Sugiyama, K.; Kawanishi, S.; Oki, Y.; Kamiya, M.; Hanada, R.; Egi, M.; Akai, S. Lipase-catalyzed Asymmetric Synthesis of Naphtho[2,3-c]furan-1(3H)-one Derivatives by a One-Pot Dynamic Kinetic Resolution/Intramolecular Diels–Alder Reaction: Total Synthesis of (–)-Himbacine *Bioorg. Med. Chem.* **2018,** *26,* 1378–1386.
40. (a) Akai, S.; Tanimoto, K.; Kanao, Y.; Omura, S.; Kita, Y. Lipase-Catalyzed Domino Kinetic Resolution of Alpha-Hydroxynitrones/Intramolecular 1,3-dipolar Cycloaddition: a Concise Asymmetric Total Synthesis of (-)-Rosmarinecine. *Chem. Commun.* **2005,** 2369–2371.
(b) Nemoto, H.; Tanimoto, K.; Kanao, Y.; Omura, S.; Kita, Y.; Akai, S. Protecting-Group-Free Catalytic Asymmetric Total Synthesis of (-)-Rosmarinecine. *Tetrahedron* **2012,** *68,* 7295–7301. See also.
(c) Akai, S.; Naka, T.; Omura, S.; Tanimoto, K.; Imanishi, M.; Takebe, Y.; Matsugi, M.; Kita, Y. Lipase-Catalyzed Domino Kinetic Resolution/Intramolecular Diels-Alder Reaction: One-Pot Synthesis of Optically Active 7-Oxabicyclo[2.2.1] heptenes from Furfuryl Alcohols and Beta-Substituted Acrylic Acids. *Chem. Eur J.* **2002,** *8,* 4255–4264.

CHAPTER 3

Organic Synthetic Reactions Under Kinetic Control

Toshiyuki Hamura
Department of Applied Chemistry for Environment, Kwansei Gakuin University, Sanda, Japan

Contents

1. INTRODUCTION

Construction of polycyclic aromatic compounds with diverse functionalities is one of the important subjects often faced in the natural and unnatural product syntheses. In this relevance, when aiming to prepare the aromatics with complex substitution pattern, conventional organic synthetic reactions with a given aromatic starting material, such as electrophilic substitution, transition-metal-catalyzed cross-coupling reaction, and directed metalation–functionalization are sometimes ineffective. This is largely due to the lack of generality of these reactions, and the introduction of various functionalities and fused rings into the specific positions of the aromatics becomes even more difficult, especially when the reactant is consisting of polycyclic structure. For that reason, new synthetic route to functionalized polycycles should be developed to overcome these problems.

Kinetic Control in Synthesis and Self-Assembly
ISBN 978-0-12-812126-9
https://doi.org/10.1016/B978-0-12-812126-9.00003-1

In this context, the reaction integration is one of the useful synthetic strategies to access the aromatic skeleton with the desired substitution pattern because only a few steps are required to prepare the target molecule, and more importantly the formation of regioisomeric mixture can be avoided by search for the suitable reaction conditions in each step. In particular, utilization of the highly reactive molecules or reactive intermediates as a reactive platform in this reaction integration is a powerful and straightforward tool in view of the rapid assembly of polycycles through the multiple annulation and/or functionalization onto a reactive core ring. In this sequential process, three kinds of activations, i.e., heat, light, and transition-metal catalyst under thermodynamic or kinetic control, permit selective activation of the starting material, which leads to the selective generation of the elaborated reactive species, allowing the construction of complex structures with high stereoselectivity, chemoselectivity, regioselectivity, and/or site-selectivity. In this manner, reaction integration of reactive intermediate, selectively generated from the precursor by controlling the reactivity under kinetic or thermodynamic conditions, would have valuable synthetic utilities for selective construction of polycyclic aromatic compounds.

In this chapter, organic synthetic reactions of reactive molecules in consideration of kinetic or thermodynamic control are summarized based on a recent progress of our synthetic approach to polyfunctionalized aromatic compounds.

2. SELECTIVE HALOGEN–LITHIUM EXCHANGE FOR SUCCESSIVE CYCLOADDITION OF ARYNE AND ISOBENZOFURAN EN ROUTE TO POLYCYCLIC AROMATIC COMPOUNDS

2.1 Isolable Isobenzofuran

Isobenzofurans are 10π electron systems with quinoid structures, which make them useful intermediates for various natural/unnatural product syntheses *(1)*. Among various possibilities, $[4+2]$ cycloaddition with dienophiles is a reliable method for the construction of a polycyclic structure *(2)*. However, isobenzofuran **I** having no substituent or alkyl groups at the C_1 and C_3 positions is unstable and regarded as a transient intermediate with a few exceptions, which have limited the availability of this attractive molecule *(3)*. In this context, Warrener and coworkers developed a new technique for stabilizing isobenzofuran by introducing an

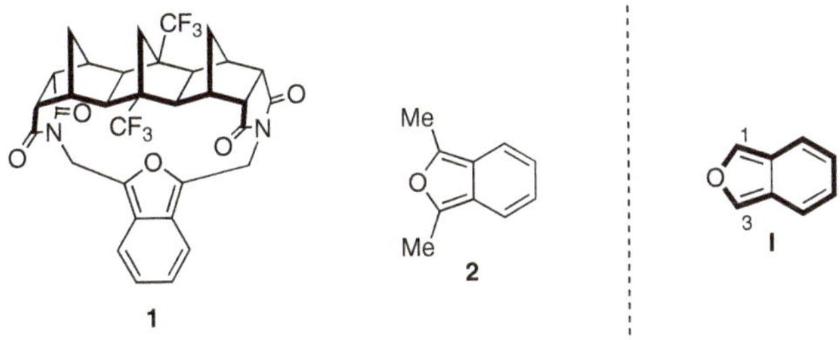

Figure 3.1 Kinetically stabilized isobenzofuran **1**.

alicyclophane macrocycle (Fig. 3.1) *(4)*. The concept depends on the rack section of the alicyclophane, acting as a steric shield to inhibit intermolecular self-dimerization or polymerization processes of isobenzofuran, whose strategy is well known as kinetic stabilization. Indeed, the new rack-isobenzofuran **1** is so stable that could be isolated, handled as regular compounds, and stored in solution for several days at 5°C and for many months in the solid state, although the structurally similar 1,3-dimethylisobenzofuran **2** is rapidly polymerized during the attempted isolation.

In contrast, we recently developed the synthetic access to dibromoisobenzofuran **5** having no substituent at the C_1 and C_3 position by reaction of epoxy-naphthalene **3** *(5,6)* with 3,6-di(2-pyridyl)-1,2,4,5-tetrazine (**4**). Essential point of this synthesis is the isolation method: purification of the crude products by silica–gel column chromatography *under Ar atmosphere* to give dibromoisobenzofuran **5** as pure form (Fig. 3.2). It is important to note that the product can be stored under Ar atmosphere in a refrigerator for

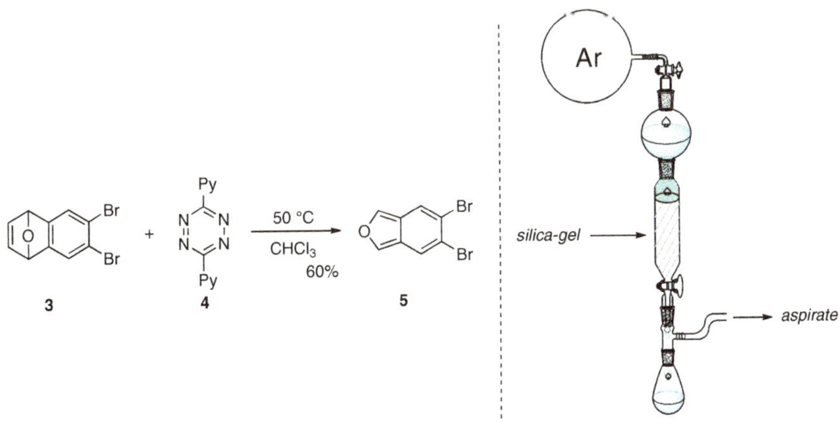

Figure 3.2 Isolability of isobenzofuran.

several months without any kinetic protection. This finding provides us great opportunities to use these potentially attractive molecules under various reaction conditions.

2.2 Selective Generation of Aryllithium Under Thermodynamic Control

To develop the synthetic utility of isobenzofurans, especially the derivative with no substitution at the C_1 and C_3 position (vide supra), we came up with an idea to use a *donor–acceptor molecule*, having electron-donating and electron-withdrawing moieties in one molecule, allowing the rapid introduction of the fused rings onto the central core ring. In particular, we are intrigued by the use of a formal equivalent to didehydroisobenzofuran **II**, possessing a highly strained 1,2-didehydroarene structure at the C_5–C_6 position in **I**, as it could serve as a reactive platform for expeditious assembly of polycycles by two-directional annulations (Fig. 3.3).

Specifically, we focus on the dual annulations of dibromoisobenzofuran **V**, a formal equivalent of didehydroisobenzofuran **II**, via successive [2 + 4] cycloadditions of benzyne and isobenzofuran (Fig. 3.4). Among three dibromides **IV**, **V**, and **VI**, if the bromine–lithium exchange occurs selectively at the side

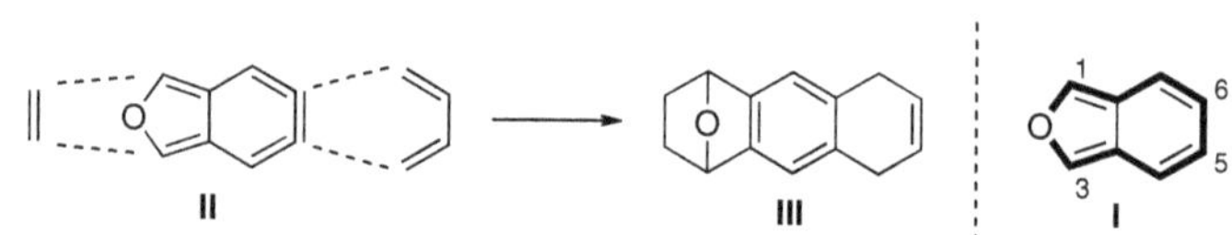

Figure 3.3 Two-directional annulations of didehydroisobenzofuran.

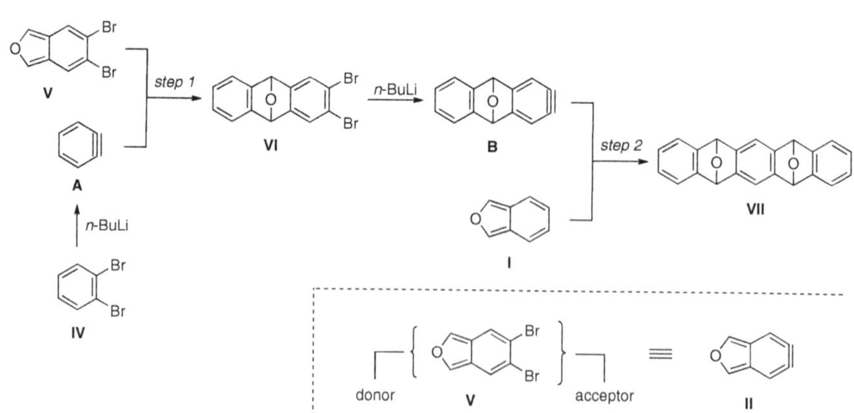

Figure 3.4 Dual annulations of dibromoisobenzofuran **V**.

of dibromoisobenzofuran **V** under kinetic or thermodynamic control *(7)*, the tandem generation of benzynes and dual cycloadditions with two different arynophiles are realized. Thus, the first [4 + 2] cycloaddition of dibromoisobenzofuran **V** and benzyne **A**, selectively generated from dibromobenzene **IV**, gives tetracyclic compound **VI** (step 1), which, in turn, serves as a benzyne donor, leading to the selective generation of benzyne **B**. Second [2 + 4] cycloaddition of benzyne **B** and isobenzofuran **I** would give polycyclic compound **VII** (step 2). Because the difference in reactivity of bromine–lithium exchange of functionalized 1,2-dibromoarenes, e.g., **IV**, **V**, and **VI**, under kinetic or thermodynamic conditions is not easily predicted, the first [4 + 2] cycloaddition was examined to check the possibility of selective generation of benzyne **A** *(8,9)*.

When 1,2-dibromobenzene (**6**) was treated with 1.0 equiv of *n*-BuLi in the presence of dibromoisobenzofuran **5** (1.0 equiv, toluene, −15 → 25°C), cycloadduct **7** was obtained only in 18% yield. In this case, a sizable amount of bis-cycloadduct **8** (25%) was produced, which indicated the similar reactivity of two bromides **6** and **7** with *n*-BuLi, resulting in the competitive generation of aryne **B** from the initially formed mono-cycloadduct **7** (Fig. 3.5). This result shows the approximate order of the reactivity among the three dibromides, **6** ≈ **7** > **5**, suggesting the difficulty in the selective generation of benzyne **A** under this condition.

On the other hand, similar reaction using excess amounts of the starting material **6** (5.0 equiv) improved the yields of the mono-cycloadduct **7** (42%). However, it is not an essential solution because presence of the large amount of the starting material **6** disturbed the second [2 + 4] cycloaddition with **7** in *one-pot process*.

Further investigation revealed that introduction of the electron-deficient fluoro group on the aromatic ring achieved selective generation of

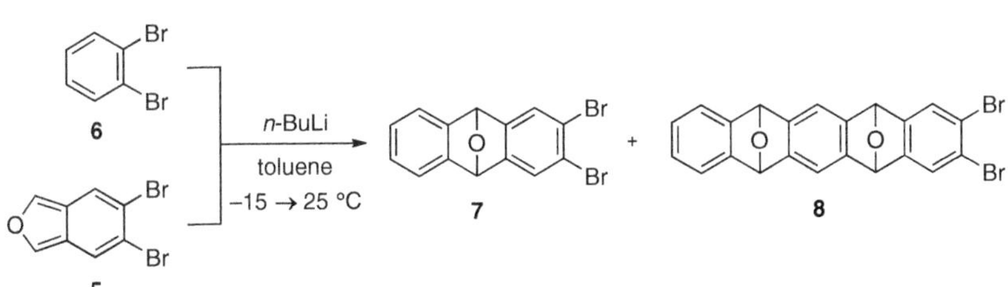

6: 1 equiv, **5**: 1 equiv 18% 25%

6: 5 equiv, **5**: 1 equiv 42% 4%

Figure 3.5 Nonselective generation of benzyne.

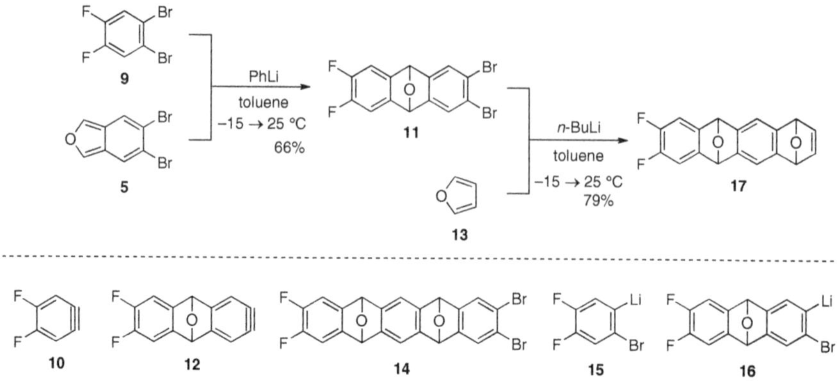

Figure 3.6 Selective generation of 4,5-difluorobenzyne **10**.

4,5-difluorobenzyne **10**, affording mono-cycloadduct **11** as almost the exclusive product. In this case, bis-cycloadduct **14**, caused by the generation of aryne **12**, was not produced at all (Fig. 3.6). This observed selectivity in the bromine–lithium exchange, leading to the selective formation of aryne **10**, is explained by the thermodynamic generation of aryllithium; (2-bromo-4,5-difluorophenyl)lithium **15** is stabilized by an electron-withdrawing fluoro atom at the *meta*-position in comparison with epoxyanthracenyllithium **16**.

The cycloadduct **11**, thus obtained, can be used for the second cycloaddition, affording the bis-cycloadduct **17** in good yield by using furan **13** (2.7 equiv toluene, $-15 \rightarrow 25^{\circ}$C) as a trapping agent.

This successive process can be performed in one-pot by sequential addition of the arynophiles. When the above mentioned first cycloaddition is conducted under the similar reaction conditions, the formation of the mono-cycloadduct is observed by tlc. The second arynophile is added to the reaction mixture at -15°C, to which is slowly added *n*-BuLi and the temperature raised to 25°C, giving the dual-cycloadduct **17**. The dual reactions with diphenylisobenzofuran **19** (first arynophile) and **18** or **20** (second arynophile) gave **21** and **22**, respectively (Fig. 3.7). Moreover, successive cycloaddition of 4,5-chlorobenzyne (**26**) is also available through the selective generation of (2-bromo-4,5-chlorophenyl)lithium (**25**) from the dibromide **24**, affording the dual cycloadduct **23** in good yield.

Moreover, 1,2,4,5-tetrabromobenzene (**27**) works well as a reactive platform *(10)*, allowing the threefold cycloadditions *in one-pot* (Fig. 3.8). Again, the key process is the selective bromine–lithium exchange of tetrabromobenzene **27** among the three bromides, **27**, **19**, and **28**, and the

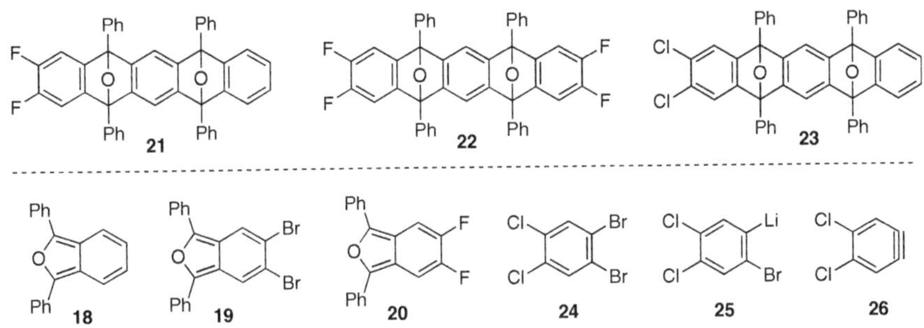

Figure 3.7 One-pot dual cycloadditions of 4,5-dihaloarynes.

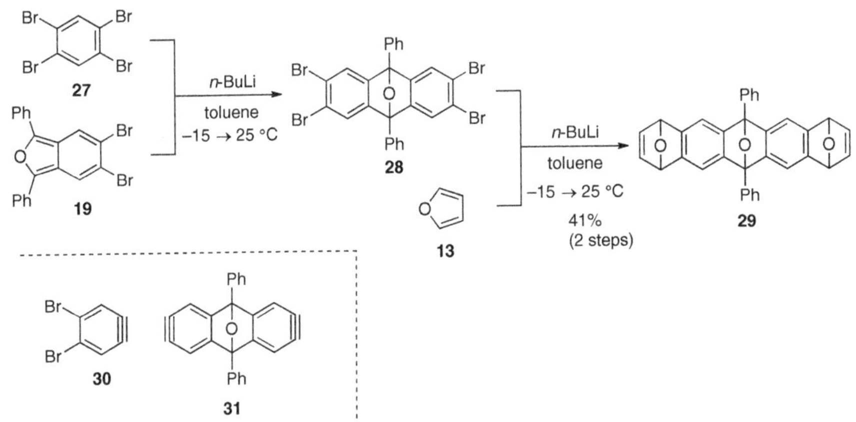

Figure 3.8 Threefold aryne cycloadditions in one-pot process.

dibromobenzyne **30**, cleanly generated, is intercepted by **19** to afford the cycloadduct **28**, which can be viewed as an equivalent to bis-aryne **31**. In fact, dual [2 + 4] cycloadditions of bis-aryne **31** by treatment of **28** with *n*-BuLi in the presence of furan gives bis-cycloadduct **29**.

2.3 Synthetic Transformation

Fig. 3.9 shows the transformations of the dual-cycloadduct **29** into a substituted pentacene *(11)*. From the synthetic point of view, the dual-cycloadduct **29** has high synthetic utility, because it is available as a precursor of bis-isobenzofuran **30**, allowing the rapid introduction of the fused rings and/or functionalization onto the isobenzofuran nucleus through the dual cycloadditions. Indeed, the transient intermediate **30**, generated by heating of **29** with tetrazine **4**, is trapped with dimethyl maleate to give the substituted pentacene **32** after acid-promoted aromatization of the dual cycloadduct **31** *(5)*.

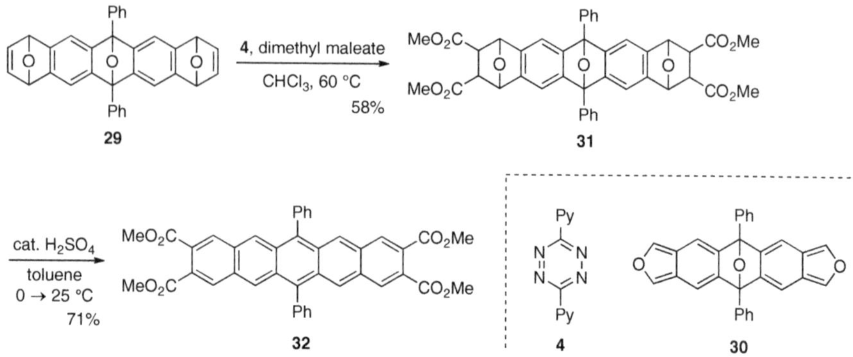

Figure 3.9 Preparation of substituted pentacene **32**.

2.4 Selective Generation of Aryllithium Under Kinetic Control

In the previous section, we described the successive cycloaddition of aryne using the selective halogen–lithium exchange of 1,2-dibromoarene under the thermodynamic control. This sequential reaction, however, has a limitation in that the introduction of electron-withdrawing groups on the benzene ring in aryne precursor (e.g., **9**) is required to restrict the competitive formation of the dual cycloadduct (Fig. 3.6).

To expand the synthetic utility of this successive process, we reexamined the aryne cycloadditions of the parent benzyne species **A** as the first cycloaddition (vide supra). The key to fulfill this sequential process is search for a suitable aryne precursor to permit the selective halogen–lithium exchange under the suitable conditions. Along these lines, we select 1,2-dihaloarene as an aryne precursor to control the reactivity of halogen–lithium exchange by taking advantage of the following two features: (1) utilization of the more electropositive halogen (type 1) or (2) tuning the reactivity by the adjacent halogen (type 2) as shown in Fig. 3.10. The first strategy depends on the relative reactivity of the halogen atom, I > Br > Cl, thus exploiting the iodoarene with higher reactivity toward *n*-BuLi in comparison with the corresponding bromoarene (kinetic control). On the other hand, the second strategy is using the strong electron–withdrawing ability of the adjacent halogen, which might reinforce the electrophilicity of the targeted halogen atom, and therefore it facilitates the halogen–lithium exchange compared to when there is less electronegative halogen.

Table 3.1 shows the result of selective generation of benzyne species **A** and its trapping using 1-halo-2-iodobenzene as a kinetically reactive precursor

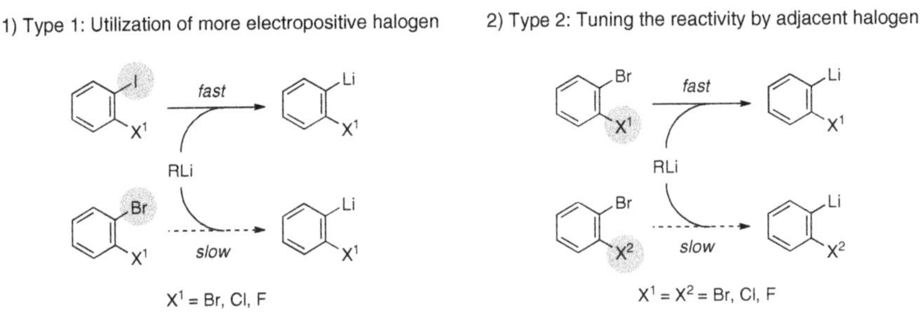

Figure 3.10 Two strategies for selective halogen–lithium exchange.

Table 3.1 Selective generation of benzyne and its trapping

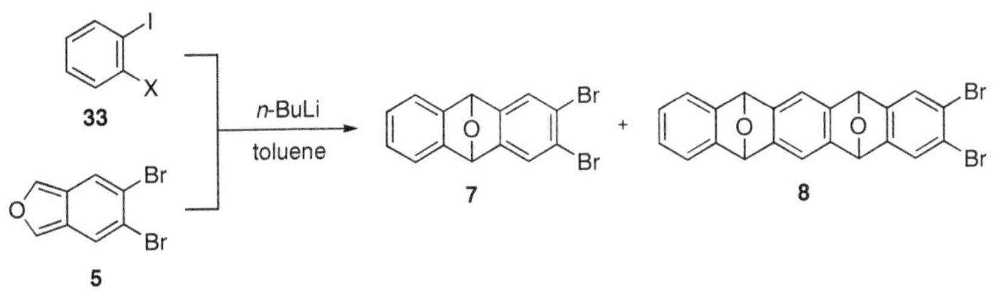

Entry	X	Temperature (°C)	Yield of 6a (%)	Yield of 7a (%)[a]
1	**33a**: Br	−78	60	9
2	**33a**: Br	−15 → 25	78	1
4	**33b**: Cl	−15 → 25	62	9

[a]The cycloadduct **8** was obtained as a mixture of diastereomers (ds: 44/56–58/42).

(type 1 reaction). As we expected, the reaction of 1-bromo-2-iodobenzene (**33a**) with 1.2 equiv of *n*-BuLi is selective to give the mono-cycloadduct **7** in 60% yield (entry 1). Importantly, the formation of the bis-cycloadduct **8** via the bromine–lithium exchange of **7** was not fully but mainly suppressed (9%) in comparison with the corresponding reaction of dibromide **6** (Fig. 3.5). Same reaction at higher reaction temperature (−15 → 25°C) gave a better yield of the desired product **7** (78%), and the bis-cycloadduct **8** was obtained only in 1% yield (entry 2). Using 1-chloro-2-iodobenzene (**33b**) as a benzyne precursor again proved feasible with *n*-BuLi (toluene, −15 → 25°C), affording **7** in 62% yield (entry 3). These results indicate that halogen–lithium exchange selectively occurred at the more electropositive iodine atom in iodohalides **33a** and **33b**, smoothly generating (2-halo)phenyllithiums, respectively, whereas the dibromoisobenzofuran **5** and the dibromo cycloadduct **7** almost untouched under these conditions. Based on these reaction outcomes, it is safe to say that use of 1,2-dihaloarenes **33a** and **33b** possessing a more electropositive iodine atom is suitable as a benzyne precursor over the bromide **6** in terms of selectivity and yield.

We next examined the another mode of selective halogen–lithium exchange of 1,2-haloarenes through the reactivity control by adjacent halogen (Fig. 3.11) *(12,13)*. As a model substrate, the tetrahalide **34** having two benzyne generation sites on each benzene ring was prepared and its site-selectivity was checked. As a result, the bromine–lithium exchange occurred at the dibromo side in **34** to selectively generate the bromo-chlorobenzyne **36**, which was intercepted by 1,3-diphenylisobenzofuran (**18**) to afford the unsymmetrical cycloadduct **35** in 65% yield, accompanied by a formation of dual cycloadduct (20%, structure not shown). This observed site-selectivity in the bromine–lithium exchange of **34** was unexpected, because (2-chlorophenyl)lithium **39** was more thermodynamically stable than (2-bromophenyl)lithium **40** by the presence of a more electron-withdrawing chlorine atom, which would suggest the desired formation of aryne **37** over that of aryne **36**. In this manner, tuning the reactivity of the bromine atom by adjacent halogen proved to be possible in site-selective bromine–lithium exchange of polyhalide **34**, although the origin of this kinetically controlled site-selectivity is not clear at this stage.

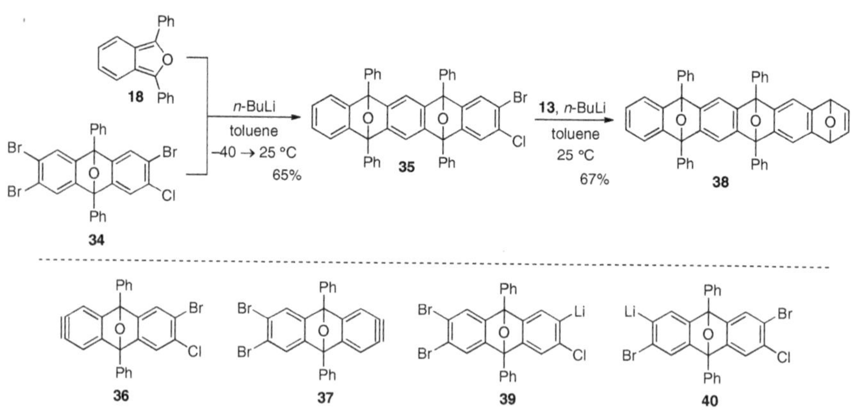

Figure 3.11 Tuning the reactivity by adjacent halogen.

In relation to the abovementioned dual cycloadditions, Suzuki and coworkers reported the stepwise generation of two benzyne species from bis-sulfonate **41**, which is sequentially trapped by ketene silyl acetal (Fig. 3.12) *(14)*. Importantly, the first cycloaddition occurs at the iodotriflate side of **41**, which arises from the selective iodine–lithium exchange. This site-selectivity is explained by the thermodynamic generation of aryllithium

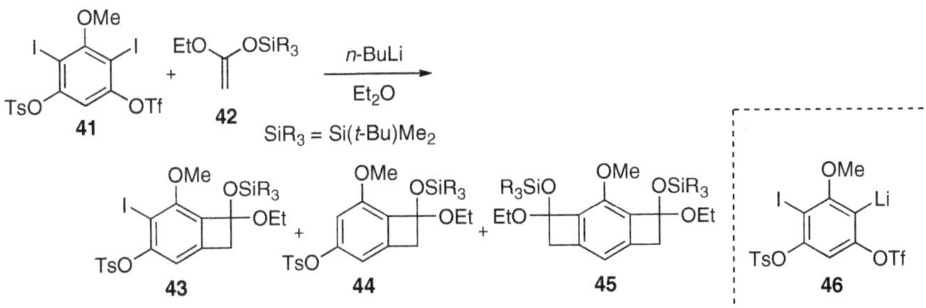

	43	44	45
−95 °C[a]	72%	4%	8%
−95 °C[b]	—	34%	24%
−95 → −78 °C[b]	—	8%	72%

a) 1.05 equiv of *n*-BuLi; b) 2.3 equiv of *n*-BuLi.

Figure 3.12 Stepwise generation of benzynes from bis-sulfonate **41** under thermodynamic control.

46, since the aryllithium **46** is strongly stabilized by higher electron-withdrawing ability of the triflate group than that of the tosylate group.

Aside from the unanticipated site-selectivity in the reaction of polyhalide **34**, threefold cycloadditions of arynes with isobenzofuran by taking advantage of these two types of kinetically controlled site-selective halogen–lithium exchange (Type 1: step 1, Type 2: step 2) can produce the highly condensed aromatic compound **50**, which is expected to be suitably converted into substituted heptacenes (Fig. 3.13).

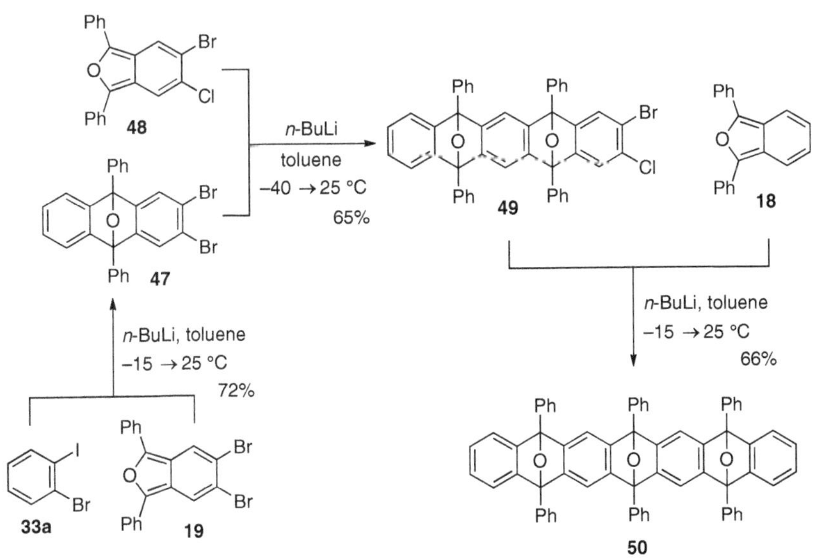

Figure 3.13 Threefold cycloadditions of arynes with isobenzofuran.

3. RING SELECTIVITY IN THE GENERATION OF ISOBENZOFURAN FOR DIVERGENT ACCESS TO POLYCYCLIC AROMATIC COMPOUNDS

3.1 Ring Selective Generation of Isobenzofuran

In the previous section, we described the two-directional annulation of dibromoisobenzofuran, a formal equivalent to didehydroisobenzofuran, for expeditious assembly of polycyles. In this successive process, the selective halogen–lithium exchange is key for selective generation of aryne and its trapping with arynophile. To develop another efficient route to polycycles with various functionalities, we paid attention to dual annulation of bis-isobenzofuran **VIII** *(15)*, which could serve as a reactive platform for linearly fused polycyclic compounds **IX** (Fig. 3.14).

In particular, we focus on the formal use of bis-isobenzofuran **VIII** by sequential generation of two isobenzofurans from diepoxyanthracene **X**. The fundamental issue that should be established in this process is the "ring selectivity," i.e., the relative susceptibilities of the two oxabicyclo rings in **X** toward the generation of isobenzofuran **XI** or **XII**, which would be realized under kinetic control (Fig. 3.15).

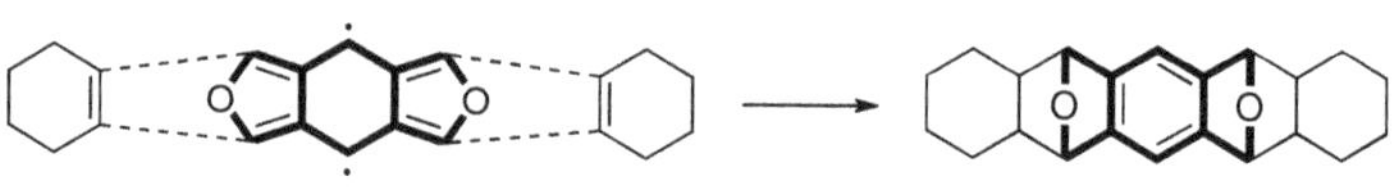

Figure 3.14 Bis-isobenzofuran as a reactive platform.

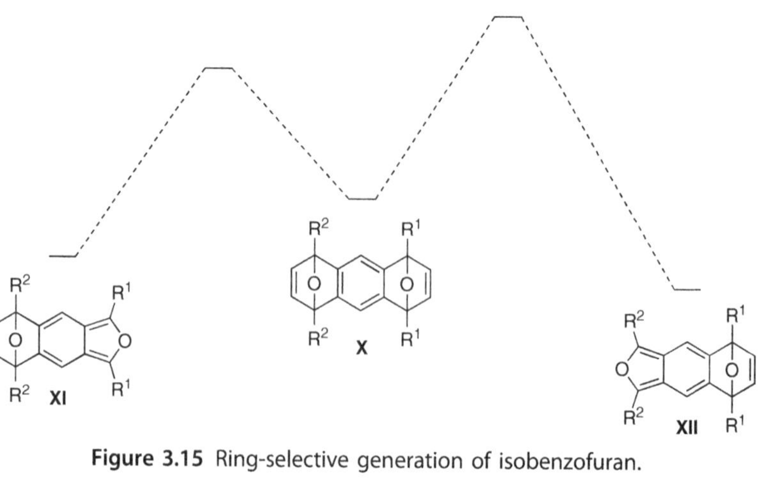

Figure 3.15 Ring-selective generation of isobenzofuran.

Thus, if the more reactive oxabicyclo ring (right ring) in **X** would undergo selective Diels–Alder, retro Diels–Alder reaction with 3,6-di(2-pyridyl)-1,2,4,5-tetrazine (vide infra), the isobenzofuran **XI**, selectively generated, can cyclize with a dienophile to give the [4 + 2] cycloadduct **XIII**. Subsequent second generation of isobenzofuran **XVI** at the remaining oxabicyclo ring (left ring) in **XIII** and trapping with a dienophile, selectively affording the bis-cycloadduct **XV**. By switching the order of the dienophiles in each cycloaddition, the isomer **XVIII** would be accessible (Fig. 3.16).

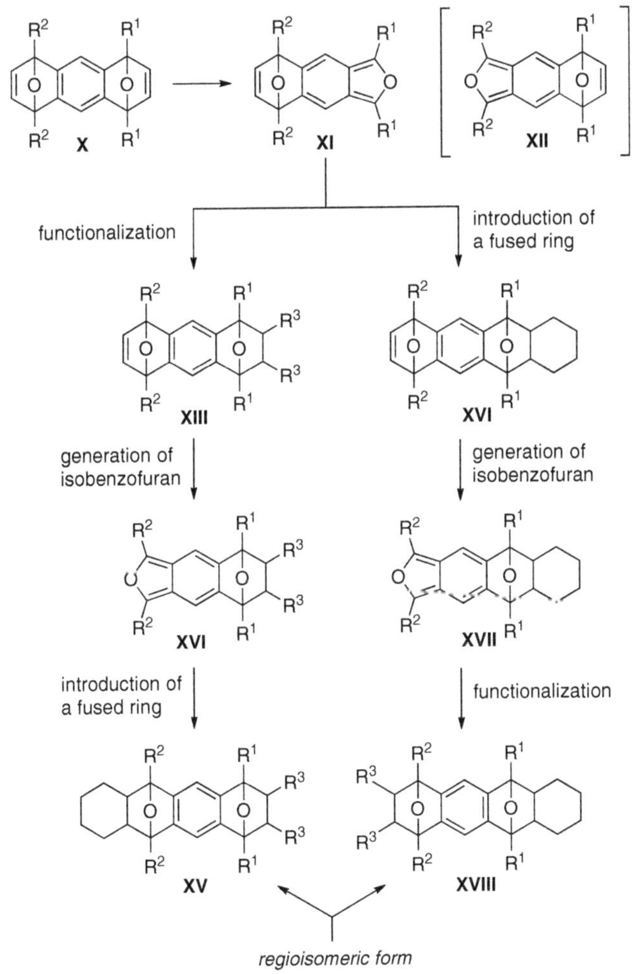

Figure 3.16 Divergent syntheses of polycycles by ring-selective generation of isobenzofuran.

On the basis of this strategy, three diepoxyanthracenes **51a–51c** *(10)* with different substitution patterns on the epoxy rings were tested for exploiting the ring-selective generation of isobenzofurans. On heating of diepoxyanthracene **51a** with tetrazine **4** in the presence of naphthoquinone **52** ($CHCl_3$, 40°C, 40 min), the less substituted oxabicyclo ring (right ring) underwent the exclusive generation of isobenzofuran **53** and subsequent cycloaddition of **53** with **52** to give the mono-cycloadduct **54** in high yield with perfect ring selectivity (Fig. 3.17). In this case, the cycloadduct **55** based on the generation of isobenzofuran **56** was not produced at all. In addition to the ring selectivity, this cycloaddition was stereoselective to give *endo* isomer **54a** as a major product (**54b**: *exo* isomer, **54a/54b** = 80:20).

This observed ring selectivity in the generation of isobenzofuran **53** from the right ring over that of the left ring can be explained by the facile nature of the interaction between the double bond in the less substituted oxabicyclo ring and the diene in tetrazine **4** (kinetic control).

Similarly, diepoxyanthracene 51b, possessing the two phenyl groups in the left ring, reacted with tetrazine **4** at the less substituted right ring to generate isobenzofuran 58, which was intercepted with 52 to give cycloadducts **57a** and 57b (57a/57b = 64:36), respectively (Fig. 3.18). Moreover, ring selectivity was observed for the reaction of tetra-substituted

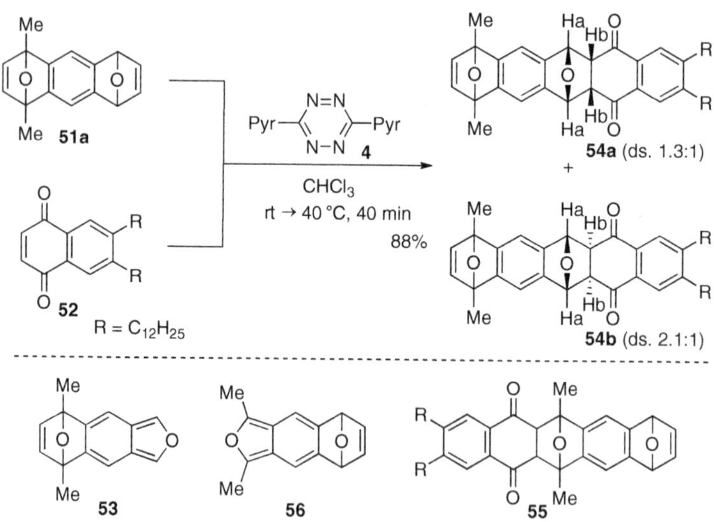

Figure 3.17 Ring selectivity in the generation of isobenzofuran **1**.

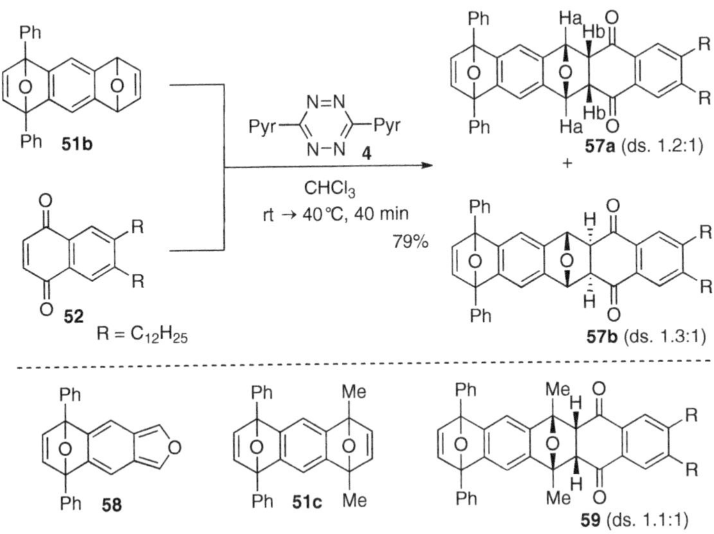

Figure 3.18 Ring selectivity in the generation of isobenzofuran **2**.

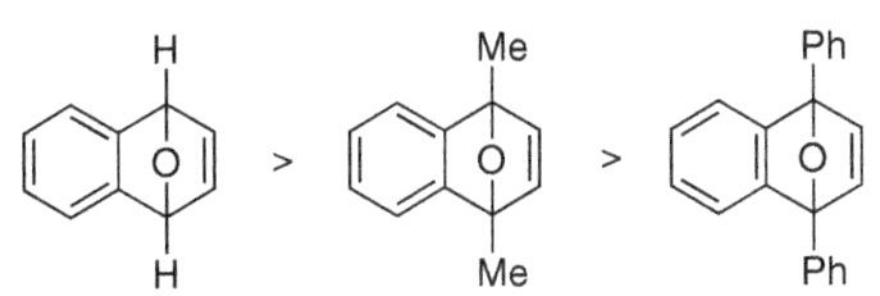

C **D** **E**

Figure 3.19 Order of the reactivity of the epoxy ring.

derivative 51c, having methyl and phenyl groups at the two epoxy rings, which led to the selective formation of the *endo*-cycloadduct 59 albeit the moderate yield (43%).

These results indicate the order of the reactivity of the epoxy ring, **C** > **D** > **E**, which is based on the steric effect of the substituents in each epoxy ring (Fig. 3.19).

3.2 Divergent Synthesis of Polyacene Derivatives

The mono-cycloadducts, thus selectively obtained, were further functionalized by the second [4 + 2] cycloaddition (Fig. 3.20). For example, treatment of *endo*-adduct **57a** with tetrazine **4** in the presence of fumaronitrile (**60**) ($CHCl_3$, 50°C) gave the [4 + 2] cycloadduct **61** as a mixture of diastereomers, which were smoothly converted into pentacenequinone **63** through the two-step sequence of aromatization.

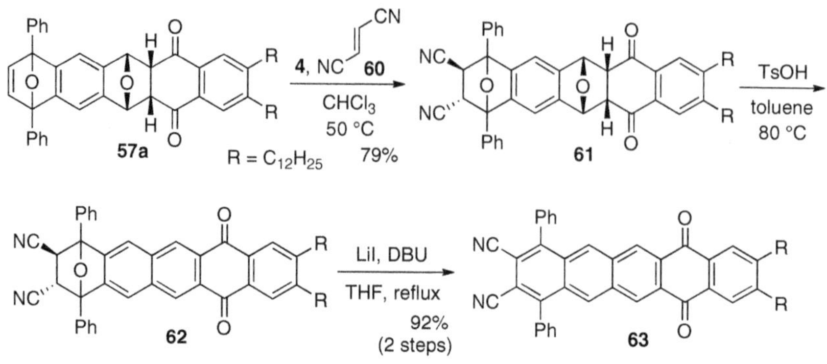

Figure 3.20 Second [4 + 2] cycloaddition of isobenzofuran.

Moreover, pentacenequinone **67**, corresponding to the isomer of quinone **63**, is selectively synthesized by switching the order of adding the trapping agents in each cycloaddition (Fig. 3.21). The first [4 + 2] cycloaddition of isobenzofuran **58**, ring-selectively generated by reaction at the less substituted right ring, and fumaronitrile (60) gave functionalized epoxyanthracene **65** after base-induced aromatization. Subsequent generation of isonaphthofuran and its trapping by naphthoquinone **52** gave pentacenequinone **67** after acid-promoted aromatization.

Lastly, an important point to emphasize is that as one of the synthetic applications, pentacenequinone **67** was converted into the pentacene **69** through the reduction of two carbonyl groups by treatment with $NaBH_4$ followed by Sn^{II}-mediated reductive aromatization (Fig. 3.22). Attaching the two long alkyl chains and the two cyano groups at the 2, 3, 9, and 10 positions sufficiently improved the solubility and the stability of the product in comparison with nonsubstituted pentacene *(16)*. Indeed, purification of

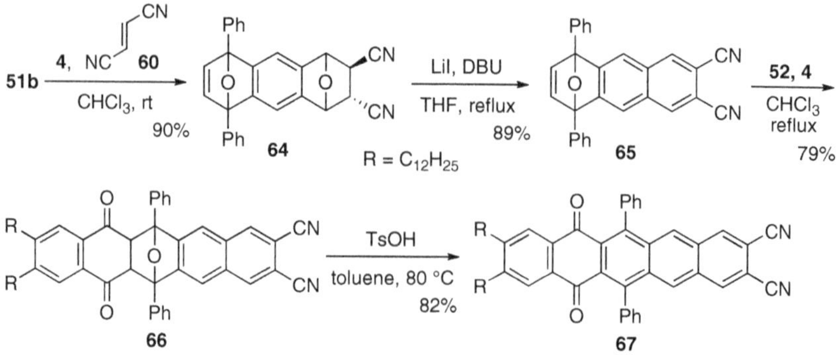

Figure 3.21 Divergent synthesis of pentacenequinone.

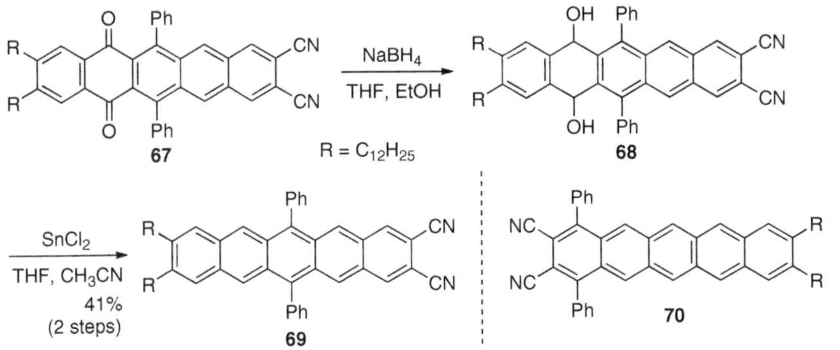

Figure 3.22 Selective synthesis of substituted pentacenes.

69 by PTLC with a stringent exclusion of air and light to give the pure product, whose half-life turned out to be around 288 min. In sharp contrast, pentacene **70**, an isomer of **69**, similarly synthesized by this two-step sequence, was fairly unstable and immediately underwent photooxidation *(17)*.

4. CONCLUSION

Synthesis of aromatic compounds has a long history of more than 100 years since the discovery of benzene. At present, however, there are a few synthetic methods enabling the introduction of various functionalities and/or fused rings into the desired positions or the desired core ring. Therefore, when we intend to prepare highly condensed aromatic compounds with multiple functionalities, the complexity of the structure makes it much more difficult to synthesize, and hence, the molecules that can be accessed are inevitably limited. In relation to this problem, we paid our attention to use highly reactive molecules, i.e., isobenzofuran and aryne, for rapid construction of polycyclic structure. Although they are generally considered to be not easy to handle, we expect developing the unique reactivity inherent in these reactive molecules would lead to exploit novel reactions utilized as a reactive platform.

Along these lines, one of our achievements on this matter is *isolability of isobenzofuran*, discovered during the purification under Ar atmosphere, which would provide great opportunities to use these potentially attractive molecules under various reaction conditions. Indeed, two-directional annulations of dibromoisobenzofuran, a formal equivalent to didehydroisobenzofuran, were realized by dual [4 + 2] cycloadditions of arynes and isobenzofurans. In this successive process, aryne could be efficiently generated by site-selective halogen—lithium exchange of dihaloarenes under thermodynamic or kinetic control. Importantly, two kind of kinetically controlled halogen—lithium exchange, newly developed, i.e., (1) utilization of the more electropositive halogen or (2) tuning the reactivity of halogen by the adjacent halogen, allowed rapid construction of highly condensed threefold cycloadducts, which were amenable to selective transformation en route to high-ordered polyacene derivatives.

Moreover, as a potentially versatile method for the rapid, selective assembly of functionalized polycycles, we also exploited the successive cycloaddition using synthetic equivalent of bis-isobenzofuran. In this reaction, isobenzofuran is ring selectively generated from diepoxyantracene in a kinetically controlled manner, and its functionalization and/or introduction of the fused ring by trapping with dienophile can selectively produce an isomeric pair of products.

Both reaction integrations with reactive intermediate, selectively generated from the precursor by careful consideration of the reactivity under kinetic or thermodynamic conditions, showed high synthetic utilities for construction of polycyclic aromatic compounds. We hope our approach described in this chapter would open a way to higher condensed aromatic compounds and provide a valuable opportunity to develop functional materials.

REFERENCES

1. (a) Friedrichsen, W. Benzo[c]furan. *Adv. Heterocyl. Chem.* **1980,** *26,* 135–241.
   (b) Friedrichsen, W. Recent Advances in the Chemistry of Benzo[c]furans and Related Compounds. *Adv. Heterocyl. Chem.* **1999,** *73,* 1–96.
2. (a) Chan, S.-H.; Yick, C.-Y.; Wong, H. N. C. 5,6-Bis(trimethylsilyl)benzo[*c*]furan: an Isolable Versatile Building Block for Linear Polycyclic Aromatic Compounds. *Tetrahedron* **2002,** *58,* 9413–9422.
   (b) Rainbolt, J. E.; Miller, G. P. 4,7-Diphenylisobenzofuran: a Useful Intermediate for the Construction of Phenyl-substituted Acenes. *J. Org. Chem.* **2007,** *72,* 3020–3030.
3. (a) Warrener, R. N. Isolation of Isobenzofuran, a Stable but Highly Reactive Molecule. *J. Am. Chem. Soc.* **1971,** *93,* 2346–2348.
   (b) Sygula, A.; Sygula, R.; Rabideau, P. W. Isocorannulenofuran: a Versatile Building Block for the Synthesis of Large Buckybowls. *Org. Lett.* **2006,** *8,* 5909–5911.
   (c) Pei, B. J.; Chan, W. H.; Lee, A. W. M. Anthracene Capped Isobenzofuran: A Synthon for the Preparations of Iptycenes and Iptycene Quinones. *Org. Lett.* **2011,** *13,* 1774–1777.
4. Warrener, R. N.; Shang, M.; Butler, D. N. A New Stabilised Form of Isobenzofuran, Rack-mounted on an Alicyclophane. *Chem. Commun.* **2001,** 1550–1551.
5. Haneda, H.; Eda, S.; Aratani, M.; Hamura, T. Dibromoisobenzofuran as a Formal Equivalent of Didehydroisobenzofuran: Reactive Platform for Expeditious Assembly of Polycycles. *Org. Lett.* **2014,** *16,* 286–289.

6. (a) Hamura, T.; Nakayama, R. A One-pot Preparation of 1,3-diarylisobenzofuran. *Chem. Lett.* **2013,** *42,* 1013–1015.
(b) Asahina, K.; Matsuoka, S.; Nakayama, R.; Hamura, T. An efficient synthetic route to 1,3-bis(arylethynyl)isobenzofuran by using alkoxybenzocyclobutenone as a reactive platform. *Org. Biomol. Chem.* **2014,** *12,* 9773–9776.
(c) Kudo, R.; Kitamura, K.; Hamura, T. 1,3-Dialkyny- and 1,3-dialkenyl-isobenzofurans: New Π-extended Congeners Prepared by Double Nucleophilic Addition of Alkynyllithiums to O-phthalaldehyde. *Chem. Lett.* **2017,** *46,* 25–28.
(d) Tozawa, H.; Kitamura, K.; Hamura, T. Water-soluble 1,3-Diarylisobenzoheterols: Syntheses and Characterization. *Chem. Lett.* **2017,** *46,* 703–706.
7. (a) Clayden, J. *Organolithiums: Selectivity for Synthesis* In: *Tetrahedron Organic Chemistry Series,* Vol. 23; Pergamon: Oxford, 2002; pp 111–147.
(b) Beak, P.; Allen, D. J. Experimental Evaluation of Transition Structure Geometry for an Aryl Bromide-alkyllithium Exchange Reaction: New Information Relevant to the Reaction Mechanism. *J. Am. Chem. Soc.* **1992,** *114,* 3420–3425.
(c) Nájera, C.; Sansano, J. M.; Yus, M. Recent Synthetic Uses of Functionalised Aromatic and Heteroaromatic Organolithium Reagents Prepared by Non-deprotonating Methods. *Tetrahedron* **2003,** *59,* 9255–9303.
(d) Dabrowski, M.; Kubicka, J.; Lulinski, S.; Serwatowski, J. Halogen–lithium Exchange between Substituted Dihalobenzenes and Butyllithium: Application to the Regioselective Synthesis of Functionalized Bromobenzaldehydes. *Tetrahedron* **2005,** *61,* 6590–6595.
8. (a) Hoffmann, R. W. *Dehydrobenzene and Cycloalkynes;* Academic Press: NewYork, 1967.
(b) Kessar, S. V. In *Comprehensive Organic Synthesis,* Vol. 4, Trost, B. M., Ed.; Pergamon: Oxford, U.K, 1991; pp 483–515.
(c) Pellissier, H.; Santelli, M. The Use of Arynes in Organic Synthesis. *Tetrahedron* **2003,** *59,* 701–730.
(d) Wenk, H. H.; Winkler, M.; Sander, W. One Century of Aryne Chemistry. *Angew. Chem. Int. Ed.* **2003,** *42,* 502–528.
(e) Yoshida, H.; Takaki, K. Aryne Insertion Reactions into Carbon–Carbon Σ-Bonds. *Synlett* **2012,** *23,* 1725–1732.
(f) Bhunia, A.; Yetra, S.; Biju, R. A. T. Recent Advances in Transition-metal-free Carbon-carbon and Carbon-heteroatom Bond-forming Reactions Using Arynes. *Chem. Soc. Rev.* **2012,** *41,* 3140–3152.
(g) Dubrovskiy, A. V.; Markina, N. A.; Larock, R. C. Use of Benzynes for the Synthesis of Heterocycles. *Org. Biomol. Chem.* **2013,** *11,* 191–218.
9. Hamura, T.; Chuda, Y.; Nakatsuji, Y.; Suzuki, K. Catalytic Generation of Arynes and Trapping by Nucleo-iodination. *Angew. Chem. Int. Ed.* **2012,** *51,* 3368–3372.
10. (a) Hart, H.; Lai, C.-Y.; Nwokogu, G.; Shamouilian, S.; Teuerstein, A.; Zlotogorski, C. Bis Annelation of Arenes with Bisaryne Equivalents. *J. Am. Chem. Soc.* **1980,** *102,* 6649–6651.
(b) Hart, H.; Ok, D. Synthesis of 1,5-diamino-1,5-dihydrobenzo[1,2-d:4,5-d']bis-triazole (DABT) and its Use as a 1,4-benzadiyne Equivalent. *J. Org. Chem.* **1986,** *51,* 979–986.
11. (a) Bendikov, M.; Wudl, F.; Perepichka, D. F. Tetrathiafulvalenes, Oligoacenenes, and Their Buckminsterfullerene Derivatives: the Brick and Mortar of Organic Electronics. *Chem. Rev.* **2004,** *104,* 4891–4946.
(b) Anthony, J. E. Functionalized Acenes and Heteroacenes for Organic Electronics. *Chem. Rev.* **2006,** *106,* 5028–5048.
(c) Anthony, J. E. The Larger Acenes: Versatile Organic Semiconductors. Angew. Chem. 2008, 120, 460 *Angew. Chem. Int. Ed.* **2008,** *47,* 452–483.

(d) Bettinger, H. F. Electronic Structure of Higher Acenes and Polyacene: The Perspective Developed by Theoretical Analyses. *Pure Appl. Chem.* **2010,** *82,* 905–915.

12. Nagaki, A.; Ichinari, D.; Yoshida, J. Three-component Coupling Based on Flash Chemistry. Carbolithiation of Benzyne with Functionalized Aryllithiums Followed by Reactions with Electrophiles. *J. Am. Chem. Soc.* **2014,** *136,* 12245–12248.
13. (a) Gilman, H.; Gorsich, R. D. Some Reactions of O-Halobromobenzenes with *N*-Butyllithium. *J. Am. Chem. Soc.* **1956,** *78,* 2217–2222.
(b) Chen, L. S.; Chen, G. J.; Tamborski, C. The Synthesis and Reactions of *Ortho* Bromophenyllithium. *J. Organomet. Chem.* **1980,** *193,* 283–292.
(c) Caster, K. C.; Keck, C. G.; Walls, R. D. Synthesis of Benzonorbornadienes: Regioselective Benzyne Formation. *J. Org. Chem.* **2001,** *66,* 2932–2936.
(d) Coe, J. W.; Wirtz, M. C.; Bashore, C. G.; Candler, J. Formation of 3-Halobenzyne: Solvent Effects and Cycloaddition Adducts. *Org. Lett.* **2004,** *6,* 1589–1592.
(e) Eda, S.; Aratani, M.; Hamura, T. Selective Halogen–lithium Exchange of 1,2-Dihaloarenes for Successive [2+4] Cycloadditions of Arynes and Isobenzofurans. *Molecules* **2015,** *20,* 19449–19462.
14. Hamura, T.; Arisawa, T.; Matsumoto, T.; Suzuki, K. Two-directional Annelation: Dual Benzyne Cycloadditions Starting from Bis(sulfonyloxy)diiodobenzene. *Angew. Chem. Int. Ed.* **2006,** *45,* 6842–6844.
15. (a) Luo, J.; Hart, H. Bisannelation with a Benzo[1,2-c:4,5-c']difuran Equivalent: a New Route to Linear Acene Derivatives. *J. Org. Chem.* **1988,** *53,* 1341–1343.
(b) Qu, H.; Cui, W.; Li, J.; Shao, J.; Chi, C. 6,13-Dibromopentacene [2,3:9,10]-Bis(dicarboximide): A Versatile Building Block for Stable Pentacene Derivatives. *Org. Lett.* **2011,** *13,* 924–927.
16. (a) Anthony, J. E.; Brooks, J. S.; Eaton, D. L.; Parkin, S. R. Functionalized Pentacene: Improved Electronic Properties from Control of Solid-state Order. *J. Am. Chem. Soc.* **2001,** *123,* 9482–9483.
(b) Payne, M. M.; Delcamp, J. H.; Parkin, S. R.; Anthony, J. E. Robust, Soluble Pentacene Ethers. *Org. Lett.* **2004,** *6,* 1609–1612.
(c) Swartz, C. R.; Parkin, S. R.; Bullock, J. E.; Anthony, J. E.; Mayer, A. C.; Malliaras, G. G. Synthesis and Characterization of Electron-deficient Pentacenes. *Org. Lett.* **2005,** *7,* 3163–3166.
(d) Miao, Q.; Chi, X.; Xiao, S.; Zeis, R.; Lefenfeld, M.; Siegrist, T.; Steigerwald, M. L.; Nuckolls, C. Organization of Acenes with a Cruciform Assembly Motif. *J. Am. Chem. Soc.* **2006,** *128,* 1340–1345.
(e) Eda, S.; Eguchi, F.; Haneda, H.; Hamura, T. A New Synthetic Route to Substituted Tetracenes and Pentacenes via Stereoselective [4+2] Cycloadditions of 1,4-dihydro-1,4-epoxynaphthalene and Isobenzofuran. *Chem. Commun.* **2015,** *51,* 5963–5966.
17. Akita, R.; Kawanishi, K.; Hamura, T. Ring Selective Generation of Isobenzofuran for Divergent Access to Polycyclic Aromatic Compounds. *Org. Lett.* **2015,** *17,* 3094–3097.

CHAPTER 4

Control of Inversion Kinetics of Bowl-Shaped Aromatic Compounds

Shuhei Higashibayashi
Faculty of Pharmacy, Keio University, Tokyo, Japan

Contents

1. INTRODUCTION

Buckybowls (geodesic polyarenes, π-bowls) are polycyclic aromatic compounds possessing bowl-shaped structures *(1–12)*. Although buckybowls are known as the partial structures of fullerenes or the cap structures of carbon nanotubes, the first synthesis of a C_{5v} symmetric buckybowl, corannulene (**1**) (Fig. 4.1), was reported in 1966 *(13)*, which was much older than the discovery of C_{60} in 1985 *(14)* or carbon nanotubes in 1991 *(15)*. However, the first synthesis of **1** took long 17-steps and bowl-shaped aromatic compound did not attract much attention until the discovery of C_{60}. The discovery of C_{60} stimulated many organic chemists to study the syntheses and properties of buckybowls. In the early stage of buckybowl chemistry, synthetic organic chemists focused on the development of synthetic methods to construct the strained conjugated bowl structures. As a result, efficient syntheses of many buckybowls in gas- or solution-phase have been developed, which supplied enough amounts of compounds and enabled the investigation of properties. After the achievements of syntheses of pristine buckybowls such as **1**, sumanene (**2**), and diindenochrysene (**3**), the syntheses of functionalized buckybowls also became the important topics for the elucidation of the

Kinetic Control in Synthesis and Self-Assembly
ISBN 978-0-12-812126-9
https://doi.org/10.1016/B978-0-12-812126-9.00004-3

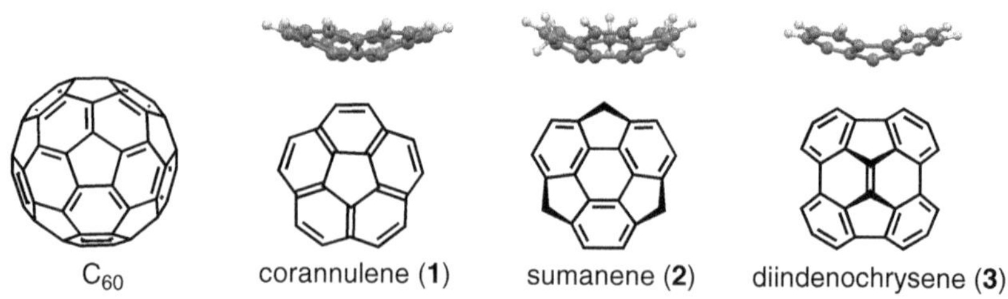

Figure 4.1 Representative pristine buckybowls.

fundamental properties of curved π-electron systems as well as the exploration of science and application. As a result, many interesting properties of buckybowls have elucidated to date. One of the unique properties of buckybowls is the bowl inversion (Fig. 4.2). The bowl structures of buckybowls are flexible enough to undergo inversion. Theoretical calculation of bowl inversion of corannulene (**1**) was first reported in 1967 *(16)* and experimental bowl inversion barrier of a corannulene derivative was determined in 1992 *(17)*. Since then, the molecular dynamics of many buckybowls have been extensively studied by means of theoretical calculations and experimental measurements. The energy barrier was found to depend on the buckybowl skeleton such as corannulene (**1**), sumanene (**2**) *(18)*, and diindenochrysene (**3**) *(19)* as well as the size of bowl structures from small to big buckybowls. Substituents were also found to affect the bowl depth and bowl inversion barrier. Extensive studies of the relationship between bowl depth and bowl inversion energy elucidated the correlation on corannulene *(20)* and sumanene skeletons *(21)*. Not only buckybowls possessing all-carbon frameworks but also heterobuckybowls containing heteroatoms in the bowl skeletons were studied *(12)*. Introduction of heteroatoms into bowl skeletons was found to affect the bowl structure and bowl inversion energy significantly. The effect of redox on bowl structure

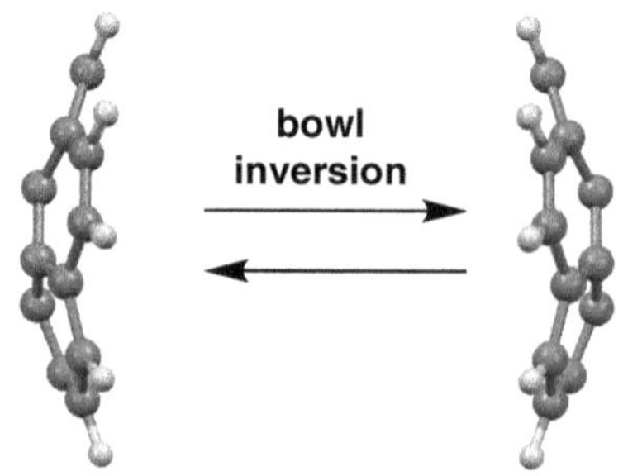

Figure 4.2 Bowl inversion of buckybowls.

and bowl inversion was also investigated by calculation and experiment. In both anionic and cationic states, the inversion barrier was found to change from that in the neutral states. Based on these extensive studies on the bowl inversion barriers of many kinds of buckybowls, it was elucidated that the bowl inversion energy barriers were varied in wide range by the structural modifications, leading to the control of the bowl inversion kinetics. Another unique feature of buckybowls related to bowl inversion is the existence of chirality. Fullerenes and carbon nanotubes are well known to possess chirality because of the three-dimensional geometry of the structures *(22)*. Buckybowls also possess inherent planar chirality because of the bowl structure and we named the chirality to be "bowl chirality" *(23)*. The chiral buckybowls possessing the bowl chirality are classified into three types: (1) the bowl structure itself possesses the chirality such as hemifullerene, (2) the introduction of substituent induces the bowl chirality into achiral bowl structure such as trimethylsumanene, and (3) the introduction of heteroatom in the bowl skeleton induces the bowl chirality into achiral bowl structure such as triazasumanene (Fig. 4.3). In fact, many kinds of buckybowls are chiral, including simple buckybowls such as substituted corannulenes and

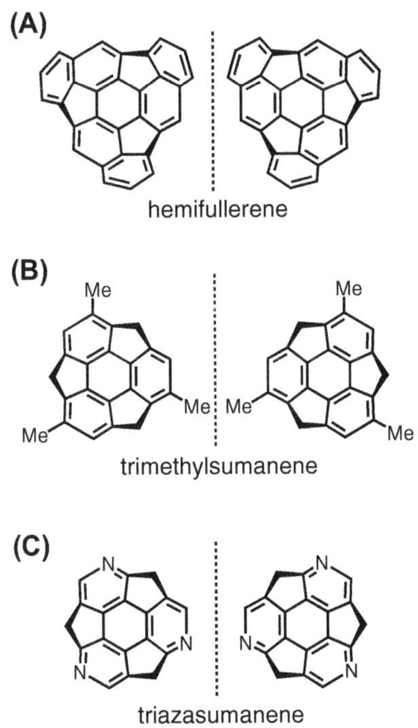

Figure 4.3 Three types of chiral buckybowls.

sumanenes. Importantly in chiral buckybowls, the bowl inversion correspond to the racemization process. Because chiral buckybowls with low bowl inversion barrier undergoes racemization quickly, the enantiomers cannot be separated in such a case. On the other hand, chiral buckybowls with high bowl inversion barriers undergo slow or no racemization. Thus, the enantiomers can be separated. In fact, enantioselective synthesis of chiral buckybowls or resolution of enantiomers by chiral HPLC was achieved in trimethylsumanenes *(21,23,24)* and triazasumanene *(25)*. Control of bowl inversion kinetics also leads to the control of bowl chirality. In this chapter, I particularly focused on the bowl inversion property of buckybowls and reviewed the inversion barriers of pristine buckybowls in Section 2, substituted buckybowls in Section 3, and heterobuckybowls in Section 4. In Section 3 and 4, the stability of bowl chirality of chiral buckybowls is also reviewed.

2. PRISTINE BUCKYBOWLS

The first synthesis of C_{5v} symmetric corannulene (**1**) was achieved in 17-steps by Barth and Lawton in 1966 *(13)*. However, the properties were not extensively studied owing to the limited supply of the material. In 1991, Scott et al. developed short-step synthesis of **1** by a flash vacuum pyrolysis (FVP) method *(26)*. Later, Siegel et al. *(27)* and Rabideau et al. *(28)* developed the solution-phase short-step syntheses of **1**, respectively. Sygula and Rabideau also reported another practical synthesis *(29)* and Siegel et al. established the kilogram-scale production of **1** *(30)*. Synthetic methods of corannulene derivatives such as Pd-catalyzed synthesis were also developed *(31)*. These developed synthetic methods of **1** and the derivatives enabled the extensive studies on the properties including bowl inversion and the application to material science. Determination of bowl inversion energy barriers have been experimentally studied by mainly NMR techniques, either 1D VT (variable temperature) NMR or 2D EXSY (exchange spectroscopy) NMR experiment. In principle, the energy barriers of buckybowls with diastereotopic protons can be determined by these NMR techniques, whereas buckybowls without diastereotopic protons such as **1** cannot be. The first experimental bowl inversion energy barrier of a buckybowl, corannulenyldimethylcarbinol (**4**) (Fig. 4.4), was first reported by Scott et al. in 1992 *(17)*. ^{1}H NMR spectrum of **4** showed a single peak for the diastereotopic methyl groups at room temperature but two resolved singlets at −90°C. The coalescence temperature for the two

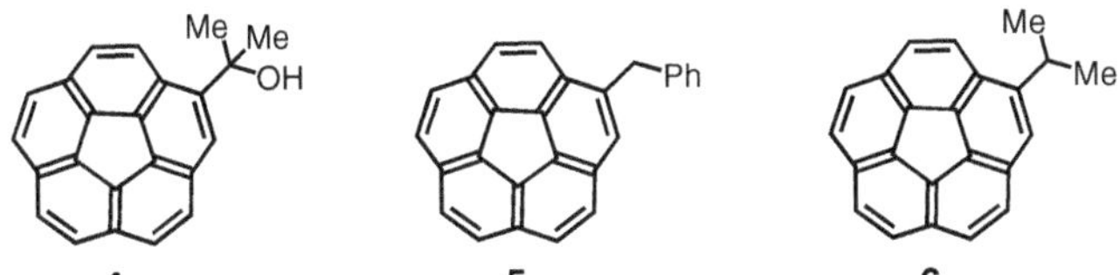

4 10.2 kcal/mol (exp.) **5** 11.2 kcal/mol (exp.) **6** 11.3 kcal/mol (exp.)

corannulene (**1**)
11.5 kcal/mol (estimated)
11.0 kcal/mol (calcd. at MP2/cc-pVDZ//B3LYP/cc-pVDZ)
10.7 kcal/mol [calcd. at B3LYP/6-311+G(d,p)]

Figure 4.4 Bowl inversion barriers of corannulene and its derivatives.

signals at $-64°C$ afforded the barrier to be $\Delta G^{\ddagger} = 10.2$ kcal/mol. This energy barrier correspond to 200,000 times inversion per second at room temperature. Rabideau et al. also determined the barriers of benzylcorannulene (**5**) and isopropylcorannulene (**6**) to be 11.2 and 11.3 kcal/mol *(32)* (Fig. 4.4). Siegel et al. estimated the barrier of corannulene (**1**) itself to be 11.5 kcal/mol *(20)* (Fig. 4.4), judging from the barriers of corannulene derivatives with bromomethyl groups (see Section 3), while the barrier of corannulene (**1**) itself cannot be obtained by NMR technique. Structures and dynamics of buckybowls also have been extensively studied by theoretical calculations. The bowl structures and bowl inversion energy barriers obtained by quantum calculations at high level of theory were well-consistent to those obtained by X-ray analyses of single crystals or NMR experiments. In addition, reliable quantum calculations at high level of theory afforded us to know the very low or high barriers or the transition state structures of inversion of many buckybowls that were not observed experimentally. Theoretical calculation revealed that the bowl inversion of **1** follows a double-well potential through a planar transition state (Fig. 4.5). Quantum calculations afforded the bowl inversion barrier of **1** to be 11.0 kcal/mol (MP2/cc-pVDZ//B3LYP/cc-pVDZ) *(20)* or 10.7 kcal/mol (B3LYP/6-311 + G[d,p])*(33)* (Fig. 4.4), which is consistent to experimental values of corannulene derivatives. Bowl inversion energy barriers of anion species of corannulene and its derivatives were also studied by theoretical calculations. Sygula et al. reported that the inversion barrier of corannulene

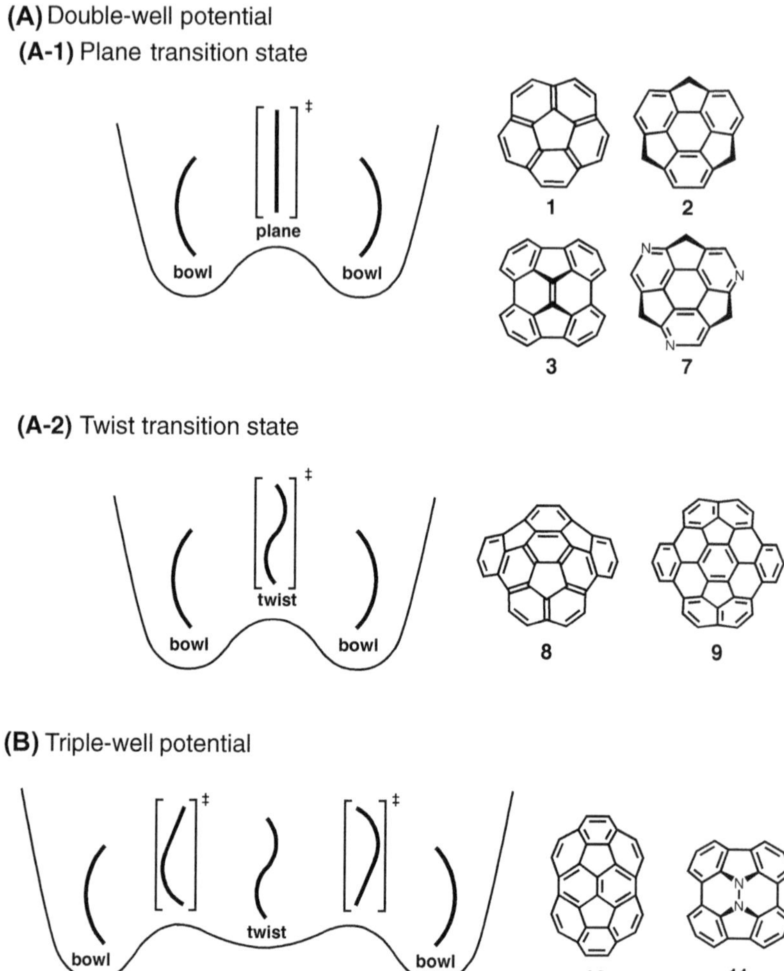

Figure 4.5 Types of bowl inversion dynamics.

decreases from 14.2 kcal/mol (neutral) to 7.9–9.2 kcal/mol for dianion and 3.2 kcal/mol for tetraanion (MP2/3-21G//HF-3-21G) *(34)*. Fontanesi, Marcaccio, Scott et al. also reported that the barrier of corannulene decreases from 6.5 kcal/mol (neutral) to 4.6, 2.8, 1.8, 1.5 kcal/mol for monoanion, dianion, trianion, tetraanion (UB3PW91/Midi) *(35)*. The inversion barrier of a corannulene derivative, phenalenyl-fused corannulene, was also reported to decrease stepwise with increasing the negative charge by Morita et al. *(36)*. In every calculation, the bowl structure is flattened with increasing the negative charge.

Based on reported theoretical studies, the bowl inversion process of buckybowls is classified into two types depending on the shape of the potential: (A) double-well potential and (B) triple-well potential (Fig. 4.5). Depending on the shape of the transition state, type A is further divided into type A-1 with a planar transition state and type A-2 with a twist (S-shaped) transition state. Most of buckybowls including corannulene (**1**), sumanene (**2**), diindenochrysene (**3**), and triazasumanene (**7**) *(25)* belong to type A-1, in which the transition state is planar. Meanwhile, two buckybowls **8** *(37)* and **9** *(38)* reported by Wu et al. belong to Type A-2 with a twist transition state. In type B, a buckybowl has two conformers: a bowl conformer with a global minimum energy and a twist conformer with a local minimum energy, and they are inverted through a nonplanar transition state. Two buckybowls, semibuckminsterfullerene (**10**) *(39,40)* and hydrazinobuckybowl (**11**) *(41)*, have been reported to be type B.

Synthesis of C_{3v} symmetric sumanene (**2**) was achieved by Sakurai, Hirao et al. in 2003 *(42)*. **2** was synthesized in 3–4 steps from norbornadiene. The experimental bowl inversion energy barrier was reported in 2008 by Hirao et al. *(18)*. **2** possesses three sets of two diastereotopic *exo* and *endo* protons in the skeleton (Fig. 4.6) and ^{1}H NMR spectrum showed two resolved doublet signals at room temperature. Thereby, determination of the coalescence temperature would afford the barrier energy. However, it was not determined by VT NMR because the coalescence was not observed even at 140°C owing to the high bowl inversion barrier of **2**. Instead, 2D EXSY NMR was applied for the determination. Trideuteriosumanene (**2**-d_3) was used for the measurement to cancel the nuclear Overhauser effect between the *gem*-protons (Fig. 4.6). From the exchange rate between the *exo* and *endo* protons through bowl inversion, the barrier was determined to be 20.3 kcal/mol in CD_2Cl_2 at 30°C. The barrier of sumanene (**2**) is approximately twice of that of corannulene (**1**). Bowl inversion of **2** was also studied by theoretical calculations, and we investigated the dependence of the inversion barrier on DFT functional (Fig. 4.6, Table 1). Among the tested DFT functionals, PBE0, ωB97XD, and M06-2X level of theory afforded a close value to the experimental value.

The experimental inversion barriers of monoanion and dianion of **2** were also determined by Hirao et al. *(18)*. In the case, *exo*-trideuteriosumanene was stereoselectively prepared from trianion of **2**, and monoanion and dianion were generated by deprotonation (Fig. 4.7). Measurement of the conversion rate from the *endo*-proton to the *exo*-proton in the VT NMR spectra allowed to determine the barriers to be 21.8 kcal/mol for monoanion

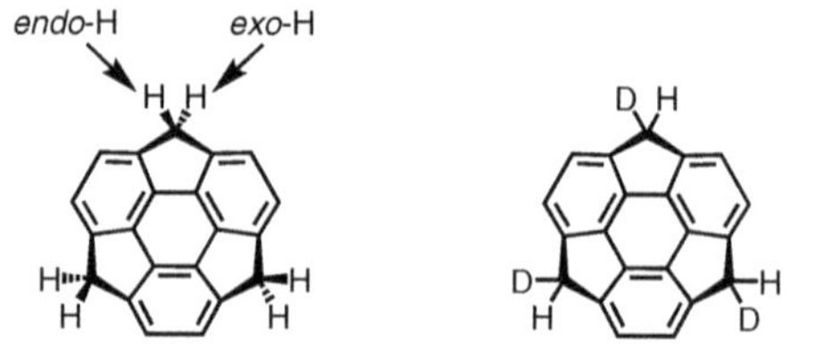

sumanene (**2**) trideuteriosumanene (**2**-d_3)
20.3 kcal/mol (exp.)

Table 1. Calculated bowl inversion barrier of **2** (kcal/mol)

	Basis set	
DFT funcional	6-31G(d)	6-311+G(d,p)
B3LYP	16.3	18.3
PBE0	18.0	19.4
ωB97XD	18.1	19.7
LC-ωPBE	17.9	-
CAM-B3LYP	16.9	-
M06	17.7	-
M06-2X	18.3	19.6

Figure 4.6 Experimental and calculated bowl inversion barriers of sumanene (**2**).

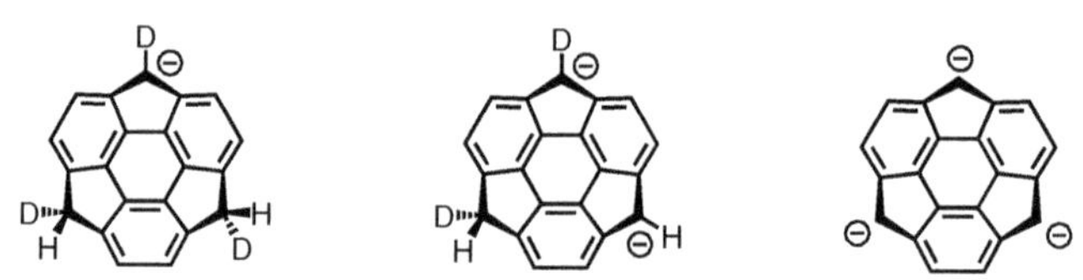

monoanion
21.8 kcal/mol (exp.)
21.6 kcal/mol (calcd.)

dianion
21.5 kcal/mol (exp.)
22.5 kcal/mol (calcd.)

trianion
20.9 kcal/mol (calcd.)

Figure 4.7 Experimental and calculated [B3LYP/6-311 + G(2d,p)//B3LYP/6-31G(d,p)] bowl inversion barriers of monoanion, dianion, and trianion of sumanene (**2**).

(10°C) and 21.5 kcal/mol for dianion (0°C) in THF-d_8, which were higher than 20.3 kcal/mol of neutral sumanene. DFT calculation (B3LYP/6-311 + G[2d,p]//B3LYP/6-31G[d,p]) was also conducted, indicating that the barriers are 19.0 kcal/mol for neutral sumanene, 21.6 kcal/mol for monoanion, 22.5 kcal/mol for dianion, and 20.9 kcal/mol for trianion. Although the higher energies of anions than that of neutral sumanene are consistent between calculation and experiment, the order between monoanion and diaion were inconsistent. This discrepancy may come from the effect of counteraction or solvent.

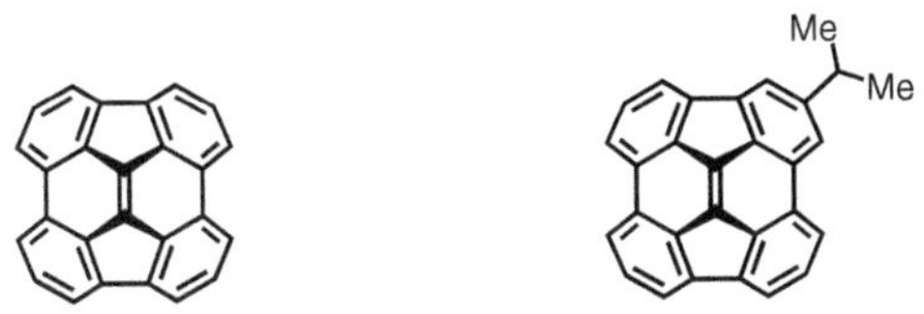

Figure 4.8 Calculated bowl inversion barrier of diindenochrysene (**3**).

First synthesis of C_{2v} symmetric diindenochrysene (**3**) was reported by Zimmerman et al. in 1995 *(43,44)*. However, the dehydrogenation method gave **3** only in very low cyclization yield (0.6%). In 1996, Scott et al. achieved the efficient synthesis of **3** in 37% by FVP *(45,46)*. In 2010, Wu et al. reported a solution-phase synthesis of **3** using Pd-catalyzed cyclization *(19)*. Wu et al. tried to determine the inversion barrier of **3** by the measurement of coalescence of diastereotopic methyl groups of isopropyldiindenochrysene in 1D VT NMR. Even at −80°C, however, the resolution of the signals was not observed, indicating that the inversion barrier is lower than 10 kcal/mol. Scott and Rabinovitz et al. reported the calculated inversion barrier of **3** to be 6.8 kcal/mol (*m*PW1PW/6-311G [d,p])*(47)* and Wu et al. reported to be 6.9 kcal/mol (M06-2X/cc-pVDZ) *(19)*, which are lower than that of corannulene (**1**) (Fig. 4.8). Scott and Rabinovitz et al. also calculated the inversion barriers of anions (*m*PW1PW/6-311G[d,p]) *(47)*. The barrier (11.4 kcal/mol) of dianion is much higher than that of neutral diindenochrysene, whereas the barrier (6.7 kcal/mol) of tetraanion is slightly lower.

Bowl inversion of bigger buckybowls with extended π-conjugated bowl structures than corannulene (**1**), sumanene (**2**), and diindenochrysene (**3**) were also studied by experimental measurements and calculations. However, only few experimental data were reported because the inversion barriers of big buckybowls are too high to be determined by NMR spectral methods. Therefore, they were mainly studied by theoretical calculations. Hirao et al. reported the synthesis and the experimental inversion barrier of mononaphthosumanene (**13**) (Fig. 4.9) *(48,49)*. For the determination of the barrier, *exo*-dideuterionaphthosumanene was prepared by selective deuteration of the dianion. Measurement of the conversion rate of *endo*-proton to *exo*-proton in 1D VT NMR afforded the barrier of **13** to be 32.2 kcal/mol in mesitylene-d_{12} (140 °C), which is much higher than

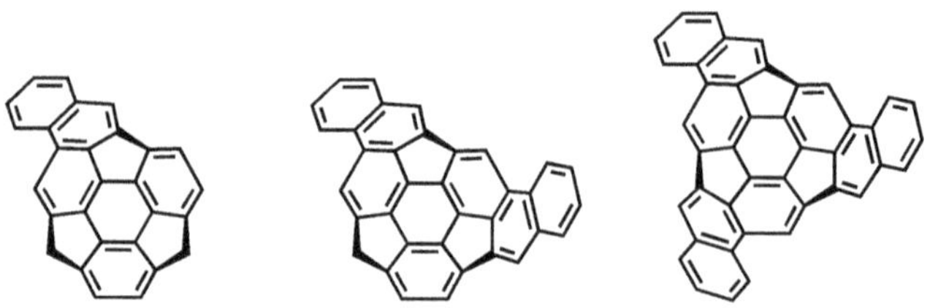

Figure 4.9 Experimental and calculated [B3LYP/6-311 + G(2d,p)//B3LYP/6-31G(d,p)] bowl inversion barriers of naphthosumanenes.

20.3 kcal/mol of sumanene (**2**). DFT calculation (B3LYP/6-311 + G [d,p]//B3LYP/6-31G[d,p]) indicated that the inversion barriers of dinaphthosumanene (**14**) and trinaphthosumanene (**15**) are 46.4 and 63.8 kcal/mol, while those of sumanene (**2**) and mononaphthosumanene (**13**) are 19.0 and 31.4 kcal/mol in the same level of theory.

Müllen et al. reported the synthesis and the inversion barrier of the buckybowl **16** (Fig. 4.10) *(50)*. By measurement of the coalescence temperature of diastereotopic protons of the substituents, the inversion barrier was determined to be 16.7 kcal/mol in $C_2D_2Cl_4$ (60°C). Siegel et al. tried to determine the inversion barrier of an indenocorannulene derivative **17** by VT NMR (Fig. 4.10) *(51)*. However, the coalescence of the diastereotopic methyl groups was not observed even at 172°C, which indicated the barrier of more than 24 kcal/mol. *Ab initio* calculation predicted the inversion barrier of indenocorannulene (**18**) to be 29.8 kcal/mol (MP2/cc-pVDZ//B3LYP/cc-pVDZ). In the same level of theory, the inversion barriers of corannulene (**1**), cyclopentacorannulene (**19**), and dihydrocyclopentacorannulene (**20**) were reported to be 11.0, 30.9, and 28.9 kcal/mol (Fig. 4.10). Sygula and Rabideau determined the inversion barrier of dihydrocyclopentacorannulene (**20**) to be 27.7 kcal/mol by the measurement of conversion rate of the *endo*-protons of dideuteriocyclopentacorannulene, prepared by the face-selective deuterogenation of **19**, to the *exo*-protons (Fig. 4.10) *(32)*. They also reported that the experimental inversion barrier of dibenzodihydrocyclopentacorannulene (**21**) was 23.1 kcal/mol, which was determined by the simulation of the coalescence of the diastereotopic protons in VT NMR *(52)*.

Wu et al. synthesized the buckybowls **22**, **8**, and **9** and investigated the bowl inversion (Fig. 4.11) *(37,38)*. They prepared the dideuterio-derivative of **22** and examined the bowl inversion by VT NMR. However, the bowl inversion was not observed even at 170°C. DFT calculation showed that

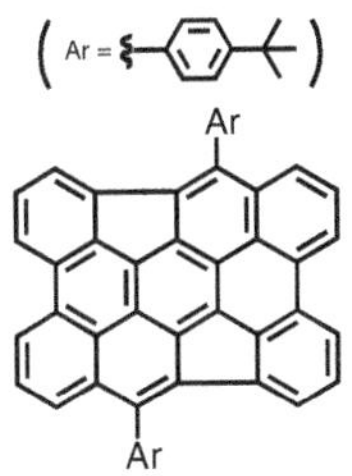

16

16.7 kcal/mol (exp.)

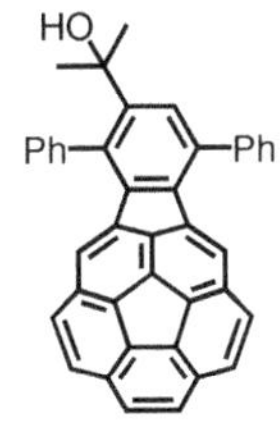

17

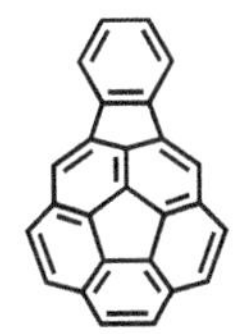

18

29.8 kcal/mol (calcd.)

19

30.9 kcal/mol (calcd.)

20

28.9 kcal/mol (calcd.)
27.7 kcal/mol (exp.)

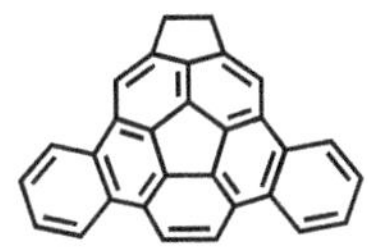

21

23.1 kcal/mol (exp.)

Figure 4.10 Experimental and calculated (MP2/cc-pVDZ//B3LYP/cc-pVDZ) bowl inversion barriers of a buckybowl **16** and corannulene derivatives.

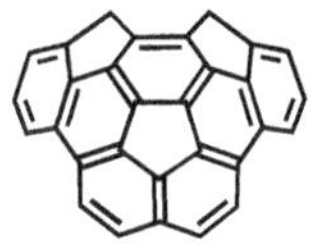

22

56.2 kcal/mol (calcd.)

8

124.3 kcal/mol (calcd.)

9

79.8 kcal/mol (calcd.)

Figure 4.11 Calculated bowl inversion barriers of buckybowls **22**, **8**, and **9** (B3LYP/cc-pVDZ).

the inversion barrier is 56.2 kcal/mol through the planar transition state (B3LYP/cc-pVDZ). They also investigated the bowl inversion of **8** and **9** by DFT calculation, which revealed that the twist (S-shaped) transition states have lower energies than the planar transition states (Fig. 4.5). The calculation indicated that the inversion barriers of **8** and **9** are 124.3 and 79.8 kcal through the twist transition states (B3LYP/cc-pVDZ). Rabideau and Agranat reported the theoretical calculation of the bowl inversion of semibuckminsterfullerene (**10**) and elucidated the bowl inversion goes through triple-well potential with the twist conformer (Fig. 4.5) *(39,40)*. The twist conformer is 46.3 kcal/mol higher than the bowl conformer and the inversion barrier is 47.7 kcal/mol (B3LYP/6-311g[d,p]) (Fig. 4.12).

10

inversion barrier 47.7 kcal/mol (calcd.)
twist conformer 46.3 kcal/mol (calcd.)
bowl conformer 0.0 kcal/mol (calcd.)

Figure 4.12 Calculated bowl inversion barrier and conformer energies of semi-buckminsterfullerene (**10**) [B3LYP/6-311g(d,p)].

3. SUBSTITUTED BUCKYBOWLS

Siegel et al. investigated the bowl inversion barriers and the bowl depths of many substituted corannulene derivatives by means of experiments and theoretical calculations *(20)*. Corannulene derivatives with acyclic substituents tend to adopt a shallower bowl depth and a lower bowl inversion barrier (class I) (Fig. 4.13). On the other hand, corannulene derivatives with tethered substituents tend to adopt a deeper bowl depth and a higher bowl inversion barrier (class II) (Fig. 4.14). From the experimental and calculated data, they proposed a correlation between the bowl depth and the inversion barrier of corannulenes in the form of a double-well potential equation $E = ax^4 - bx^2$, where E is the energy, x is the reaction coordinate (bowl depth), and a and b are coefficients. The bowl inversion energy ΔE is given by the equation $\Delta E = E_0 - E_1 = ax_1^4$, where E_0 is the energy for the transition planar structure and E_1 and x_1 are the energy and reaction coordinate, respectively, for a bowl structure with a minimum energy. The experimental and calculated values for corannulenes were well fitted by this model.

Because of the three-dimensional bowl structure, buckybowls can possess chirality. The bowl chirality can be induced by the π-conjugated skeleton itself (as, for example, in hemifullerene), by the introduction of substituents, or by the replacement of skeletal carbons by heteroatoms (as in heterobuckybowls) (Fig. 4.3). Although many buckybowls possess this bowl chirality, one of the challenging tasks in buckybowl chemistry was the control of the bowl chirality, the enantioselective synthesis of chiral buckybowls. The primary problem of enantioselective synthesis of buckybowls is that of introducing homochirality. The secondary problem that occurs with buckybowls with relatively low bowl inversion barriers is that of racemization through bowl inversion. To isolate an enantiomer of a chiral buckybowl, the chiral buckybowl must have an enough high

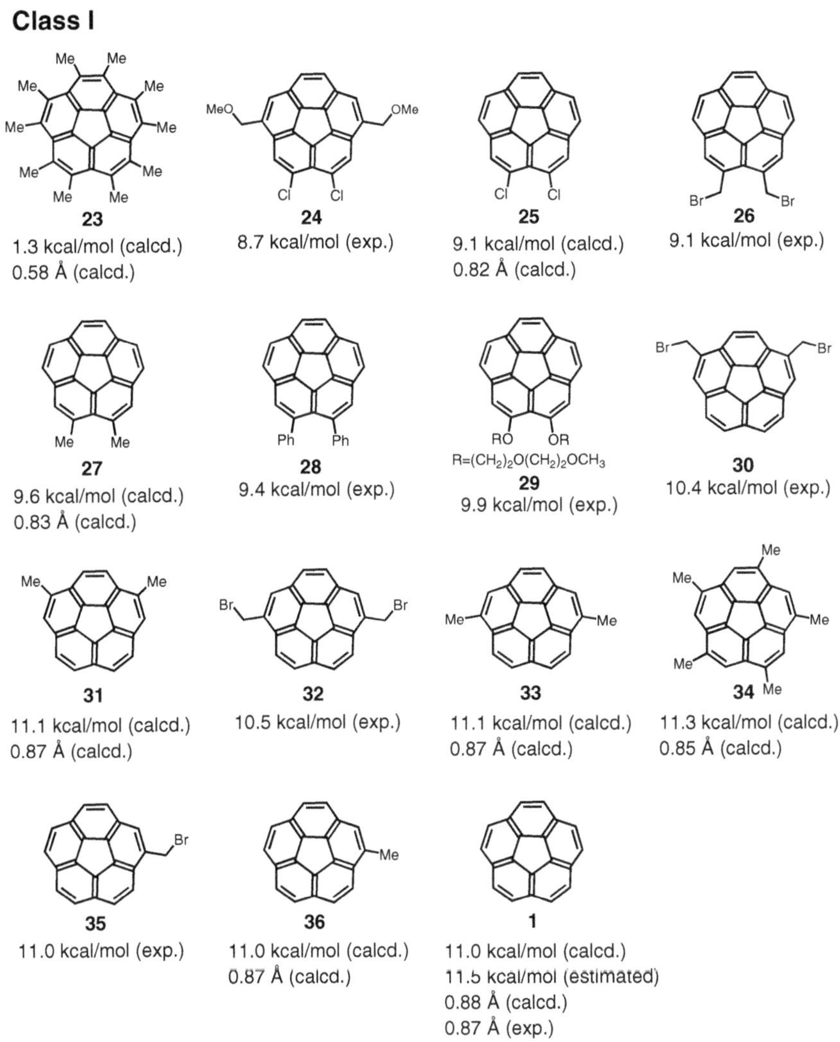

Figure 4.13 Experimental and calculated bowl inversion barriers (MP2/cc-pVDZ//B3LYP/cc-pVDZ) and bowl depths (B3LYP/cc-pVDZ) of corannulene derivatives with acyclic substituents.

inversion barrier. Given the problems, we tried and achieved the first enantioselective synthesis of chiral trimethylsumanene (**50**) *(23)*. Because sumanene (**2**) has a bigger bowl inversion barrier (20.3 kcal/mol) than corannulene (**1**) (11.5 kcal/mol), a chiral sumanene derivative was more suitable to overcome the secondary racemization problem. Concerning the primary problem, we employed a chiral transmission method in which the

Class II

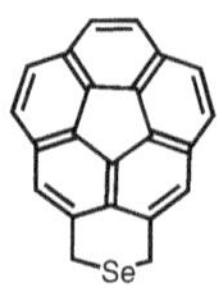

37
13.0 kcal/mol (exp.)

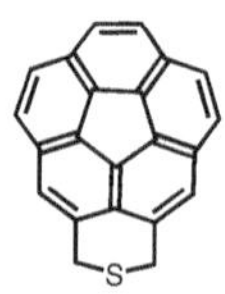

38
13.9 kcal/mol (exp.)
14.5 kcal/mol (calcd.)
0.90 Å (calcd.)

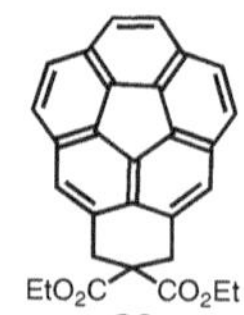

39
15.5 kcal/mol (exp.)

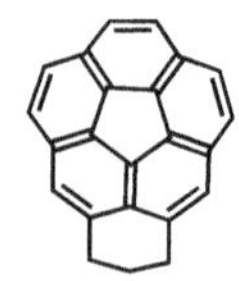

40
15.7 kcal/mol (exp.)
0.93 Å (calcd.)

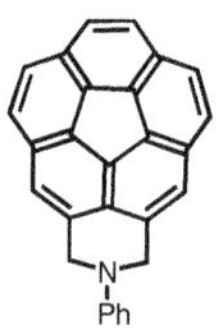

41
16.7 kcal/mol (exp.)

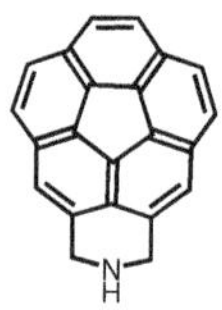

42
16.1 kcal/mol (exp.)
16.4 kcal/mol (calcd.)
0.94 Å (calcd.)

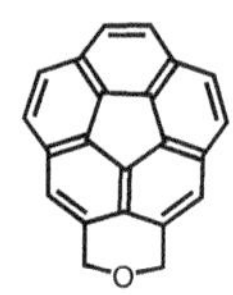

43
17.3 kcal/mol (exp.)
16.9 kcal/mol (calcd.)
0.95 Å (calcd.)

44
27.7 kcal/mol (exp.)
28.9 kcal/mol (calcd.)
1.07 Å (calcd.)

45
30.9 kcal/mol (calcd.)
1.08 Å (calcd.)

Figure 4.14 Experimental and calculated bowl inversion barriers (MP2/cc-pVDZ//B3LYP/cc-pVDZ) and bowl depths (B3LYP/cc-pVDZ) of corannulene derivatives with tethered substituents.

chirality of the sp^3 stereogenic center was transformed into bowl chirality (Fig. 4.15). The C_3 symmetric *syn*-tris(oxonorborneno)benzene **8** was prepared by Pd-catalyzed *syn*-selective cyclotrimerization of enantiomerically pure iodonorbornenone (1*S*,4*S*)-**46**. **47** was converted to trimethylated *syn*-tris(norbornadieno)benzene **48** and the ring-opening/closing metathesis reaction using Grubbs second generation catalyst afforded hexahydrotrimethylsumanene **49**. In the final chirality-transmission process, the bowl inversion barrier of **50** must be taken into account, because bowl inversion of (*C*)-**50** would afford the racemic compound. The bowl inversion energy of **50** was estimated to be ca. 21 kcal/mol by comparison between the experimental value (20.3 kcal/mol) of **2** and the calculated values (18.3 and 19.2 kcal/mol at B3LYP/6-311 + G[d,p]) of **2** and **50**. Because this energy

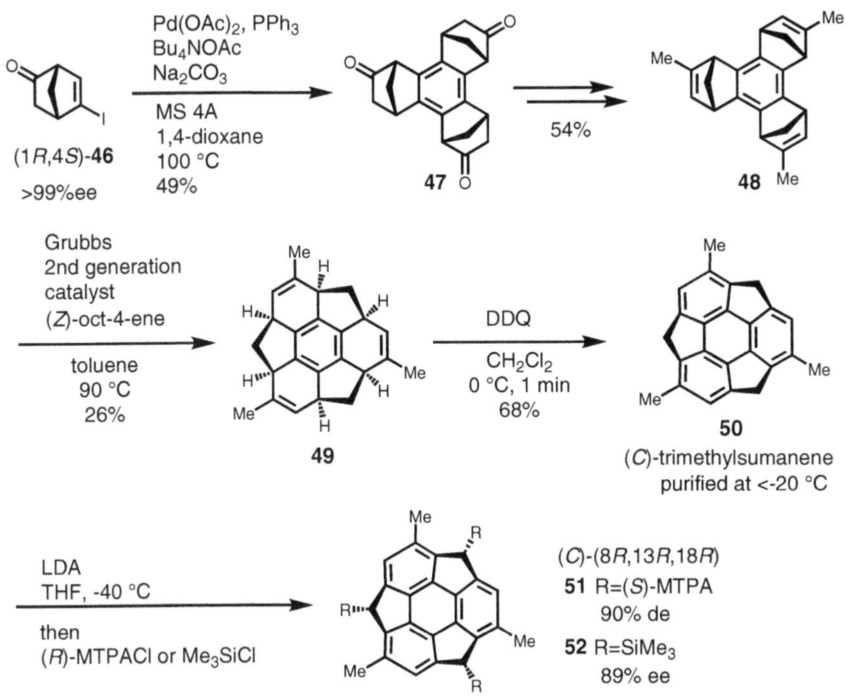

Figure 4.15 Enantioselective synthesis of chiral trimethylsumanene.

correspond to a half-life time for racemization of about 2 h at 0°C, the final aromatization of **49** was carried out with an excess of 2,3-dichloro-5,6-dicyano-1,4-benzoquinone (DDQ) at 0°C for a short time (1 min) to prevent the racemization. The resulting reaction mixture was quickly purified by silica gel chromatography at a low temperature (below −20°C) to afford the chiral buckybowl **50**. The chirality of **50** was confirmed by circular dichroism (CD) spectroscopy. The CD spectrum of **50** observed at −40°C scarcely changed at −40°C (Fig. 4.16A). In contrast, a gradual decrease in the intensity of the CD spectrum was observed at 10°C (Fig. 4.16B). By following the decay of the CD spectra at 10°C, the racemization/bowl inversion barrier of **50** was determined to be 21.6 kcal/mol (283 K), which is 1.3 kcal/mol higher than 20.3 kcal/mol of sumanene (**1**) *(8a)*.

Because it was difficult to determine the enantiomeric excess (ee) of (*C*)-**50** directly due to the racemization derived from the relatively low bowl inversion energy, the ee of (*C*)-**50** was estimated by derivatization to compounds **51** and **52**. Trianion of (*C*)-**50** was generated at low temperature and (*S*)-MTPA (PhC[CF_3] [OMe]CO) groups or $SiMe_3$

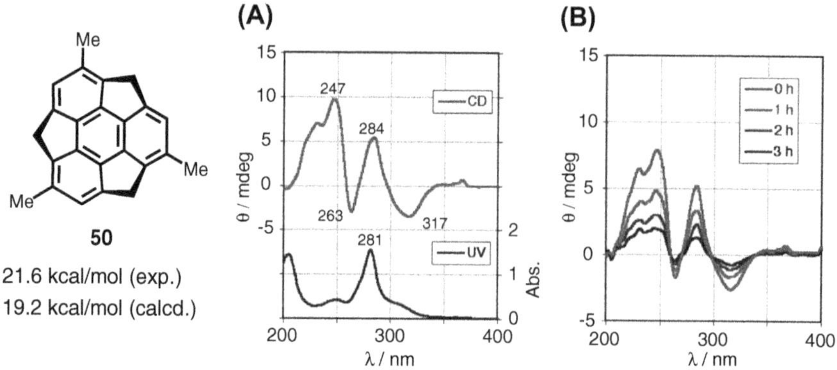

Figure 4.16 Experimental and calculated [B3LYP/6-311 + G(d,p)] bowl inversion barriers and (A)(B) circular dichroism spectra of chiral trimethylsumanene (**50**).

groups were introduced at the benzylic positions with *exo* orientation (Fig. 4.15). The introduction of the substituents induced new stereogenic centers at the benzylic positions. As a result, the bowl inversion cannot cause the racemization but result instead in the diastereomerization between the *exo*-orientation and the *endo*-orientation of substituents (Fig. 4.17). The diastereomerization through the bowl inversion, however, was negligible because the bulky substituents significantly favor the *exo*-orientation than the *endo*-orientation. Thus, the ratio of (*C*)-(8*R*, 13*R*, 18*R*)-**51** or **52** from (*C*)-**50** and (*A*)-(8*S*, 13*S*, 18*S*)-**51** or **52** from (*A*)-**50** bearing *exo*-substituents can be analyzed without racemization or diastereomerization through the bowl inversion. In the case of **51**, the formation of (*S*)-MTPA-(*C*)-(8*R*,13*R*,18*R*)-**51** from (*C*)-**50** and (*S*)-MTPA-(*A*)-(8*S*,13*S*,18*S*)-**51** from the contaminant (*A*)-**50**, which correspond to the relevant diastereomers, permits the determination of the diastereomeric excess (de) of **51** to be 90% by means of ^{1}H NMR analysis. In contrast, (*C*)-(8*R*,13*R*,18*R*)-**52** and (*A*)-(8*S*,13*S*,18*S*)-**52** are enantiomers that can be discriminated by means of chiral HPLC analysis. The ee of **52** was determined to be 89% by chiral HPLC analysis. From the results of **51** and **52**, the ee of (*C*)-**50** was estimated to be >89%. Because the ee of the starting material was >99%, only a relatively small amount of racemization had occurred during the aromatization from **49** to **50** and subsequent purification step of **50.**

Chiral HPLC resolution of **52** is possible because the presence of the sp^3 stereogenic centers in addition to the bowl chirality prevents racemization of the compound by bowl inversion. In contrast, the resolution of a

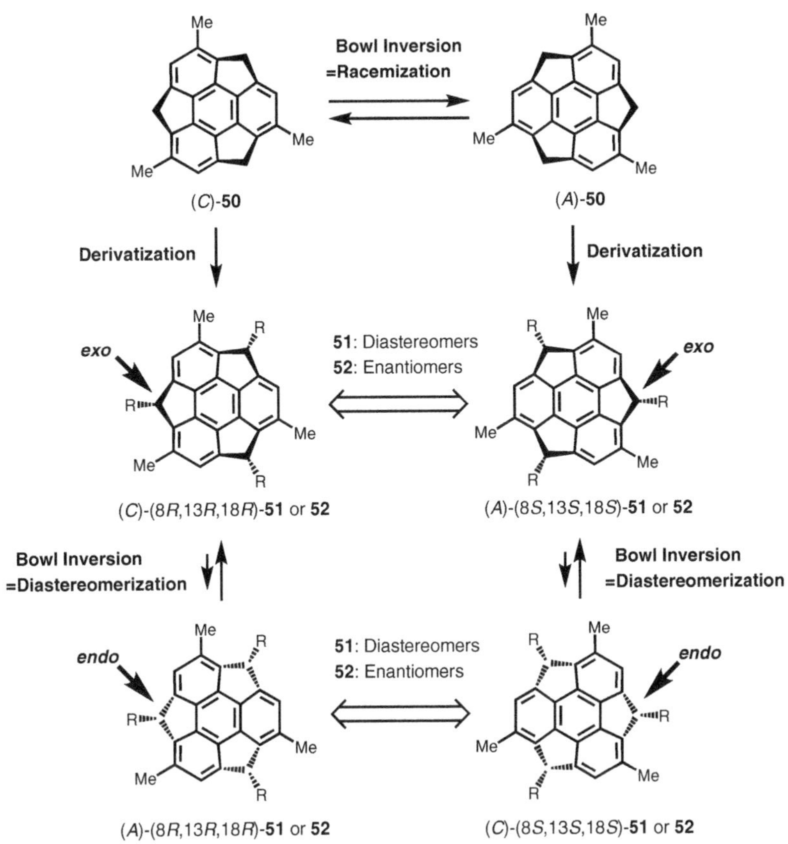

Figure 4.17 Relationship of stereoisomers of **51** and **52** through bowl inversion.

buckybowl with exclusive bowl chirality, such as trimethylsumanene (**50**), is difficult to achieve because of the rapid racemization of the compound. However, if a buckybowl has a high bowl inversion energy, the slow rate of racemization on the HPLC time scale might permit the resolution of the racemic buckybowl by means of chiral HPLC. Separation of racemic chiral buckybowls might therefore be another effective strategy for obtaining enantiomerically pure or enantiomerically enriched compounds. (±)-Trimethylsumanenetrione (**53**) *(21,23)*, prepared by oxidation of (±)-**50**, was a candidate for this process because its bowl inversion energy was estimated to be ca. 23.5 kcal/mol, by comparison between the experimental value (20.3 kcal/mol) of **2** and the calculated values (B3LYP/6-311 + G(d, p):18.3 and 21.5 kcal/mol) of **2** and **53**, which corresponds to a half-life of about 44 h, even at 10°C. This value could be high enough

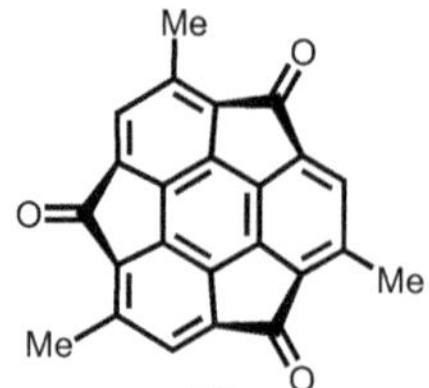

53

23.3 kcal/mol (exp.)

21.5 kcal/mol (calcd.)

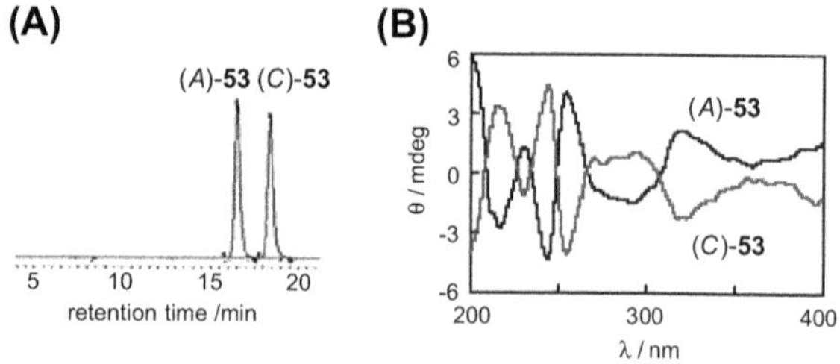

Figure 4.18 Experimental and calculated [B3LYP/6-311 + G(d,p)] bowl inversion barriers, HPLC separation, and circular dichroism (CD) spectra of chiral trimethylsumanenetrione (**53**).

to permit the isolation of enantiomerically enriched **53** by means of chiral HPLC. Indeed, resolution of **53** was achieved by chiral HPLC at 9°C, giving samples of each of the enantiomerically enriched forms of **53** (Fig. 4.18). By using the resulting enantiomerically enriched samples of **53**, we determined the rate constant of the racemization/bowl inversion and the bowl inversion energy to be 23.4 kcal/mol (303 K) (MeCN) from the time-dependent decay of the intensity of the CD spectra. This value was higher than the corresponding value for **50** (21.6 kcal/mol).

We also investigated the substituents effects on the bowl inversion barriers of substituted sumanenes by experiments and theoretical calculations (Figs. 4.19 and 4.20) *(21,53,54)*. In contrast to corannulene (**1**), Me and OH groups caused higher bowl inversion energies than that of sumanene (**2**). In the theoretical analyses, Me and OH groups destabilize the planar transition state rather than the bowl structure by the steric effect. On the other hand, CO_2H, CHO, CN, NO_2 groups stabilized the planar transition state rather than the bowl structure by the π-conjugation, giving lower energy than that of sumanene (**2**).

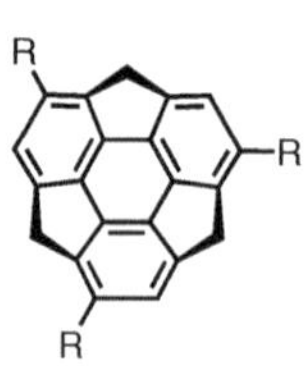

54	R=CO_2H	16.6 kcal/mol (calcd.)
55	R=CHO	16.8 kcal/mol (calcd.)
56	R=CN	17.5 kcal/mol (calcd.)
57	R=F	18.1 kcal/mol (calcd.)
2	R=H	18.3 kcal/mol (calcd.), 20.3 kcal/mol (exp.)
59	R=Br	18.4 kcal/mol (calcd.)
60	R=Cl	18.6 kcal/mol (calcd.)
61	R=OH	18.8 kcal/mol (calcd.)
62	R=Me	19.2 kcal/mol (calcd.), 21.6 kcal/mol (exp.)

Figure 4.19 Experimental and calculated [B3LYP/6-311 + G(d,p)] bowl inversion barriers of substituted sumanene derivatives.

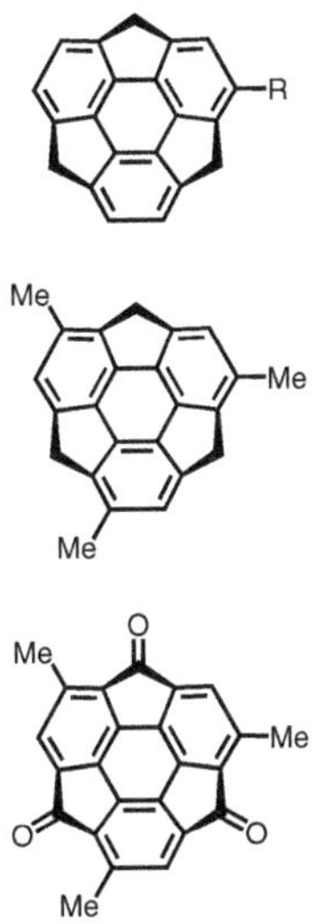

62 R=NO_2 18.5 kcal/mol (calcd.), 19.8 kcal/mol (exp.)
63 R=CHO 19.2 kcal/mol (calcd.), 20.2 kcal/mol (exp.)
2 R=H 19.7 kcal/mol (calcd.), 20.5 kcal/mol (exp.)
64 R=I 19.6 kcal/mol (calcd.), 20.8 kcal/mol (exp.)
65 R=OH 20.0 kcal/mol (calcd.), 21.2 kcal/mol (exp.)

50 21.5 kcal/mol (calcd.), 21.3 kcal/mol (exp.)

53 22.9 kcal/mol (calcd.), 23.3 kcal/mol (exp.)

Figure 4.20 Experimental and calculated [ωB97XD/6-311 + G(d,p)] bowl inversion barriers of substituted sumanene derivatives.

4. HETEROBUCKYBOWLS

After the achievements of pristine buckybowls such as corannulene (**1**), or sumanene (**2**), the heterobuckybowls containing heteroatoms in the bowl skeleton have been attracting much attention owing to the expected unique properties. Introduction of heteroatoms in the bowl-shaped skeleton causes significant change on the bowl geometry, the bowl inversion, and the various physicochemical properties. Reported heterobuckybowls are summarized in Figs. 4.21 and 4.22. Most of synthesized heterobuckybowls are heterosumanenes **66–73** *(55–57,59–65)* in which three methylenes of **2** were replaced by heteroatoms (Fig. 4.21). These heteroatoms are bigger atoms possessing longer bond lengths than carbon atom. These bigger heteroatoms introduced at the peripheral position of bowl cause shallower bowl depth and lower bowl inversion barrier than pristine sumanene (**2**). The bowl depths of trithiasumanene **66a** (0.79 Å) *(55–57)* and triselenasumanene **71a** (0.47 Å) *(56,57,64)* were reported to be shallower than that of sumanene (**2**) (1.11 Å). The calculated bowl inversion barrier of **66a** was reported to be only 1.8 kcal/mol at B3LYP/6-31G(d) level of theory by Sastry et al. *(58)*, which is due to the substantially shallow bowl structure. Furthermore, trisilasumanenes **67** *(57,59,60–62)*, mixed heterosumanenes **68**, **69** *(63)*, trigermanosumanenes **70** *(62,63)*, triphosphasumanenes **72** *(65)*, and triiodonium-sumanene **73** *(57)* were almost planar. Reported

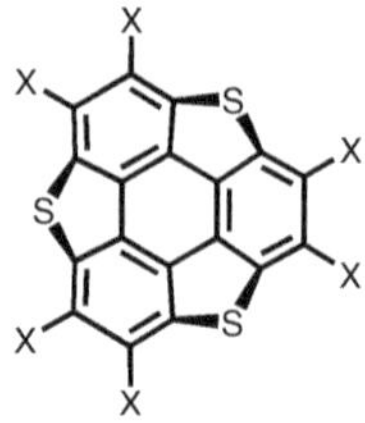

66a X=H
66b X=OBu
1.8 kcal/mol
[calcd. at B3LYP/6-31G(d)]

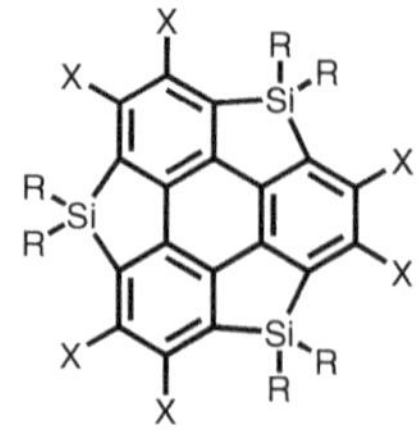

67a X=OBu R=Ph
67b X=H R=Bu, Me, ^{i}Bu

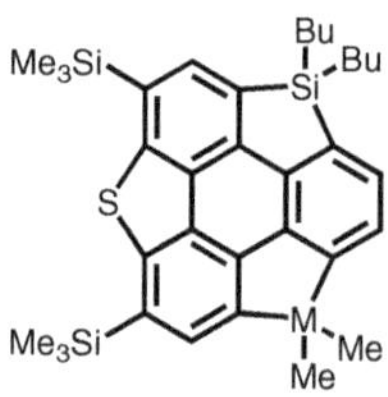

68 M=Si
69 M=Sn

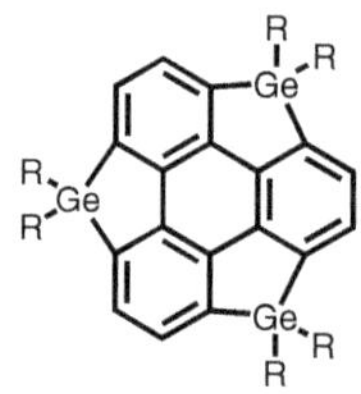

70 R =Bu, Me

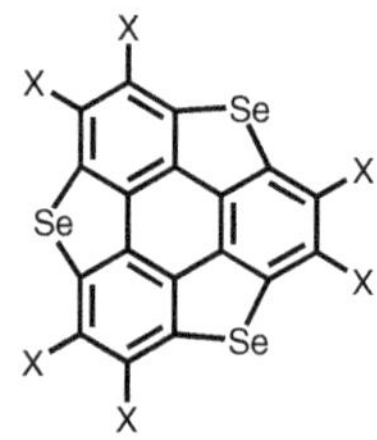

71a X=OBu
71b X=H

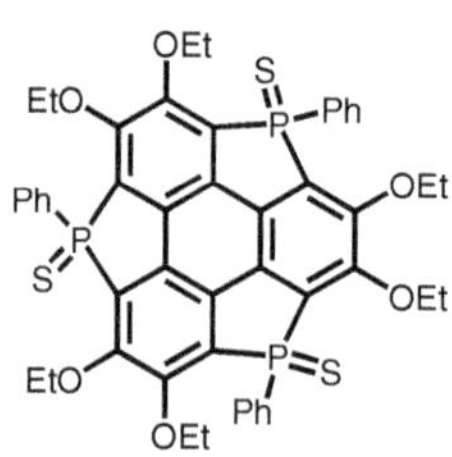

72

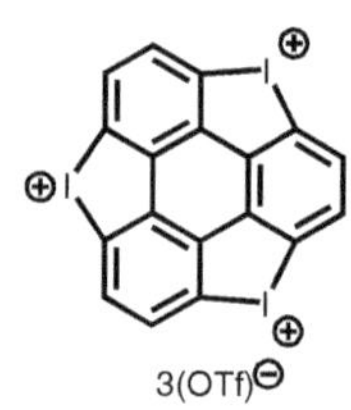

73

Figure 4.21 Heterosumanenes.

azabuckybowls possessing nitrogen atoms in the bowl skeleton are shown in Fig. 4.22. Only three kinds of azabuckybowls **74**, **75**, **77** possessing sumanene, corannulene, and diindenochrysene skeletons were reported to date. Triazasumanene (**74**) is a first azabuckybowl reported by us *(25)*. Unlike heterosumanenes in Fig. 4.21, nitrogen atoms were introduced in the benzene rings of **2** as pyridine rings. As a second azabuckybowl and a first heterocorannulene buckybowl, Ito and Nozaki et al. and Hiroto and Shinokubo et al. independently reported pentabenzoazacorannulene derivatives **75a** and **75b** *(66,67)*. Third azabuckybowl **77** was reported by us *(41,68)*, which is a first heterobuckybowl possessing diindenochrysene skeleton. The bowl inversion barrier of pentabenzocorannulene was

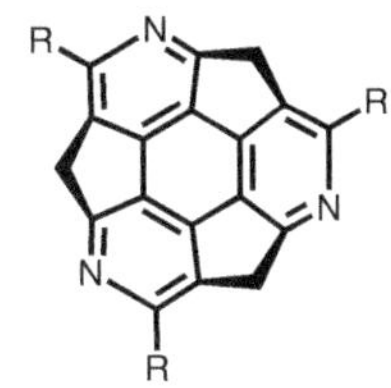

74a R=SMe
74b R=SO_2Me

42.2 kcal/mol (exp.)
39.9 kcal/mol [calcd. at B3LYP/6-311+G(d,p)]

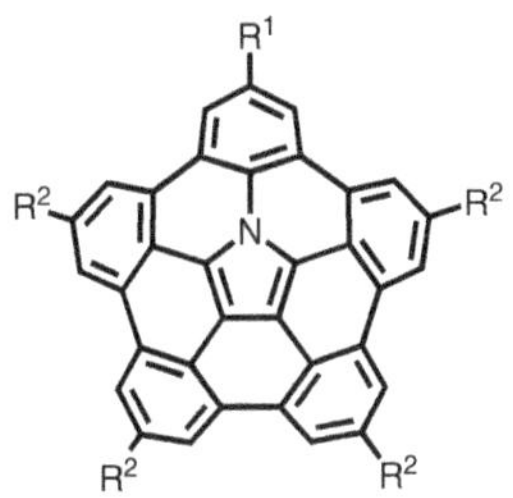

75a R^1=^{t}Bu, R^2=H
75b R^1=H, R^2=^{t}Bu

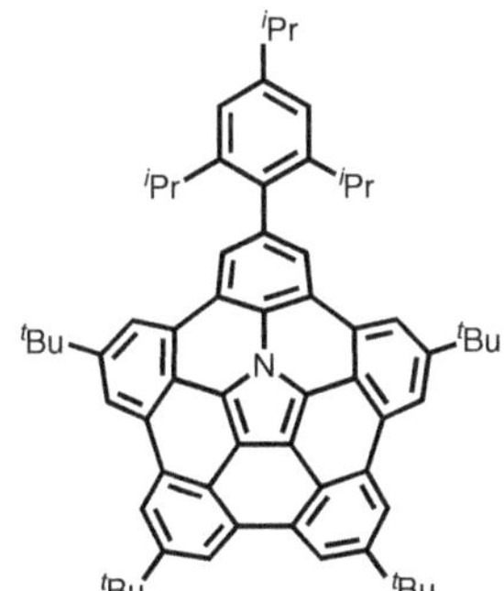

76

23.3 kcal/mol (exp.)
19.9 kcal/mol (calcd. at B3LYP/cc-pVDZ)

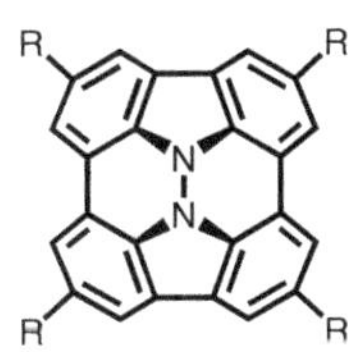

77a R=^{t}Bu
77b R=H

2.0 kcal/mol [calcd. at ωB97XD/6-311+G(d,p)]

Figure 4.22 Azabuckybowls.

determined to be 23.3 kcal/mol by measurement of 2D EXSY NMR spectra of **76** with diastereotopic protons, which is higher than the barrier (20.3 kcal/mol) of sumanene (**2**). Because the corresponding all-carbon framework of **75** is not a closed-shell structure, the effect of the introduction of nitrogen atom on the bowl inversion barrier is not clear in this skeleton.

As described in Section 1, the bowl chirality of chiral buckybowls can be induced by the π-conjugated skeleton itself, by the introduction of substituents, or by the replacement of skeletal carbons by heteroatoms. In Section 3, as chiral buckybowls induced by substituents, the enantioselective synthesis of trimethylsumanene (**50**) and its derivatives by the chiral transmission method and the determination of bowl inversion barriers by the measurement of racemization rate were described. Although trimethylsumanene (**50**) or trimethylsumanenetrione (**53**) possess the bowl chirality, they underwent racemization at room temperature owing to the low bowl inversion barriers (21.6 and 23.4 kcal/mol). In contrast, triazasumanene (**74**) was expected to possess high bowl inversion energies because of the deep bowl structure as a result of the doping effect of the nitrogen atoms. Chiral triazasumanenes (**74**) were enantioselectively synthesized by the chiral transmission method as same as chiral trimethylsumanene (**50**) (Fig. 4.23). Cyclotrimerization of enantiopure azanorbornene derivative (1*S*,4*R*)-**78** was converted to *syn*-**79** exclusively under Pd-catalyzed cyclotrimerization conditions. Compound **79** was converted to **81** through **80** with a bowl structure by hydrolysis of amides, condensation, and following transformations. Although the final aromatization of **49** in the enantioselective synthesis of chiral trimethylsumanene (**50**) was carried out with an excess of DDQ at 0°C for 1 min and **50** was

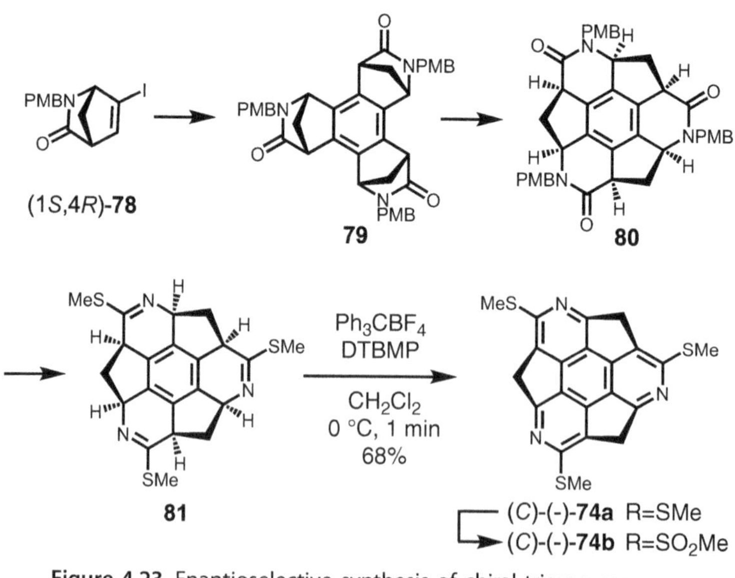

Figure 4.23 Enantioselective synthesis of chiral triazasumanene.

purified below −20°C to prevent the racemization, aromatization of **81** was carried out at room temperature for 8 h by Ph_3CBF_4 and DTBMP (2,6-*t*-butyl-4-methylpyridine) and (*C*)-(−)-**74a** was purified and isolated at room temperature because DFT calculation predicted that the racemization of **74a** did not take place at room temperature. (*C*)-(−)-**74b** was also prepared from (*C*)-(−)-**74a** by oxidation at room temperature. DFT calculation (B3LYP/6-311 + G[d,p]) indicated that the bowl inversion barrier of **74a** was 39.9 kcal/mol, which correspond to 1.1 billion years for a half-life time at 20°C. In fact, ee of both enantiomers of **74a** were determined to be >99% by chiral HPLC analysis at room temperature. The CD spectra of **74a** were recorded at room temperature and no decay was observed after 1 week (Fig. 4.24). These results indicated that compound **74a** should have much higher bowl inversion energy, and it does not racemize at room temperature. The racemization of enantiopure **74a** in tetralin solution was observed at 215°C and the bowl inversion barrier was determined to be 42.2 kcal/mol from the racemization rate by chiral HPLC analysis, which correspond to 54 billion years for the half-life time at 20°C. X-ray analysis of **74b** showed the deeper bowl depth (1.30 Å) than sumanene (**2**) (1.1 Å). The extraordinarily high bowl inversion energy, deep bowl depth, and the stable bowl chirality were induced by the introduction of nitrogen atoms, which is due to the shorter bond length of carbon—nitrogen bond than carbon—carbon bond at the peripheral positions of bowl structure (Fig. 4.25).

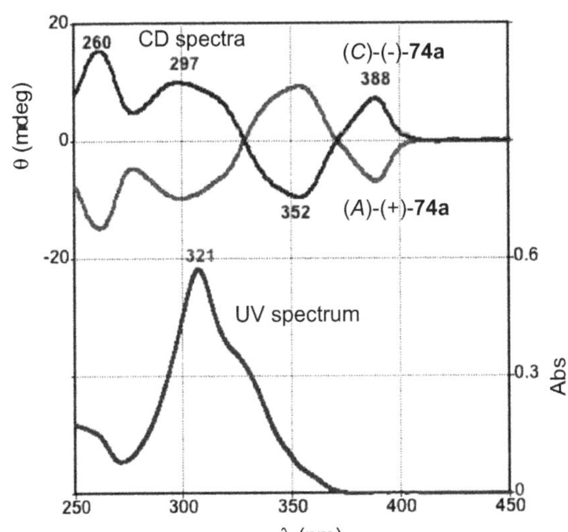

Figure 4.24 Circular dichroism (CD) and UV-Vis spectra of compound **74a** at room temperature.

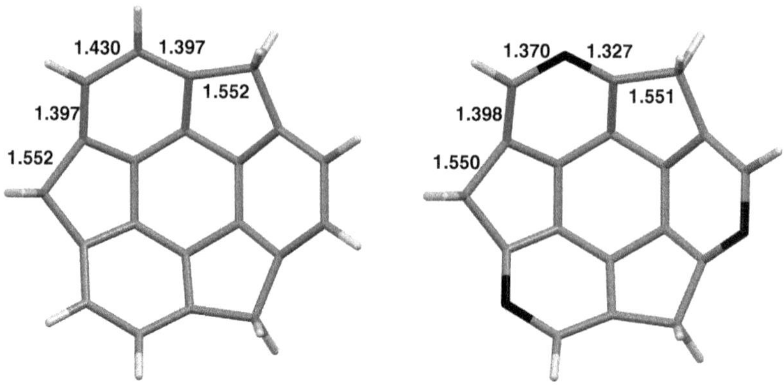

2

74c R=H

Figure 4.25 Calculated structures of **2** and **74c** (R = H) [B3LYP/6-311 + G(d,p)] and the bond lengths (Å).

In contrast to triazasumanene **74**, introduction of two nitrogens atoms as a hydrazine structure in the central position of diindenochrysene (**3**) skeleton caused lower bowl inversion barrier. The calculated bowl inversion barrier of **77b** was only 2.0 kcal/mol, which is much lower than 7.3 kcal/mol of **3** (Figs. 4.8 and 4.22). The effect of the introduced nitrogen atoms on the bowl inversion energy in **77b** is not as simple as the effect in triazasumanene **74**. Although the carbon–nitrogen bond length of **77b** is shorter than the corresponding carbon–carbon bond of **3** in calculation (Fig. 4.26), the nitrogen–nitrogen bond length of **77b** is longer than the length of the corresponding carbon–carbon double bond of **3**. The lengths

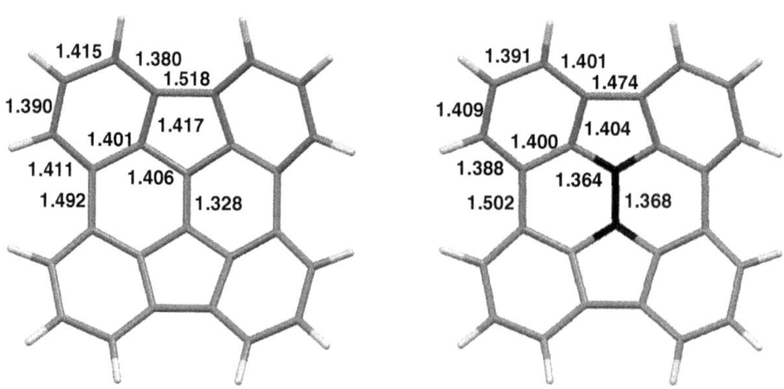

3

77b R=H

Figure 4.26 Calculated structures of **3** and **77b** [ωB97XD/6-311 + G(d,p)] and the bond lengths (Å).

of other bonds were also changed between **77b** and **4**. Nevertheless, as a result of introduction of the hydrazine structure, the bowl depth (1.22 Å) of **77b** became shallower than that (1.40 Å) of **3**, leading to the lower bowl inversion barrier of **77b** than that of **3**. In addition to the change of the bowl depth, **77b** has another different property in the bowl inversion dynamics from **3**. Analysis of the bowl inversion dynamics by theoretical calculation elucidated that **77b** has a minor twist conformation in addition to the major bowl conformation and the bowl inversion goes through nonplanar transition states on a triple-well potential, whereas the bowl inversion of **3** goes through the planar transition state on a double-well potential (Fig. 4.27). Because calculated planar structure of **77b** has higher energy than the nonplanar transition structure (Fig. 4.27), the change of the transition state between **3** and **77b** also contributes to the lower bowl inversion barrier of **77b**.

77 also possesses an interesting property concerning the geometry and the bowl inversion associated with the redox. The cyclic voltammogram of **77a** showing two reversible redox waves (Fig. 4.28) indicated the reversible generation of the stable monocation and dication of **77a**. Monocation $\mathbf{77a}^{\bullet+}\cdot SbF_6^-$ and dication $\mathbf{77a}^{2+}\cdot 2SbF_6^-$ were also

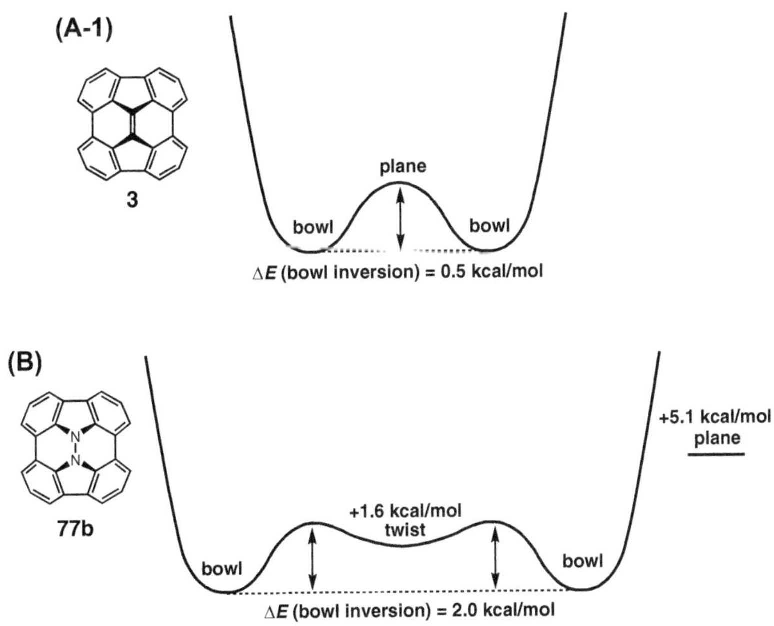

Figure 4.27 Calculated energy diagram of bowl inversion of **3** (A-1) on a double-well potential and **77b** (B) on a triple-well potential [ωB97XD/6-311 + G(d,p)].

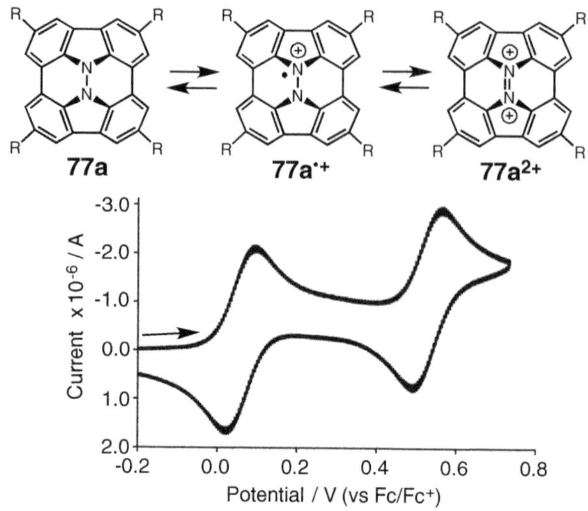

Figure 4.28 Cyclic voltammogram of **77a**.

synthesized and isolated by chemical oxidation of **77a**. X-ray crystallographic analysis of $\mathbf{77a}^{\bullet+}$ and $\mathbf{77a}^{2+}$ and DFT calculation of $\mathbf{77b}^{\bullet+}$ and $\mathbf{77b}^{2+}$ revealed that $\mathbf{77}^{\bullet+}$ possesses a shallower bowl structure than **77** and $\mathbf{77}^{2+}$ possesses a completely planar structure (Fig. 4.29). The calculated bowl depth and bowl inversion barrier of $\mathbf{77b}^{\bullet+}$ was 0.81 Å and 0.5 kcal/mol, which are smaller and lower than those (1.22 Å and 2.0 kcal/mol) of **77b**. In addition, the bowl inversion of $\mathbf{77b}^{\bullet+}$ changed to double-well potential from the triple-well potential of **77b**. The comparison of the bond lengths of the calculated structures indicated significant shortening of the N—N bond length in the order of **77b** (bowl: 1.37 Å), $\mathbf{77b}^{\bullet+}$ (1.31 Å), and $\mathbf{77b}^{2+}$ (1.27 Å), which is responsible for the flattening of the structure. The geometry of **77** can be reversibly changed between bowl and plane shapes by the redox and the bowl inversion dynamics was also controlled (Fig. 4.30).

5. SUMMARY

Among the research on buckybowls including the synthesis, property, and applications, bowl inversion property of many kinds of reported buckybowls were particularly focused in this chapter. The magnitude of bowl inversion barrier was altered in very wide range by the kinds of buckybowl skeleton, the extension of bowl π-conjugation, the substituents, the heteroatoms, and the redox. In other words, the bowl inversion kinetics

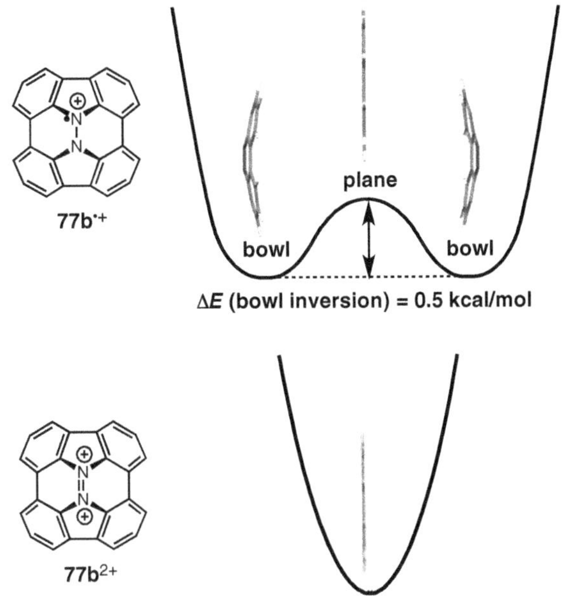

Figure 4.29 Calculated structures and dynamics of **77b**$^{\bullet+}$ and **77b**$^{2+}$ [ωB97XD/6-311 + G(d,p)].

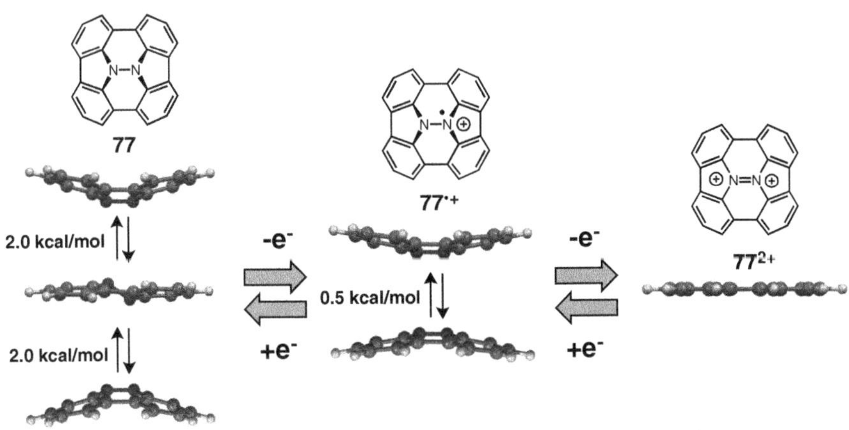

Figure 4.30 Bowl inversion dynamics of **77**, **77**$^{\bullet+}$ and **77**$^{2+}$.

can be controlled by these factors. Particularly, introduction of substituents and heteroatoms is an effective method to control the bowl inversion barrier of pristine buckybowls without changing the bowl size. The correlation between the bowl inversion barriers and bowl depths was elucidated in corannulene and sumanene skeletons. Introduction of heteroatoms has drastic effect on inversion barrier changing the order of energy as demonstrated in heterobuckybowls **66**, **67**, **74**, and **77**. The control of bowl inversion barrier by heteroatom even realized chiral buckybowl with high stability at room temperature. Introduction of substituents is useful to adjust the inversion barrier in a few kcal/mol range. Although only bowl inversion property of a single buckybowl molecule in gas- or solution-phase is described in this chapter, bowl inversion properties on metal surface *(69,70)* or in supramolecular complex *(71)* were also investigated. Based on the basic research results, the bowl inversion property is well understood and the kinetics is controllable. In the next stage, buckybowls and the bowl inversion are expected to be applied to the field of molecular machine.

REFERENCES

1. Tsefrikas, V. M.; Scott, L. T. Geodesic Polyarenes by Flash Vacuum Pyrolysis. *Chem. Rev.* **2006,** *106,* 4868–4884.
2. Wu, Y.-T.; Siegel, J. S. Aromatic Molecular-bowl Hydrocarbons: Synthetic Derivatives, Their Structures, and Physical Properties. *Chem. Rev.* **2006,** *106,* 4843–4867.
3. Sygula, A. Chemistry on a Half-shell: Synthesis and Derivatization of Buckybowls. *Eur. J. Org Chem.* **2011,** 1611–1625.
4. Higashibayashi, S.; Sakurai, H. Synthesis of Sumanene and Related Buckybowls. *Chem. Lett.* **2011,** *40,* 122–128.
5. Amaya, T.; Hirao, T. A Molecular Bowl Sumanene. *Chem. Commun.* **2011,** *47,* 10524–10535.
6. Schmidt, B. M.; Lentz, D. Syntheses and Properties of Buckybowls Bearing Electron-withdrawing Groups. *Chem. Lett.* **2014,** *43,* 171–177.
7. Hou, X.-Q.; Sun, Y.-T.; Liu, L.; Wang, S.-T.; Geng, R.-L.; Shao, X.-F. Bowl-shaped Conjugated Polycycles. *Chin. Chem. Lett.* **2016,** *27,* 1166–1174.
8. Li, X.; Kang, F.; Inagaki, M. *Small* **2016,** *12,* 3206–3223.
9. Sygula, A.; Rabideau, P. W. In *Synthesis and Chemistry of Polycyclic Aromatic Hydrocarbons with Curved Surfaces: Buckybowls; Carbon-rich Compounds;* Haley, M. M., Tykwinski, R. R., Eds.; Wiley-VCH, 2006; pp 529–565.
10. Petrukhina, M. A., Scott, L. T., Eds. *Fragments of Fullerenes and Carbon Nanotubes;* Wiley: Hoboken, 2012.
11. Wu, Y.-T.; Siegel, J. S. In *Synthesis, Structures, and Physical Properties of Aromatic Molecular-bowl Hydrocarbons; Topics in Current Chemistry 349, Polyarenes I;* Siegel, J. S., Wu, Y.-T., Eds.; Springer, 2014; pp 63–120.
12. Higashibayashi, S.; Sakurai, H. In *Growing Buckybowl Chemistry; Polycyclic Arenes and Heteroarenes: Synthesis, Properties, and Applications;* Miao, Q., Ed.; Wiley-VCH, 2015; pp 61–84.
13. Barth, W. E.; Lawton, R. G. Dibenzo[*ghi,mno*]fluoranthene. *J. Am. Chem. Soc.* **1966,** *88,* 380–381.
14. Kroto, H. W.; Heath, J. R.; O'Brien, S. C.; Curl, R. F.; Smalley, R. E. C_{60}: Buckminsterfullerene. *Nature* **1985,** *318,* 162–163.
15. Iijima, S. Helical Microtubules of Graphitic Carbon. *Nature* **1991,** *354,* 56–58.

16. Gleicher, G. J. Calculations on the Corannulene System. *Tetrahedron* **1967,** *23,* 4257–4263.
17. Scott, L. T.; Hashemi, M. M.; Bratcher, M. S. Corannulene Bowl-to-bowl Inversion Is Rapid at Room Temperature. *J. Am. Chem. Soc.* **1992,** *114,* 1920–1921.
18. Amaya, T.; Sakane, H.; Muneishi, T.; Hirao, T. Bowl-to-bowl Inversion of Sumanene Derivatives. *Chem. Commun.* **2008,** 765–767.
19. Chang, H.-I.; Huang, H.-T.; Huang, C.-H.; Kuo, M.-Y.; Wu, Y.-T. Diindeno [1,2,3,4-*defg*;1′,2′,3′,4′-*mnop*]chrysenes: Solution-phase Synthesis and the Bowl-to-bowl Inversion Barrier. *Chem. Commun.* **2010,** *46,* 7241–7243.
20. Seiders, T. J.; Baldridge, K. K.; Grube, G. H.; Siegel, J. S. Structure/energy Correlation of Bowl Depth and Inversion Barrier in Corannulene Derivatives: Combined Experimental and Quantum Mechanical Analysis. *J. Am. Chem. Soc.* **2001,** *123,* 517–525.
21. Higashibayashi, S.; Tsuruoka, R.; Soujanaya, Y.; Purushotham, U.; Sastry, G. N.; Seki, S.; Ishikawa, T.; Toyota, S.; Sakurai, H. Trimethylsumanene: Enantioselective Synthesis, Substituent Effect on Bowl Structure, Inversion Energy, and Electron Conductivity. *Bull. Chem. Soc. Jpn.* **2012,** *85,* 450–467.
22. Thilgen, C.; Diederich, F. Structural Aspects of Fullerene Chemistry - a Journey through Fullerene Chirality. *Chem. Rev.* **2006,** *106,* 5049–5135.
23. Higashibayashi, S.; Sakurai, H. Asymmetric Synthesis of a Chiral Buckybowl, Trimethylsumanene. *J. Am. Chem. Soc.* **2008,** *130,* 8592–8593.
24. Tsuruoka, R.; Higashibayashi, S.; Ishikawa, T.; Toyota, S.; Sakurai, H. Optical Resolution of Chiral Buckybowls by Chiral HPLC. *Chem. Lett.* **2010,** *39,* 646–647.
25. Tan, Q.; Higashibayashi, S.; Karanjit, S.; Sakurai, H. Enantioselective Synthesis of a Chiral Nitrogen-doped Buckybowl. *Nat. Commun.* **2012,** *3,* 891.
26. Scott, L. T.; Hashemi, M. M.; Meyer, D. T.; Warren, H. B. Corannulene. A Convenient New Synthesis. *J. Am. Chem. Soc.* **1991,** *113,* 8592–8593.
27. Seiders, T. J.; Elliott, E. L.; Grube, G. H.; Siegel, J. S. Synthesis of Corannulene and Alkyl Derivatives of Corannulene. *J. Am. Chem. Soc.* **1999,** *121,* 7804–7813.
28. Sygula, A.; Rabideau, P. W. Non-pyrolytic Syntheses of Buckybowls: Corannulene, Cyclopentacorannulene, and a Semibuckminsterfullerene. *J. Am. Chem. Soc.* **1999,** *121,* 7800–7803.
29. Sygula, A.; Rabideau, P. W. A Practical, Large Scale Synthesis of the Corannulene Systems. *J. Am. Chem. Soc.* **2000,** *122,* 6323–6324.
30. Butterfield, A. M.; Gilomen, B.; Siegel, J. S. Kilogram-scale Production of Corannulene. *Org. Process Res. Dev.* **2012,** *16,* 664–676.
31. Reisch, H. A.; Bratcher, M. S.; Scott, L. T. Imposing Curvature on a Polyarene by Intramolecular Palladium-catalyzed Arylation Reactions: A Simple Synthesis of Dibenzo[*a,g*]corannulene. *Org. Lett.* **2000,** *2,* 1427–1430.
32. Sygula, A.; Abdourazak, A. H.; Rabideau, P. W. Cyclopentacorannulene: Π-Facail Stereoselective Deuterogenation and Determination of the Bowl-to Bowl Inversion Barrier for a Constrained Buckybowl. *J. Am. Chem. Soc.* **1996,** *118,* 339–343.
33. Priyakumar, U. D.; Punnagai, M.; Mohan, G. P. K.; Sastry, G. N. A Computational Study of Cation–π Interactions in Polycyclic Systems: Exploring the Dependence on the Curvature and Electronic Factors. *Tetrahedron* **2004,** *60,* 3037–3043.
34. Sygula, A.; Rabideau, P. W. Structure and Inversion Barriers of Corannulene, its Dianion and Tetraanion. An Ab Initio Study. *J. Mol. Struct.* **1995,** *333,* 215–226.
35. Bruno, C.; Venassi, R.; Passalacqua, A.; Paolucci, F.; Fontanesi, C.; Marcaccio, M.; Jackson, E. A.; Scott, L. T. Electrochemical and Theoretical Investigation of Corannulene Reduction Processes. *J. Phys. Chem. B* **2009,** *113,* 1954–1962.

36. Nishida, S.; Morita, Y.; Ueda, A.; Kobayashi, T.; Fukui, K.; Ogasawara, K.; Sato, K.; Takui, T.; Nakasuji, K. Curve-structured Phenalenyl Chemistry: Synthesis, Electronic Strucure, and Bowl-inversion Barrier of a Phenalenyl-fused Corannulene Anion. *J. Am. Chem. Soc.* **2008,** *130,* 14954–14955.
37. Wu, T.-C.; Hsin, H.-J.; Kuo, M.-Y.; Li, C.-H.; Wu, Y.-T. Synthesis and Structural Analysis of a Highly Curved Buckybowl Containing Corannulene and Sumanene Fragments. *J. Am. Chem. Soc.* **2011,** *133,* 16319–16321.
38. Wu, T.-C.; Chen, M.-K.; Lee, Y.-W.; Kuo, M.-Y.; Wu, Y.-T. Bowl-shaped Fragments of C_{70} or Higher Fullerenes: Synthesis, Structural Analysis, and Inversion Dynamics. *Angew. Chem. Int. Ed.* **2013,** *52,* 1289–1293.
39. Rabideau, P. W.; Abdourazak, A. H.; Folsom, H. E.; Marcinow, Z.; Sygula, A.; Sygula, R. Buckybowls: Synthesis and Ab Initio Calculated Structure of the First Semibuckminsterfullerene. *J. Am. Chem. Soc.* **1994,** *116,* 7891–7892.
40. Biedermann, P. U.; Pogodin, S.; Agranat, I. Inversion Barrier of Corannulene. A Benchmark for Bowl-to-bowl Inversions in Fullerene Fragments. *J. Org. Chem.* **1999,** *64,* 3655–3662.
41. Higashibayashi, S.; Pandit, P.; Haruki, R.; Adachi, S.; Kumai, R. Redox-dependent Transformation of a Hydrazinobuckybowl between Curved and Planar Geometries. *Angew. Chem. Int. Ed.* **2016,** *55,* 10830–10834.
42. Sakurai, H.; Daiko, T.; Hirao, T. A Synthesis of Sumanene, a Fullerene Fragment. *Science* **2003,** *301,* 1878.
43. Hagen, S.; Nuechter, U.; Nuechter, M.; Zimmermann, G. High Temperature Synthesis of Bowl-shaped Subunits of Fullerenes. II. From 9,9′-bi-9*H*-fluorenylidene to Diindeno[l,2,3,4-*defg*; 1′,2′,3′,4′-*mnop*]chrysene. *Polycycl. Aromat. Comp.* **1995,** *4,* 209–217.
44. Hagen, S.; Christoph, H.; Zimmermann, G. High Temperature Synthesis to Bowl-shaped Subunits of Fullerenes - IV. From 4-[9H-fluorenylidene-(9)]-4H-cyclopenta[def]phenanthrene to Fluoreno[1,9,8-abcd]corannulene and Difluoreno [1,9,8,7-*cdefg*;2′,1′,9′,8′-*klmno*]anthracene. *Tetrahedron* **1995,** *51,* 6961–6970.
45. Scott, L. T. Fragments of Fullerenes: Novel Syntheses, Structures and Reactions. *Pure Appl. Chem.* **1996,** *68,* 291–300.
46. Bronstein, H. E.; Choi, N.; Scott, L. T. Practical Synthesis of an Open Geodesic Polyarene with a Fullerene-type 6:6-Double Bond at the Center: Diindeno[1,2,3,4-*defg*;1′,2′,3′,4′-*mnop*]chrysene. *J. Am. Chem. Soc.* **2002,** *124,* 8870–8875.
47. Shenhar, R.; Beust, R.; Hoffman, R. E.; Willner, I.; Bronstein, H. E.; Scott, L. T.; Rabinovitz, M. Lithium Reduction of the Bowl-shaped C60 Fragment Diindeno [1,2,3,4-*defg*;1',2',3',4'-*mnop*]chrysene: an Interplay between Experiment and Calculation. *J. Org. Chem.* **2001,** *66,* 6004–6013.
48. Amaya, T.; Nakata, T.; Hirao, T. Synthesis of Highly Strained Π-Bowls from Sumanene. *J. Am. Chem. Soc.* **2009,** *131,* 10810–10811.
49. Amaya, T.; Hirao, T. Bowl-to-bowl Inversion of Sumanene Derivatives. *Pure Appl. Chem.* **2012,** *84,* 1089–1100.
50. Liu, J.; Osella, S.; Ma, J.; Berger, R.; Beljonne, D.; Schollmeyer, D.; Feng, X.; Müllen, K. Fused Dibenzo[*a*,*m*]rubicene: A New Bowl-shaped Subunit of C_{70} Containing Two Pentagons. *J. Am. Chem. Soc.* **2016,** *138,* 8364–8367.
51. Wu, Y.-T.; Hayama, T.; Baldridge, K. K.; Linden, A.; Siegel, J. S. Synthesis of Fluoranthenes and Indenocorannulenes: Elucidation of Chiral Stereoisomers on the Basis of Static Molecular Bowls. *J. Am. Chem. Soc.* **2006,** *128,* 6870–6884.
52. Marcinow, Z.; Sygula, A.; Ellern, A.; Rabideau, P. W. Lowering Inversion Barriers of Buckybowls by Benzannelation of the Rim: Synthesis and Crystal and Molecular Structure of 1,2-Dihydrocyclopenta[*b*,*c*]dibenzo[*g*,*m*]corannulene. *Org. Lett.* **2001,** *3,* 3527–3529.

53. Shrestha, B. B.; Karanjit, S.; Higashibayashi, S.; Sakurai, H. Correlation between Bowl-inversion Energy and Bowl Depth in Substituted Sumanenes. *Pure Appl. Chem.* **2014,** *86,* 747–753.
54. Niti, N.; Panda, G.; Higashibayashi, S.; Sakurai, H. Synthesis of Hydroxysumanene and Substituent Effect of Hydroxy Group on Bowl Inversion Dynamics and Electronic Structure. *J. Org. Chem.* **2016,** *81,* 11978–11981.
55. Imamura, K.; Takimiya, K.; Otsubo, T.; Aso, Y. Triphenyleno[1,12-*bcd*:4,5-*b'c'd'*:8,9-*b''c''d''*]trithiophene: the First Bowl-shaped Heteroaromatic. *Chem. Commun* **1999,** 1859–1860.
56. Li, X.; Zhu, Y.; Shao, J.; Wang, B.; Zhang, S.; Shao, Y.; Jin, X.; Yao, X.; Fang, R.; Shao, X. Non-pyrolytic, Large-scale Synthesis of Trichalcogenasumanene: A Two-step Approach. *Angew. Chem. Int. Ed.* **2014,** *53,* 535–538.
57. Tan, Q.; Zhou, D.; Zhang, T.; Liu, B.; Xu, B. Iodine-doped Sumanene and its Application for the Synthesis of Chalcogenasumanenes and Silasumanenes. *Chem. Commun.* **2017,** *53,* 10279–10282.
58. Sastry, G. N.; Priyakumar, U. D. The Role of Heteroatom Substitution the Rigidity and Curvature of Buckybowls. A Theoretical Study. *J. Chem. Soc. Perkin Trans.* **2001,** *2,* 30–40.
59. Furukawa, S.; Kobayashi, J.; Kawashima, T. Development of a Sila-friedel-crafts Reaction and its Application to the Synthesis of Dibenzosilole Derivatives. *J. Am. Chem. Soc.* **2009,** *131,* 14192–14193.
60. Furukawa, S.; Kobayashi, J.; Kawashima, T. Development of a Sila-friedel-crafts Reaction and its Application to the Synthesis of Dibenzosilole Derivatives. *Dalton Trans.* **2010,** *39,* 9329–9336.
61. Tanikawa, T.; Saito, M.; Guo, J. D.; Nagase, S. Synthesis, Structures and Optical Properties of Trisilasumanene and its Related Compounds. *Org. Biomol. Chem.* **2011,** *9,* 1731–1735.
62. Zhou, D.; Gao, Y.; Liu, B.; Tan, Q.; Xu, B. Synthesis of Silicon and Germanium-containing Heterosumanenes via Rhodium-catalyzed Cyclodehydrogenation of Silicon/germanium-hydrogen and Carbon-hydrogen Bonds. *Org. Lett.* **2017,** *19,* 4628-4531.
63. Tanikawa, T.; Saito, M.; Guo, J. D.; Nagase, S.; Minoura, M. Synthesis, Structures, and Optical Properties of Heterasumanenes Containing Group 14 Elements and Their Related Compounds. *Eur. J. Org Chem.* **2012,** 7135–7142.
64. Sun, Y.; Li, X.; Sun, C.; Shen, H.; Hou, X.; Lin, D.; Zhang, H.-L.; Di, C.; Zhu, D.; Shao, X. Trichalcogenasumanene Ortho-Quinones: Synthesis, Properties, and Transformation into Various Heteropolycycles. *Angew. Chem. Int. Ed.* **2017,** *56,* 13470–13474.
65. Furukawa, S.; Suda, Y.; Kobayashi, J.; Kawashima, T.; Tada, T.; Fujii, S.; Kiguchi, M.; Saito, M. Triphosphasumanene Trisulfide: High Out-of-plane Anisotropy and Janus-type Π-Surfaces. *J. Am. Chem. Soc.* **2017,** *139,* 5787–5792.
66. Ito, S.; Tokimaru, Y.; Nozaki, K. Benzene-fused Azacorannulene Bearing an Internal Nitrogen Atom. *Angew. Chem. Int. Ed.* **2015,** *54,* 7256–7260.
67. Yokoi, H.; Hiraoka, Y.; Hiroto, S.; Sakamaki, D.; Seki, S.; Shinokubo, H. Nitrogen-embedded Buckybowl and its Assembly with C_{60}. *Nat. Commun.* **2015,** *6,* 8215.
68. Yamamoto, K.; Palash, P.; Higashibayashi, S. Non-planar [*n*]Cyclo-1,8-carbazolylenes (*N* = 3,4,6) and [3]Cyclo-1,8-carbazolylenyl B, P, PO, SiPh Complexes. *Chem. Eur J.* **2017,** *23,* 14011–14016.
69. Jaafar, R.; Pignedoli, C. A.; Bussi, G.; Ait-Mansour, K.; Groening, O.; Amaya, T.; Hirao, T.; Fasel, R.; Ruffieux, P. Bowl Inversion of Surface-adsorbed Sumanene. *J. Am. Chem. Soc.* **2014,** *136,* 13666–13671.

70. Fujii, S.; Ziatdinov, M.; Higashibayashi, S.; Sakurai, H.; Kiguchi, M. Bowl Inversion and Electronic Switching of Buckybowls on Gold. *J. Am. Chem. Soc.* **2016,** *138,* 12142−12149.
71. Juricek, M.; Strutt, N. L.; Barnes, J. C.; Butterfield, A. M.; Dale, E. J.; Baldridge, K. K.; Stoddart, J. F.; Siegel, J. S. Induced-fit Catalysis of Corannulene Bowl-to-bowl Inversion. *Nature Chem.* **2014,** *6,* 222−2228.

CHAPTER 5

Asymmetric Hydrogenation of Aromatic Carbocycles

Yusuke Makida, Ryoichi Kuwano
Department of Chemistry, Faculty of Science, Kyushu University, Fukuoka, Japan

Contents

1. INTRODUCTION

Asymmetric hydrogenation of arenes, including heteroarenes, is a promising method for preparing optically active five- or six-membered rings. However, it is not easy to precisely control the dearomative hydrogenation because large energy is generally required for the dearomatization process. As a matter of fact, the asymmetric hydrogenation of arenes had been an unsettled issue in organic chemistry two decades ago. In 1998, Bianchini reported the enantioselective hydrogenation of quinoxaline with a chiral iridium catalyst *(1)*. This is the first successful achievement of high enantioselectivity (90% ee) with asymmetric catalysis in the dearomative hydrogenation. However, the substrate scope of the asymmetric reaction is limited to only one substrate, 2-methylquinoxaline. Kuwano and Ito had successfully developed the enantioselective hydrogenation of various indoles by a chiral rhodium catalyst in 2000 *(2)*. Since then, many research groups have devoted extensive efforts to the progress of the enantioselective hydrogenation of heteroarenes. A broad range of heteroarenes can nowadays be converted to the corresponding chiral heterocycles in high enantioselectivity through the asymmetric catalysis *(3)*.

Carbocyclic arenes are more undesirable substrates for the hydrogenation than heterocyclic ones because the benzene ring is highly stabilized by

Kinetic Control in Synthesis and Self-Assembly
ISBN 978-0-12-812126-9
https://doi.org/10.1016/B978-0-12-812126-9.00005-5

its own aromaticity. Various homogeneous as well as heterogeneous catalysts are known to enable the hydrogenation of benzene rings, but the catalyst commonly needs harsh conditions for promoting the reaction. The harsh conditions, particularly high temperature, may hamper a high level of selectivity control in the hydrogenation of arenes *(4)*.

Most asymmetric reactions involve two competitive stereochemical pathways. One gives the desired stereoisomer, and the other leads to formation of its antipode. The difference of the two pathways correlates with the stereoselectivity of the asymmetric reaction. In enantioselective reactions, the selectivity is commonly expressed in the enantiomeric excess (ee) or enantiomer ratio (er) of the chiral product. The ee is defined as the absolute difference between the mole fractions of both enantiomers (where $F_{(\mathrm{R})} + F_{(\mathrm{S})} = 1$) and can be described using er through Eq. (5.1):

$$\mathrm{ee} = \left|F_{(\mathrm{R})} - F_{(\mathrm{S})}\right| = \left|\frac{\left(F_{(\mathrm{R})}/F_{(\mathrm{S})}\right) - 1}{\left(F_{(\mathrm{R})}/F_{(\mathrm{S})}\right) + 1}\right| = \left|\frac{\mathrm{er} - 1}{\mathrm{er} + 1}\right| \tag{5.1}$$

The stereoselectivity is directly correlated to the ratio of the reaction rates of the two pathways, when the reaction creates a new stereogenic center on the substrate before or during its rate determining step. Considering an enantioselective single-step reaction (Fig. 5.1) to simplify the following discussion, the er of chiral product is equal to the ratio of the two rate constants, k_R and k_S (Eq. 5.2), which are the rate constants for the pathway leading to the formation of *R*- and *S*-product, respectively.

$$\mathrm{er} = \frac{k_\mathrm{R}}{k_\mathrm{S}} \left(\mathrm{or}\ \frac{k_\mathrm{S}}{k_\mathrm{R}}\right) \tag{5.2}$$

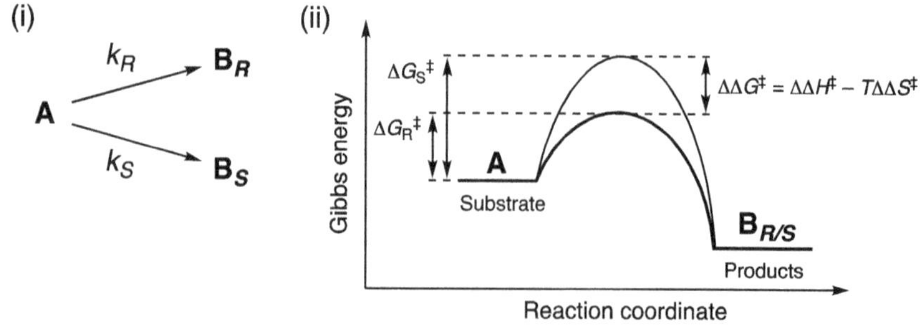

Figure 5.1 Simplified reaction model. (i) A transformation contains two competive pathways; both displayed a first-order dependence on [**A**] with rate constant (*k*). (ii) Energy diagram of the two diastereomeric pathways.

When the former constant is larger than the latter (er $\geq$ 1), the major enantiomer possesses *R*-configuration. The gaps between the activation enthalpies $\Delta\Delta H^{\ddagger}$ and entropies $\Delta\Delta S^{\ddagger}$ can be defined as follows:

$$\Delta\Delta H^{\ddagger} = \Delta H_{\mathrm{S}}^{\ddagger} - \Delta H_{\mathrm{R}}^{\ddagger} \tag{5.3}$$

$$\Delta\Delta S^{\ddagger} = \Delta S_{\mathrm{S}}^{\ddagger} - \Delta S_{\mathrm{R}}^{\ddagger} \tag{5.4}$$

Each rate constant is expressed with the function of reaction temperature *T*, each enthalpy $\Delta H^{\ddagger}$ and entropy $\Delta S^{\ddagger}$ of activation as shown in the following equations, in which k_{B}, h, and R are Boltzmann, Planck, and gas constants, respectively:

$$k_{\mathrm{R}} = \frac{k_{\mathrm{B}}T}{h}e^{-\frac{\Delta G_{\mathrm{R}}^{\ddagger}}{RT}} = \frac{k_{\mathrm{B}}T}{h}e^{-\frac{\Delta H_{\mathrm{R}}^{\ddagger}}{RT}}\cdot e^{\frac{\Delta S_{\mathrm{R}}^{\ddagger}}{R}} \tag{5.5}$$

$$k_{\mathrm{S}} = \frac{k_{\mathrm{B}}T}{h}e^{-\frac{\Delta G_{\mathrm{S}}^{\ddagger}}{RT}} = \frac{k_{\mathrm{B}}T}{h}e^{-\frac{\Delta H_{\mathrm{S}}^{\ddagger}}{RT}}\cdot e^{\frac{\Delta S_{\mathrm{S}}^{\ddagger}}{R}} \tag{5.6}$$

From Eqs. (5.1)–(5.6), the enantioselectivity, er or ee, is expressed with the following equations:

$$\mathrm{er} = e^{-\frac{\Delta H_{\mathrm{R}}^{\ddagger} - \Delta H_{\mathrm{S}}^{\ddagger}}{RT}}\cdot e^{\frac{\Delta S_{\mathrm{R}}^{\ddagger} - \Delta S_{\mathrm{S}}^{\ddagger}}{R}} = e^{\frac{\Delta\Delta H^{\ddagger}}{RT}}\cdot e^{-\frac{\Delta\Delta S^{\ddagger}}{R}} \tag{5.7}$$

$$\mathrm{ee} = \frac{e^{\frac{\Delta\Delta H^{\ddagger}}{RT}}\cdot e^{-\frac{\Delta\Delta S^{\ddagger}}{R}} - 1}{e^{\frac{\Delta\Delta H^{\ddagger}}{RT}}\cdot e^{-\frac{\Delta\Delta S^{\ddagger}}{R}} + 1} \tag{5.8}$$

To understand dependence of the stereoselectivity on reaction temperature, the ee of product is plotted against *T* at various $\Delta\Delta H^{\ddagger}$ by using Eq. (5.8) as shown in Fig. 5.2. The five lines indicate the relationships between the ee and *T* at $\Delta\Delta H^{\ddagger} = 1.0$, 1.5, 2.0, 2.5, 3.0 kcal/mol and $\Delta\Delta S^{\ddagger} = 0.0$ kcal/mol·K. The graph indicates that higher reaction temperature is disadvantageous for the highly stereoselective reaction. At −100°C, 1.0 kcal/mol of $\Delta\Delta H^{\ddagger}$ is enough for yielding the desired product with 90% ee. Meanwhile, the asymmetric reaction requires over 2.5 kcal/mol of the energy gap for achieving the high enantioselectivity at 150°C. Therefore, chiral catalyst or auxiliary must be carefully and precisely designed with steric repulsion and/or van der Waals interaction to make the large energy gap for the stereoselective reaction at high temperature.

As described above, the dearomatization of benzene rings requires high reaction temperature for producing the desired cyclohexane with adequate

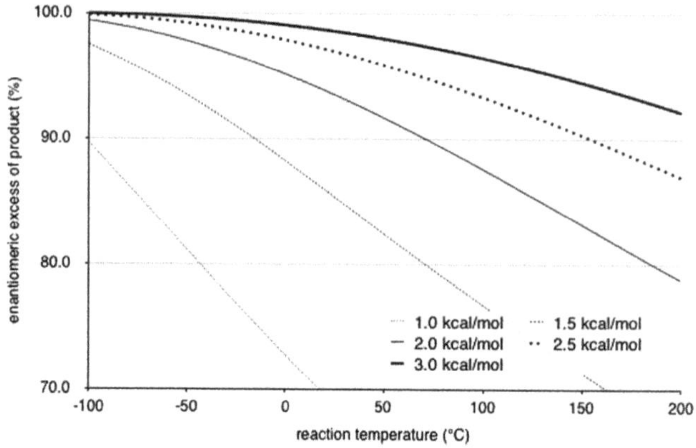

Figure 5.2 Temperature versus enantiomeric excess profiles in several activation enthalpy gaps.

efficiency. The high temperature would obstruct high stereoselectivity in the asymmetric reduction of carbocyclic arenes. Overcoming this difficulty, some successful reports have been made on the highly stereoselective hydrogenation of arenes, although the asymmetric reaction is one of the underdeveloped issues in synthetic organic chemistry. This chapter overviews the asymmetric hydrogenations of the carbocyclic arenes with the combinational use of chiral auxiliary and achiral metal-supported catalyst as well as through homogeneous asymmetric catalysis.

2. ASYMMETRIC HYDROGENATION OF AROMATIC CARBOCYCLES WITH CHIRAL AUXILIARY

Metal-supported catalysts modified with enantiomerically pure compounds have been successfully applied for the asymmetric hydrogenation of alkenes and ketones *(5)*. The hydrogenation of arenes has been also attempted by using the chiral heterogeneous catalyst. However, this strategy resulted in significantly low enantioselectivity *(6)*. As another approach toward successful use of metal-supported catalyst for the asymmetric hydrogenation, some studies have been focused on the use of chiral auxiliary, which can be installed on the achiral substrate and enables a sufficient chiral discrimination to give a diastereo-enriched product *(7)*. In 1994, Lemaire tried to open up the possibility of the combination of chiral auxiliary and heterogeneous catalysis in the asymmetric hydrogenation of carbocyclic arenes (Scheme 5.1) *(8)*. The hydrogenation of *o*-cresol was attempted by using

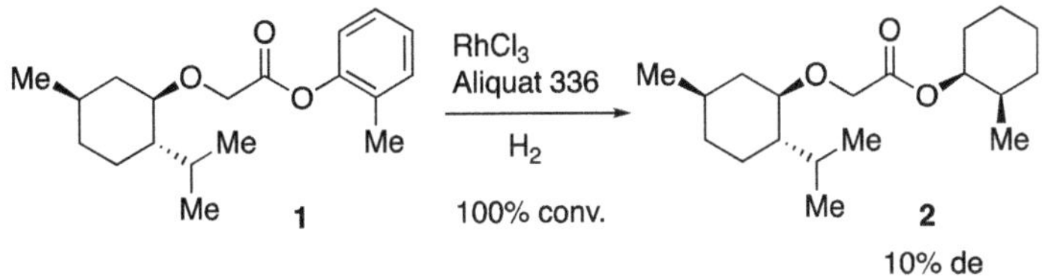

1 2

Scheme 5.1 Attempt for the stereoselective hydrogenation of covalently modified *o*-cresol with chiral auxiliary through heterogeneous catalysis.

colloidal rhodium particles as the catalyst and (−)-menthoxyacetyl group as the auxiliary. The reaction proceeded with complete conversion, but the desired cyclohexanol derivative **2** was obtained with very low diastereomeric excess (de). The chiral auxiliary may locate too far from the reactive site to achieve high stereoselectivity.

By using the combination of an appropriate chiral auxiliary and metal-supported catalyst, Besson successfully achieved high stereoselectivity in the hydrogenation of *o*-toluic acid (Scheme 5.2) *(9)*. When the substrate is modified with (*S*)-pyroglutamic acid methyl ester, its hydrogenation with Rh/Al_2O_3 catalyst afforded the desired product **4b** with 95% de. The carbonyl group on the 5-membered ring in the auxiliary might be crucial for the high stereoselectivity. Indeed, the hydrogenation of proline-installed substrate **3a** proceeded with moderate stereoselectivity.

Prins attempted the asymmetric hydrogenation of *o*-toluidine with (*S*)-proline (Scheme 5.3) *(10)*. The auxiliary was connected to *o*-toluidine through an amide bond. The hydrogenation of proline-2-methylanilide **5** afforded the cyclohexylamine derivative **6** with up to 57% de.

The benzene ring of anthranic acid was reduced to the chiral cyclohexane with high stereoselectivity by using the proline auxiliary (Scheme 5.4) *(11)*. The auxiliary was installed in anthranic acid through two amide bonds to form the rigid cyclic structure. The substrate **7** was transformed

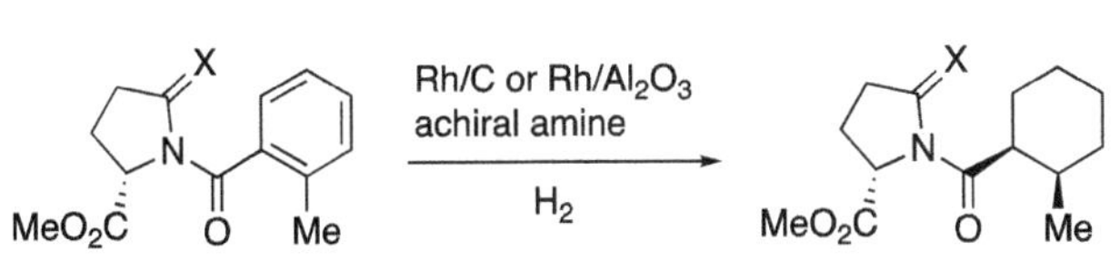

3a (X = 2H)
3b (X = O)

4a (X = 2H), 68% de
4b (X = O), 95% de

Scheme 5.2 Impact of auxiliary on diastereoselectivity for the hydrogenation of *o*-toluic acid.

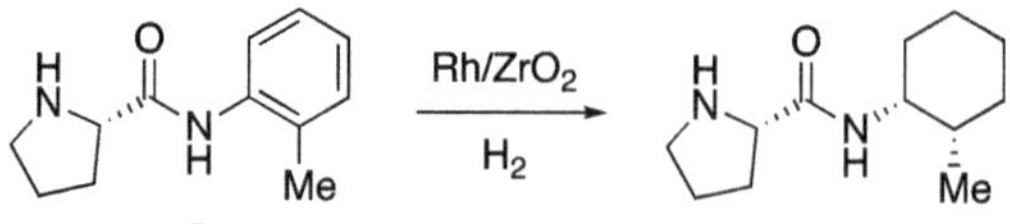

5 **6**
57% de

Scheme 5.3 Diastereoselective hydrogenation of *o*-toluidine with (*S*)-proline.

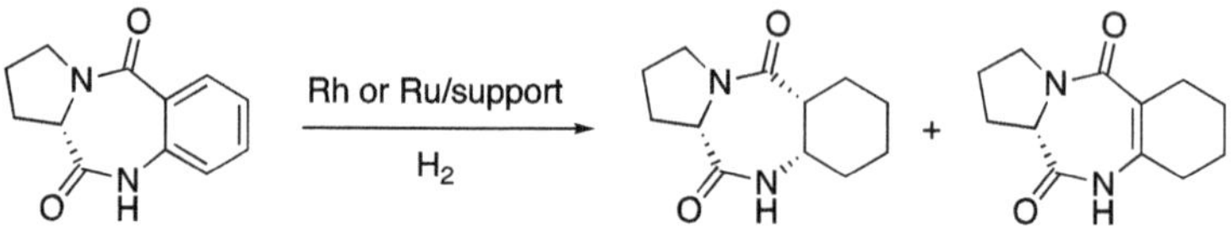

7 **8** **9**
96% de *major product* (60–70%)

Scheme 5.4 Rigid cyclic structure enabled high diastereoselectivity for the hydrogenation of anthranic acid.

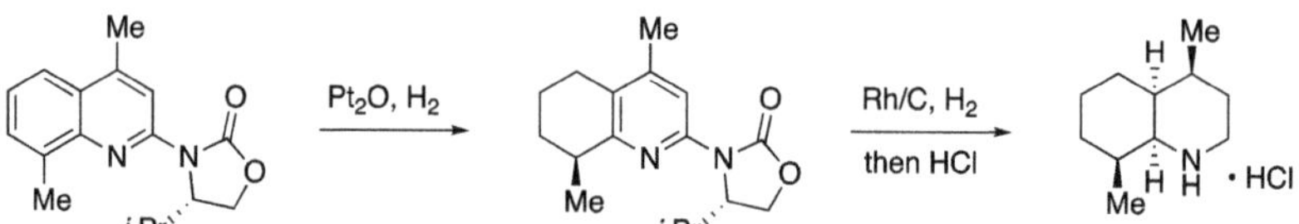

10 **11** **12**
95%, 78% de 81%, 74% de
99% ee (for major diastereomer)

Scheme 5.5 Sequential stereoselective hydrogenation of quinoline carbocycle and remaining pyridine ring, followed by hydrolytic cleavage of chiral auxiliary.

into the fully hydrogenated product **8** with 96% de with Rh/Al_2O_3 catalyst, but partly saturated **9** was obtained as the major product.

In 2010, chemo- and stereoselective hydrogenation of quinoline carbocycles has been reported by Glorius and coworkers (Scheme 5.5) *(12)*. The pyridine ring of quinoline core is exclusively reduced to give 1,2,3,4-tetrahydroquinoline in general, while the hydrogenation selectively takes place on the carbocycle to afford 5,6,7,8-tetrahydroquinoline in this report. The chiral substrate **10**, which has the oxazolidinone auxiliary at its 2-position, was converted to the desired product **11** with 78% de. Furthermore, the remaining pyridine ring of **11** was fully saturated with H_2 through Rh/C catalyst with good all *cis*-selectivity. The aminal moiety was hydrolyzed by treating hydrochloric acid to give the optically active decahydroquinoline **12**.

3. CATALYTIC ASYMMETRIC HYDROGENATION OF AROMATIC CARBOCYCLES

In 2011, Glorius reported an intriguing ligand effect in the hydrogenation of 2,3-diphenylquinoxaline **13a** with N-heterocyclic carbene (NHC) —ruthenium catalyst, which is generated in situ from [Ru(η^2-methallyl)$_2$(cod)], an imidazolinium/imidazolium salt, and potassium *tert*-butoxide *(13)*. The pyrazine moiety of quinoxaline is selectively reduced with H_2 to afford 1,2,3,4-tetrahydroquinoxaline **14a** as the sole product, when the NHC ligand has *N*-aryl substituents (Scheme 5.6, Eq. 5.1). By stark contrast, use of ICy (**17**), which has secondary alkyl groups on its nitrogens, leads to the selective formation of the 5,6,7,8-tetrahydroquinoxaline **15a** (Scheme 5.6, Eq. 5.2).

Based on the above findings, the ruthenium-catalyzed hydrogenation of **13a** was attempted with a series of optically active NHCs, which have chiral secondary *N*-alkyl groups. As a result of the ligand screening, the NHC ligand generated from **20** was found to be the best ligand for the chemo- and enantioselective hydrogenation of quinoxaline carbocycles (Scheme 5.7). With the chiral NHC—ruthenium catalyst, 6-alkyl- and 6-aryl-2,3-diphenylquinoxalines were transformed into the products **15** with good or moderate enantioselectivities. Although the large steric hindrance of the two phenyl groups at the 2- and 3-positions may obstruct the undesired

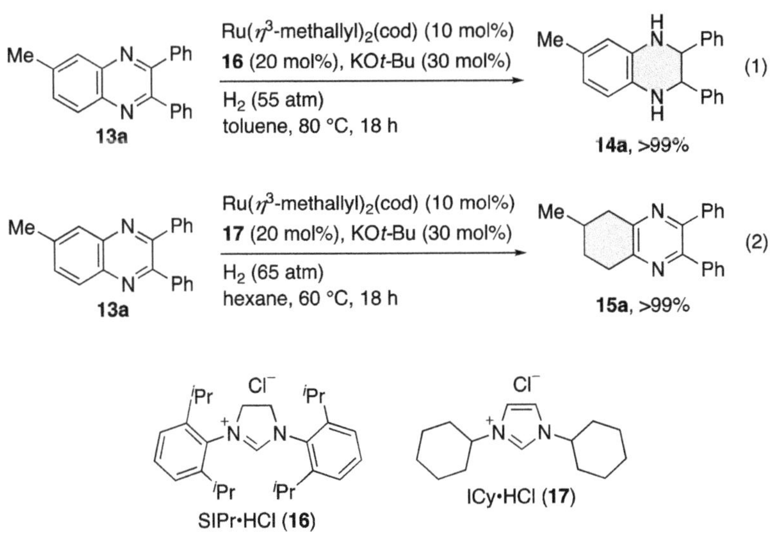

Scheme 5.6 Ligand-controlled chemoselective hydrogenation of quinoxaline.

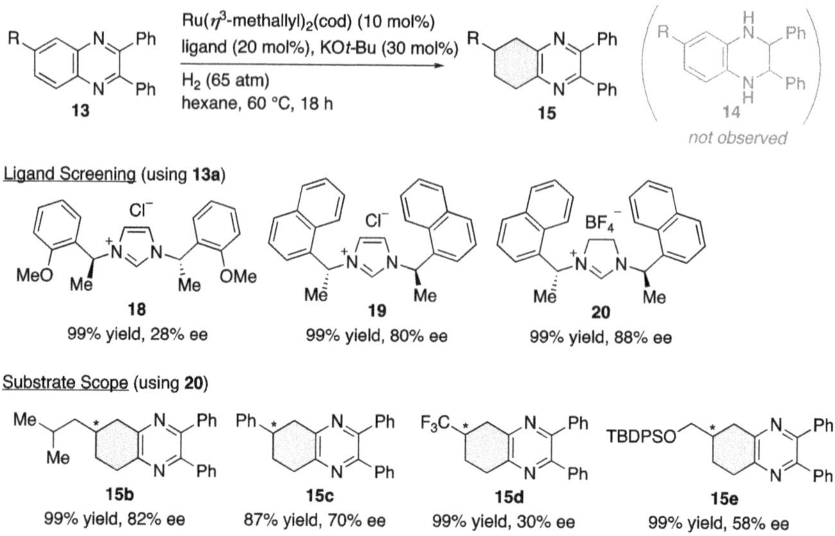

Scheme 5.7 Asymmetric hydrogenation of quinoxaline carbocycles with NHC–ruthenium catalyst.

reduction of the pyrazine ring, it is noteworthy that the reactive heteroaromatic ring remained intact during the hydrogenation of the carbocycles.

After the above pioneering work, Kuwano reported the asymmetric hydrogenation of the carbocycles of quinolines **21** and isoquinolines **24** (Scheme 5.8) *(14)*. The benzene ring is selectively reduced by using the catalyst *in situ*-generated from Ru(η^3-methallyl)$_2$(cod) and chiral bisphosphine, PhTRAP (**27**), which can form a *trans*-chelate complex with various transition metals. The chemoselectivity is affected by the choice of the phosphine ligand. The ligand effect in the hydrogenation of **24a** is summarized in Table 5.1. The chemoselectivity, ratio of **25a** and **26a**, strongly

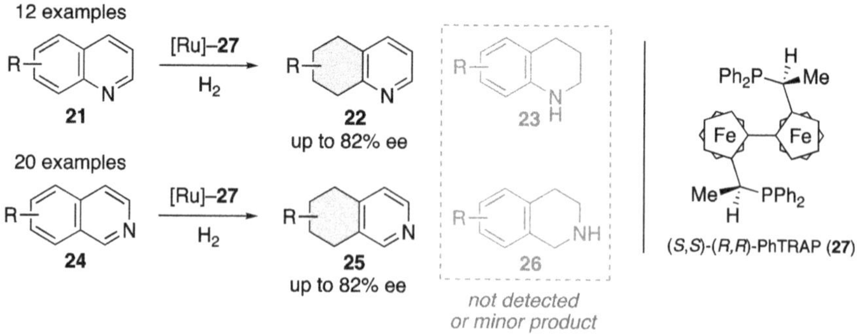

Scheme 5.8 Chemo- and enantioselective hydrogenation of quinoline and isoquinoline carbocycles with PhTRAP–ruthenium catalyst.

Table 5.1 Effect of ligand on chemoselectivity for the asymmetric hydrogenation of isoquinoline carbocycles

Ligand	Bite angle (°)	Conv. (%)	Yield 25a (%)	Yield 26a (%)	25a: 26a	ee 25a (%)
PPh_3	–	21	0	21	0:100	–
28	88.3	21	0	21	0:100	–
rac-**29**	92.7	94	2	92	2:98	–
30	99.1	82	5	77	6:94	–
31	101.2	58	9	49	16:84	–
(*R*,*R*,*R*)-**32**	160.1	98	54	44	55:45	39
(*S*,*S*)-(*R*,*R*)-**27**	163.6	>99	>99	0	100:0	58

ee, enantiomeric excess.

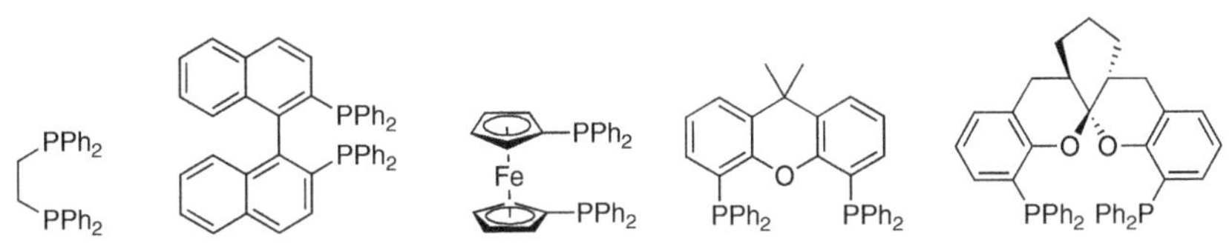

DPPE (**28**) *rac*-BINAP (**29**) DPPF (**30**) Xantphos (**31**) (*R*,*R*,*R*)-SKP (**32**)

depends on the bite angle of the bisphosphine ligand. Larger bite angle is favorable for the hydrogenation of the benzene ring. Indeed, remarkable formation of 5,6,7,8-tetrahydroisoquinoline **25a** was observed in the reaction using SKP (**32**), which chelates a metal atom in large bite angle as with **27**. Only **27**–ruthenium catalyst exhibits high chemoselectivity to exclusively produce **25a**. The chiral ruthenium catalysis allows the selective formation of various 5- or 8-substituted products **22** or **25** with little formation of **23** or **26**. The hydrogenation products were obtained with moderate to good enantioselectivities. However, the reactions of 6- or 7-substituted **21** or **24** were accompanied by a considerable amount of reduction of their pyridine ring.

The **27**–ruthenium catalyst is useful for the enantioselective hydrogenation of naphthalenes *(15)*. A variety of 2,6- and 2,7-disubstituted naphthalenes **33** are transformed into the corresponding chiral tetralines **34** with high enantioselectivites (up to 92% ee) (Scheme 5.9). Furthermore, the chiral ruthenium catalyst also allows the regioselective hydrogenation of an unsymmetrically-substituted naphthalene with high enantioselectivities.

In 2018, Zhou reported a highly enantioselective hydrogenation of 9-acetamidophenanthrenes **35**, including chrysenes, with a chiral ruthenium complex (Scheme 5.10) *(16)*. The asymmetric hydrogenation proceeded

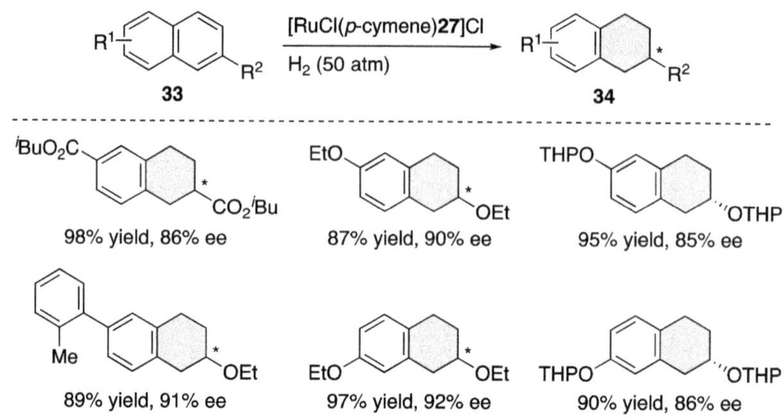

Scheme 5.9 Asymmetric hydrogenation of 2,6- and 2,7-disubstituted naphthalenes.

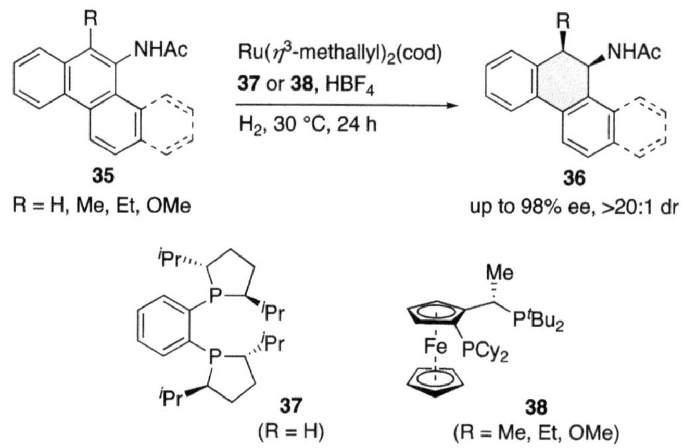

Scheme 5.10 Ruthenium-catalyzed highly enantioselective hydrogenation of 9-acetamidephenanthrenes.

with the ruthenium catalyst generated from Ru(η^3-methallyl)$_2$(cod) and **37**, yielding the chiral 9,10-dihydrophenanthrenes with high enantiomeric excesses (up to 98% ee). Each ring of phenanthrene possesses aromaticity, while the hydrogenation exclusively takes place at the K-region of phenanthrene. The C—C double bond is known to have the most olefinic (decreased aromatic) character. The ruthenium-catalyzed hydrogenation is applicable to the asymmetric hydrogenation of 9,10-disubstituted phenanthrenes. The *cis*-disubstituted dihydrophenanthrenes are formed in high diastereoselectivity as well as in high enantioselectivity using **38** as the ligand. It should be kept in mind that the stereoselective hydrogenation of tetrasubstituted alkenes still remains a challenging task in asymmetric catalysis.

4. SUMMARY

Higher reaction temperature is commonly unfavorable for the development of a new asymmetric reaction. To achieve high stereoselectivity, the asymmetric reaction at high temperature requires large activation energy gap of the transition states between two competitive stereochemical pathways. Therefore, chiral auxiliary or catalyst must be carefully and precisely designed to make the large energy gap.

The dearomative reduction of carbocyclic arenes requires high reaction temperature in most cases because the unsaturated bonds in the substrate are highly stabilized with its own aromaticity. The carbocyclic arenes remain a formidable target for asymmetric hydrogenation. Overcoming this difficulty, some highly stereoselective hydrogenations have been successfully developed by means of chiral auxiliary or asymmetric catalysis. With the former approach, the asymmetric hydrogenations of several nonfused *o*-disubstituted benzenes have been attempted with the combinational use of chiral auxiliaries and metal-particle catalyst. High stereoselectivities were successfully achieved in the reduction of *o*-toluic and anthranic acids. Furthermore, some asymmetric catalyses have been proved useful for the asymmetric hydrogenation of some (hetero)arene-fused benzene rings, e.g., quinoxalines and quinolines as well as naphthalenes. However, only limited arene substrates can be converted to the corresponding chiral cyclohexanes with high enantioselectivity as described above.

REFERENCES

1. Bianchini, C.; Barbaro, P.; Scapacci, G.; Farnetti, E.; Graziani, M. Enantioselective Hydrogenation of 2-Methylquinoxaline to (−)-(2*S*)-2-Methyl-1,2,3,4-tetrahydroquinoxaline by Iridium Catalysis. *Organometallics* **1998,** *17,* 3308−3310.
2. Kuwano, R.; Sato, K.; Kurokawa, T.; Karube, D.; Ito, Y. Catalytic Asymmetric Hydrogenation of Heteroaromatic Compounds, Indoles. *J. Am. Chem. Soc.* **2000,** *122,* 7614−7615.
3. (a) Glorius, F. Asymmetric Hydrogenation of Aromatic Compounds. *Org. Biomol. Chem.* **2005,** *3,* 4171−4175.
(b) Zhou, Y.-G. Asymmetric Hydrogenation of Heteroaromatic Compounds. *Acc. Chem. Res.* **2007,** *40,* 1357−1366.
(c) Kuwano, R. Catalytic Asymmetric Hydrogenation of 5-Membered Heteroaromatics. *Heterocycles* **2008,** *76,* 909−922.
(d) Wang, D. S.; Chen, Q. A.; Lu, S. M.; Zhou, Y.-G. Asymmetric Hydrogenation of Heteroarenes and Arenes. *Chem. Rev.* **2012,** *112,* 2557−2590.
(e) Chen, Q. A.; Ye, Z. S.; Duan, Y.; Zhou, Y.-G. Homogeneous Palladium-catalyzed Asymmetric Hydrogenation. *Chem. Soc. Rev.* **2013,** *42,* 497−511.
(f) Luo, Y.-E.; He, Y.-M.; Fan, Q.-H. Asymmetric Hydrogenation of Quinoline Derivatives Catalyzed by Cationic Transition Metal Complexes of Chiral Diamine Ligands: Scope, Mechanism and Catalyst Recycling. *Chem. Rec.* **2016,** *16,* 2693−2707.

4. (a) Gual, A.; Godard, C.; Castillon, S.; Claver, C. Soluble Transition-metal Nanoparticles-catalysed Hydrogenation of Arenes. *Dalton Trans.* **2010,** *39,* 11499–11512.
(b) Giustra, Z. X.; Ishibashi, J. S. A.; Liu, S.-Y. Homogeneous Metal Catalysis for Conversion between Aromatic and Saturated Compounds. *Coord. Chem. Rev.* **2016,** *314,* 134–181.
5. (a) Blaser, H.-U. Enantioselective Synthesis Using Chiral Heterogeneous Catalysts. *Tetrahedron Asymmetry* **1991,** *2,* 843–866.
(b) Heitbaum, M.; Glorius, F.; Escher, I. Asymmetric Heterogeneous Catalysis. *Angew. Chem. Int. Ed.* **2006,** *45,* 4732–4762.
6. Gual, A.; Godard, C.; Philippot, K.; Chaudret, B.; Denicourt-Nowicki, A.; Roucoux, A.; Castillon, S.; Claver, C. Carbohydrate-derived 1,3-Diphosphite Ligands as Chiral Nanoparticle Stabilizers: Promising Catalytic Systems for Asymmetric Hydrogenation. *ChemSusChem* **2009,** *2,* 769–779.
7. Gualandi, A.; Savoia, D. Substrate Induced Diastereoselective Hydrogenation/Reduction of Arenes and Heteroarenes. *RSC Adv.* **2016,** *6,* 18419–18451.
8. Nasar, K.; Fache, F.; Lemaire, M.; Béziat, J.-C.; Besson, M.; Gallezot, P. Stereoselective Reduction of Disubstituted Aromatics on Colloidal Rhodium. *J. Mol. Catal.* **1994,** *87,* 107–115.
9. (a) Besson, M.; Blanc, B.; Champelet, M.; Gallezot, P.; Nasar, K.; Pinel, C. Diastereoselective Hydrogenation of Substituted Aromatics on Supported Metal Catalysts. *J. Catal.* **1997,** *170,* 254–264.
(b) Besson, M.; Neto, S.; Pinel, C. Diastereoselective Hydrogenation of *O*-Toluic Acid Derivatives over Supported Rhodium and Ruthenium Heterogeneous Catalysts. *Chem. Commun.* **1998,** 1431–1432.
(c) Barbosa, L. A. M. M.; Sautet, P. Pyroglutamic Acid as a Chiral Auxiliary in the Diastereoselective Hydrogenation of Disubstituted Aromatic Rings on Rh(111): a Periodic Density Functional Theory Approach. *J. Catal.* **2003,** *217,* 23–29.
10. Ranade, V. S.; Prins, R. Diastereoselective Hydrogenation of (*S*)-Proline-2-methylanilide. *J. Catal.* **1999,** *185,* 479–486.
11. Ranade, V. S.; Consiglio, G.; Prins, R. Catalytic Diastereoselective Hydrogenation of (*S*)-Proline-modified Anthranilic Acid. *Catal. Lett.* **1999,** *58,* 71–74.
12. Heitbaum, M.; Fröhlich, R.; Glorius, F. Diastereoselective Hydrogenation of Substituted Quinolines to Enantiomerically Pure Decahydroquinolines. *Adv. Synth. Catal.* **2010,** *352,* 357–362.
13. Urban, S.; Ortega, N.; Glorius, F. Ligand-controlled Highly Regioselective and Asymmetric Hydrogenation of Quinoxalines Catalyzed by Ruthenium N-Heterocyclic Carbene Complexes. *Angew. Chem. Int. Ed.* **2011,** *50,* 3803–3806.
14. (a) Kuwano, R.; Ikeda, R.; Hirasada, K. Catalytic Asymmetric Hydrogenation of Quinoline Carbocycles: Unusual Chemoselectivity in the Hydrogenation of Quinolines. *Chem. Commun.* **2015,** *51,* 7558–7561.
(b) Jin, Y.; Makida, Y.; Uchida, T.; Kuwano, R. Ruthenium-catalyzed Chemo- and Enantioselective Hydrogenation of Isoquinoline Carbocycles. *J. Org. Chem.* **2018,** *83,* 3829–3839.
15. Kuwano, R.; Morioka, R.; Kashiwabara, M.; Kameyama, N. Catalytic Asymmetric Hydrogenation of Naphthalenes. *Angew. Chem. Int. Ed.* **2012,** *51,* 4136–4139.
16. Yan, Z.; Xie, H.-P.; Shen, H.-Q.; Zhou, Y.-G. Ruthenium-catalyzed Hydrogenation of Carbocyclic Aromatic Amines: Access to Chiral Exocyclic Amines. *Org. Lett.* **2018,** *20,* 1094–1097.

CHAPTER 6

Kinetic Control in Natural Product Synthesis

Kosuke Namba, Eisaku Ohashi
Graduate School of Pharmaceutical Science, Tokushima University, Tokushima, Japan

Contents

1. INTRODUCTION

Advances in organic synthetic chemistry have been remarkable, and in a simple comparison of chemical yields and stereoselectivity, there are many excellent reactions that are considered to be quite difficult to find further progress. However, even when they are combined, synthesis of natural products having complex carbon frameworks and many functional groups is usually extremely difficult. That is, although "multistep synthesis" that combines various chemical transformation reactions is required for the synthesis of such complex natural products, it is still difficult to synthesize by the simple combinations of existing chemical reactions. To synthesize complex natural products, it is necessary to develop innovative and truly practical synthetic methods based on a deep understanding and consideration of target molecule, in addition to constructing a reasonable and efficient route. Furthermore, it is also required to set reaction substrates, reagents, and reaction conditions in consideration of the selectivity generated by kinetic control and thermodynamic control in all reaction steps. In other words, by taking the advantage of kinetic conditions and thermodynamic conditions, desired selectivities such as stereoselectivity,

Kinetic Control in Synthesis and Self-Assembly
ISBN 978-0-12-812126-9
https://doi.org/10.1016/B978-0-12-812126-9.00006-7

functional group selectivity, regioselectivity, site selectivity, and so on would be formed. For this reason, reaction design based on the concept of kinetic control and thermodynamic control is always performed in the field of total synthesis. In this chapter, we will introduce the examples of the effective use of kinetic control and thermodynamic control in the total synthesis, focusing the total synthesis of palau'amine that was recently accomplished in our laboratory.

2. PALAU'AMINE

Palau'amine (**1**) was originally isolated from a sponge, *Stylotella agminata*, in 1993 by Scheuer as a novel class of pyrrole–imidazole alkaloids *(1)*, and its proposed structure was revised in 2007 *(2)*. Since the initial structure elucidation and the later revision, palau'amine (**1**) has been an attractive synthetic target due to its intriguing molecular architecture and significant biological properties including antifungal, antitumor, and immunosuppressive activities. Its noteworthy structural features include two guanidine moieties, a fused polycyclic system containing a spirocycle, a fully substituted complex cyclopentane ring, eight contiguous stereogenic centers including a nitrogen-substituted quaternary carbon center, and the *trans*-bicyclo[3.3.0]octane skeleton (D/E ring junction). Not surprisingly, many attempts to synthesize palau'amine and related compounds have been reported, and numerous reviews of these different approaches have been published *(3)*. However, until our report in 2015 *(4)*, there had been only one report of a total synthesis, by Baran's group in 2010 *(5)*, which was followed by the development of an asymmetric version in 2011 *(6)*. The most difficult challenge in the total synthesis of **1** is the construction of a *trans*-bicyclo[3.3.0]octane system that corresponds to the D/E ring junction. Thus, we were looking to establish an efficient method for constructing the unique polycyclic ring system including a *trans*-bicyclo [3.3.0]octane skeleton that makes up the core structure of **1**. In 2015, we finally accomplished the total synthesis of palau'amine **1** through the construction of the ABDE tetracyclic ring system by taking advantage of kinetic control *(4)*.

Our plan for the construction of tetracyclic ring core **2** is outlined in Fig. 6.1. Palau'amine **1** would be obtained by the transformation of functional groups of **2**. Because the acylimine moiety of **3** is highly electron-deficient, we expected that the nucleophilic addition of amide anion to the C10 carbon center would occur to form the D ring.

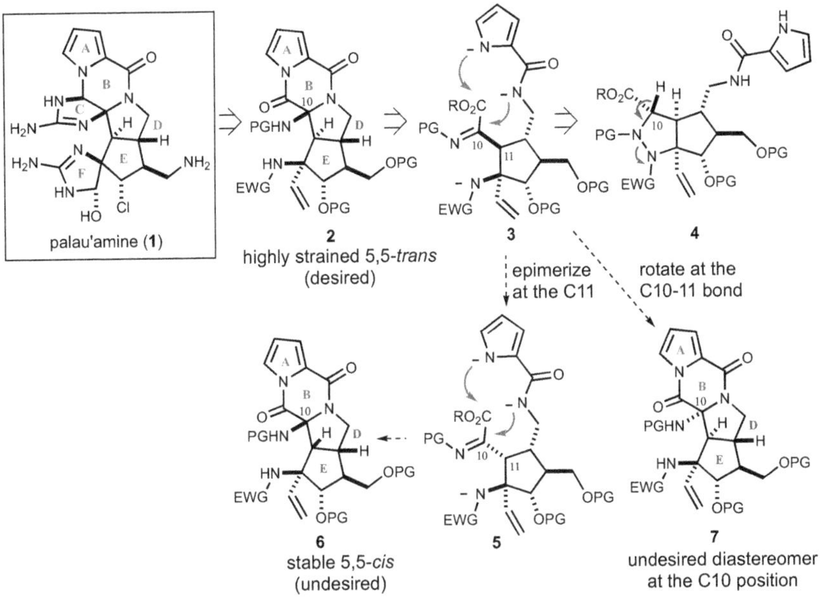

Figure 6.1 Structure of palau'amine (**1**) and its synthetic plan.

The iminoester moiety of **3** would be generated from hydrazide **4** by elimination to form an imine at the C10 position with concomitant N−N bond cleavage. Overman and coworkers have also employed a hydrazine fragmentation in their efforts toward the palau'amine core, in which reductive cleavage of the N−N bond occurred spontaneously due to the high strain of a pyrazolidine ring *(7)*. On the other hand, we planned to adopt an E1cB eliminative cleavage of the N−N bond of **4** to obtain iminoester **3** directly, although, unlike the case of reductive cleavage, there have been very few examples of such eliminative cleavage of N−N bonds *(8)*. The direct formation of **3** under basic conditions was expected to induce further cascade cyclization reactions leading to **2**, i.e., the ABDE tetracyclic ring core **2** might be obtained from **4** in a single step. However, as the *trans*-bicyclo[3.3.0]octane skeleton is highly strained, there is concern that a stable *cis*-bicyclo[3.3.0]octane skeleton is formed via an intramolecular cyclization of **5** that is derived from an epimerization at the C11 position. In addition, the formation of undesired diastereomer **7** is also a concern, if rotation of the C10-11 bond occurs before the nucleophilic addition of amide anion of **3** (Fig. 6.1). Therefore, a suitable kinetic control was required for the cyclization of **3** to occur in prior to the epimerization at the C11 position and the rotation of C10-11 bond, overcoming the steric strain of the *trans*-bicyclo[3.3.0]octane skeleton (D/E ring junction). To implement our plan, we therefore first attempted to synthesize the cyclization precursor **4** by means of $Hg(OTf)_2$-catalyzed cyclization reaction developed in our laboratory *(9)*.

3. SYNTHESIS OF CYCLIZATION PRECURSOR 4

Our synthesis started with a Morita–Baylis–Hillman reaction of 2-cyclopenten-1-one **8** and methoxyacetaldehyde followed by a sequential operation of acetylation, a Luche reduction, and a TBS protection to afford **9** as a 3:1 diastereomeric mixture in 49% four-step yield (Fig. 6.2). Without separation, the diastereomeric mixture of the acetates **9** was subjected to an Ireland–Claisen rearrangement. After treatment of **9** with LHMDS/TBSCl/HMPA in THF at −78°C, refluxing in toluene induced the desired reaction to afford the cyclopentylidene carboxylic acid **10** in good yield. The double bond geometry of **10** was determined to be *Z* by an NOE experiment on its amide derivative. Next, we attempted to prepare an acyltosylhydrazide by coupling **10** with *N*-tosylhydrazide under the combined action of 1-ethyl-3-(3-dimethylaminopropyl)carbodiimide (EDCI) and 4-dimethylaminopyridine (DMAP) in dichloromethane. Surprisingly, the nitrogen atom masked with a tosyl group participated in the condensation to give a diastereomeric mixture of **11** in 78% yield. After the examination of various reaction conditions, we found that the regioselectivity of acylation of *N*-tosylhydrazine depends on the presence of DMAP catalyst, i.e., *N,N*-acyltosylhydrazines are the sole products in the presence of DMAP, whereas *N*-acyl-*N'*-tosylhydrazines are obtained without DMAP catalyst *(10)*. Although the origin of regioselectivity with

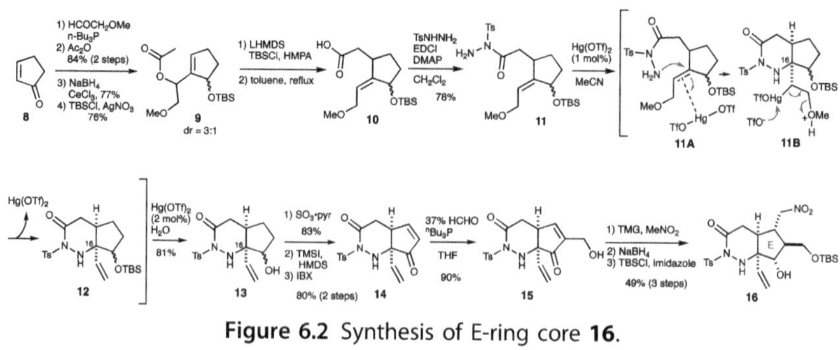

Figure 6.2 Synthesis of E-ring core **16**.

DMAP catalyst is still unclear, investigations to elucidate the mechanism are currently underway in our laboratory. Next, the $Hg(OTf)_2$-catalyzed cyclization as a key reaction for construction of the E-ring system was examined. Treatment of **11** with 1 mol % of $Hg(OTf)_2$ in acetonitrile at room temperature induced a hydrazidomercuration reaction to give pyridazine intermediate (**11A** → **11B**). Subsequent protodemercuration of **11B** afforded desired vinyl-heterocycle **12** and methanol while regenerating the $Hg(OTf)_2$ catalyst. After a thin-layer chromatography (TLC) check of conversion of **12** into **13**, the reaction mixture was directly treated with an additional 2 mol% of $Hg(OTf)_2$ and H_2O to remove the TBS group at the C17 position, and these sequential operations gave **13** in 81% yield from **11**. The stereochemical outcome at the ring junction was completely controlled to be *cis* regardless of the stereochemistry of the secondary alcohol at C17. The structure of each diastereomer of **13** was unambiguously confirmed by an X-ray diffraction study and NOE studies. SO_3·pyridine oxidation of the diastereomeric mixture of **13** gave the corresponding ketone in 83% yield. Direct oxidation of the resulting ketone to enone **14** using IBX or selenium dioxide was not successful, nor was oxidation of the ketone via its enolate using selenium halide, sulfinimidoyl chloride, or *N*-bromosuccinimide (NBS). Although the preparation of the silyl enol ether was initially difficult, a combination of TMSI and hexamethyldisilazane in dichloromethane was later found to give the corresponding trimethylsilylenolether in quantitative yield. In contrast to the direct oxidation of the ketone, IBX oxidation of this silylenolether provided enone **14** in good yield. A Morita–Baylis–Hillman reaction of **14** with formaldehyde gave alcohol **15** in excellent yield. Subsequent 1,4-addition of nitromethane in the presence of a catalytic amount of 1,1,3,3-tetramethylguanidine afforded the desired adduct. We found that this readily underwent dehydration during column chromatography purification to give the exomethylene product. Therefore, the crude 1,4-adduct was directly subjected to reduction with $NaBH_4$ followed by TBS protection of the primary alcohol to afford **16** in 49% yield from **15**. The stereochemistry of **16** was confirmed by an NOE experiment *(11)*, and it was the result of kinetic controls that derived from a convex attack of nitromethane followed by trans-orientation of α-side chain to avoid steric repulsion with the resulting nitromethyl group.

In this way, we accomplished to construct the tetrasubstituted carbon center possessing nitrogen at the C16 position by means of $Hg(OTf)_2$-catalyzed cyclization. On the other hand, the similar cyclization of **17** by using NBS afforded not hydropyridazine **19** but lactone **18** (Fig. 6.3) *(12)*. The reaction initially afforded multiproducts, and the products slowly converged on lactone **18**. Thus, with regard to the formation of **18**, we speculated that the *O*-cyclization occurred predominantly to give an unstable **17A**, which was converted into **17B** through sequential reactions that involved the migration of nitrogen atoms and the hydrolysis of

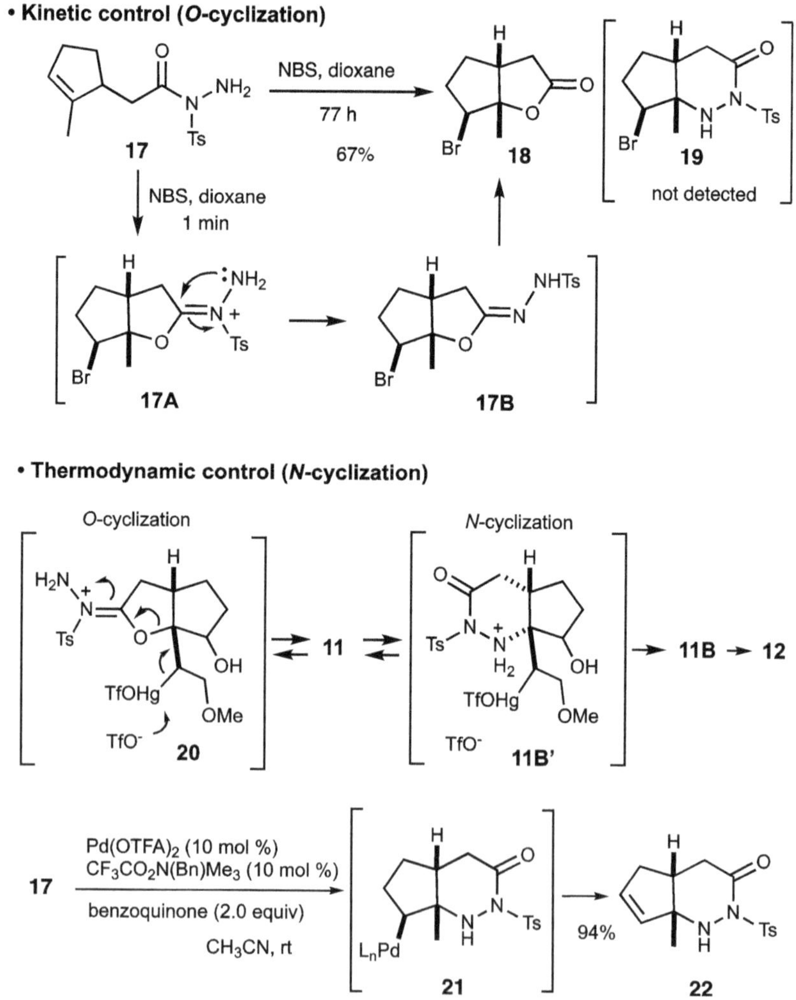

Figure 6.3 *N*,*O*-selectivity in the cyclization of *N*,*N*-acyltosylhydrazine according to the thermodynamics and kinetic controls.

resulting hydrazonether **17B** (Fig. 6.3). This result revealed that the bromohydrazidation reaction predominantly induces the *O*-cyclization to give lactone, although γ,δ-unsaturated analogues can possibly afford both *O*-cyclized 5-membered lactone and *N*-cyclized 6-membered hydropyridazine. In clear contrast, the $Hg(OTf)_2$-catalyzed cyclization reaction of γ,δ-unsaturated *N,N*-acyltosylhydrazide **11** afforded the *N*-cyclized hydropyridazine **12**, and *O*-cyclized lactone was not observed (Fig. 6.2). Based on the results of the bromohydrazidation reaction of **17**, the $Hg(OTf)_2$-catalyzed cyclization reaction of **11** was also suggested to predominantly give the *O*-cyclized product **20**. However, because the starting materials and mercury-cyclized intermediate are in equilibrium state, unlike the bromoxygenation intermediate **17A**, *(13)* the unstable oxymercuration product **20** immediately reverts to **11** before the migration of nitrogen atoms. On the other hand, N-cyclized product **11B′** led to an alternative deprotonation step to generate the stable hydropyridazine and trifluoromethanesulfonic acid, the latter of which induced a protodemercuration step as shown in Fig. 6.3. Similarly, the cyclization reaction of **17** with a palladium(II) catalyst as a divalent transition metal also afforded *N*-cyclized product **22** in 94% yield (Fig. 6.3) *(12)*. Therefore, we found that the N,O-selectivity of the cyclization of *N,N*-acyltosylhydrazine can be switched by the electrophiles, i.e., the electrophiles leading to reversible or irreversible cyclization intermediates afford *N*-cyclized hydropyridazines or O-cyclized lactones, respectively. In other words, the cyclization precursor *N,N*-acyltosylhydrazine **17** predominantly gave the *O*-cyclized **18** under kinetic control and the *N*-cyclized hydropyridazines under thermodynamics control. Although we used the thermodynamics condition for the construction of C16 position of palau'amine, kinetic control condition should be adopted if the construction of tetrasubstituted carbon center possessing oxygen is required. Thus, this is one example of the proper use of the kinetic control and thermodynamics control in the total synthesis.

Having established the construction of E-ring core **18** that includes the quaternary carbon center at the C16 position, transformation of **18** into cyclization precursor **4** was next examined. Treatment of **18** with SmI_2 induced removal of the tosyl group and reduction of the nitro group to amine. The resulting primary amine was directly protected with the Fmoc group, after which further protections of the secondary alcohol and acylhydrazide proceeded smoothly to give **23** after treatment with *tert*-butyldimethylsilyl trifluoromethanesulfonate as silylating reagent followed by di-*tert*-butyl dicarbonate (Boc_2O) in the presence of a catalytic amount of DMAP. As we expected, Boc protection occurred on the nitrogen possessing the acyl group due to the relative acidity of the NH protons and steric hindrance. Treatment of **23** with triethylsilyl trifluoromethanesulfonate and N,N-diisopropylethylamine at −78°C readily afforded a silyl ketene aminal, and this crude product was treated with NBS in a mixed solvent of THF and methanol to give bromide **24** in an 82% two-step yield along with 7%−14% of starting material **23**. The stereochemistry of **24** was determined by a nuclear Overhauser effect spectroscopy (NOESY) experiment, which proved the β-configuration of bromide. It is likely that the attack of bromide on the concave face was due to steric hindrance by the vinyl group. This selectivity is also a kind of kinetic control. Subsequent methanolysis of **24** afforded an amide anion **25** that immediately induced an intramolecular S_N2 reaction leading to **26**. At this stage, we considered **26** would readily epimerize to **27** to avoid unfavorable steric interactions on the more crowded concave face. The stereochemistry of **27** was in fact confirmed by a NOESY experiment. Next, we attempted to introduce a strong electron-withdrawing group onto the nitrogen on the tetrasubstituted carbon center. After various attempts, we found that only a trifluoroacetyl group could be introduced to afford **28** by treatment with an excess amount of trifluoroacetic anhydride. Although over-acetylated product **29** was also obtained, the extra trifluoroacetyl group on the carbamate was easily removed in methanol at 40°C in quantitative yield. Finally, removal of the Fmoc group and condensation of the resulting primary amine with pyrrole trichloromethyl ketone afforded **30**, the precursor of the key cascade reaction, in an 83% three-step yield (Fig. 6.4).

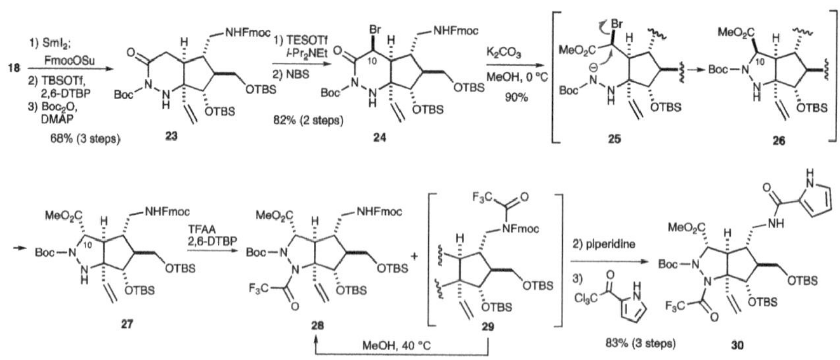

Figure 6.4 Synthesis of cyclization precursor **30**.

4. SINGLE-STEP CONSTRUCTION OF ABDE TETRACYCLIC RING CORE USING KINETIC CONTROL CONDITION

Having prepared the cyclization precursor **30**, we attempted its single-step conversion into an ABDE tetracyclic ring system. To induce E1cB eliminative cleavage of the N—N bond, **30** was treated with 3.0 equiv of lithium hexamethyldisilazide (LHMDS) acting as a strong base at −78°C and was then allowed to warm to room temperature (condition I). This reaction afforded the expected tetracyclic compound **31** that corresponds to the ABDE ring of palau'amine, including the *trans*-bicyclo[3.3.0]octane skeleton, in modest yield (Fig. 6.5). The *trans*-fused D/E ring junction and the stereochemistry of **31** were confirmed by an NOE experiment, and by comparison of the coupling constant at the C11 proton with that of the natural product. Its characteristic value ($J = 14.5$ Hz) is typical of a *trans*-bicyclo[3.3.0]octane skeleton. The reaction pathway leading to the ABDE ring system **31** is explained below. Treatment of **30** with 3.0 equiv of a strong base would abstract two NH protons and the hydrogen at the C10 position, thereby inducing a β-elimination of the nitrogen of **30A** that

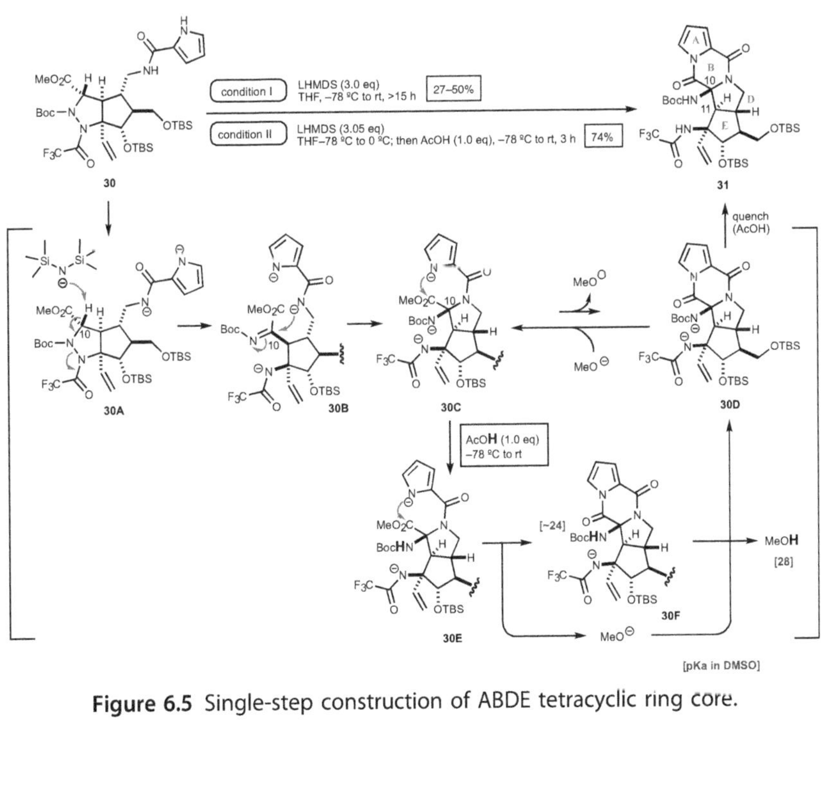

Figure 6.5 Single-step construction of ABDE tetracyclic ring core.

possesses an electron-withdrawing group, as shown in Fig. 6.5. This would lead simultaneously to both cleavage of the N—N bond and imine formation at the C10 position to give **30B**. The nucleophilic addition of amide anion to the electron-deficient C10 carbon center would occur immediately to give **30C** possessing a *trans*-bicyclo[3.3.0]octane skeleton (D/E ring junction). Furthermore, the cascade reaction would not stop at **30C** due to the remaining pyrrole anion, and subsequent condensation of this with the methyl ester would form the B ring to give **31** (Fig. 6.5). These results encouraged us to scale up this reaction from a few milligrams to 100 mg with the goal of completing the total synthesis of palau'amine. However, the reaction was found to suffer from poor reproducibility, often leading to poor yields even after a prolonged reaction time. After careful consideration of the reaction mechanism based on the kinetic control, we established an improved method involving the partial protonation of the anionic intermediates with acetic acid (condition II). On treatment with 3.0 equiv of LHMDS, substrate **30** undergoes stepwise lithiation of the two NH protons at −78°C and then the C10-proton at around 0°C. Once the C10 proton is abstracted, the formation of **30B** and the subsequent cyclization of amide anion leading to **30C** would proceed rapidly and completely. Indeed, at this stage all that could be detected on TLC was the protonated form of **30C**, indicating fast conversion of **30A** into **30C** as well as slow formation of **30D** from **30C**. The smaller p*K*a value (in DMSO) of the pyrrole moiety (considered to be less than 23) than of methanol (28) predicts that the equilibrium between **30C** and **30D** should favor the former, which led us to remove the methoxide ion without quenching the pyrrole anion. Thus, after checking the conversion of **30** into **30C** by TLC, the mixture was cooled to −78°C again, and an exact 1.0 equiv of acetic acid was slowly added. On warming to room temperature, the desired product **31** was obtained in up to 74% yield with good reproducibility on acceptable scales for total synthesis (Fig. 6.5).

The mechanism underlying the cascade reaction under "condition B" is explained as follows. The intermediate **30C** possesses three nitrogen anions, that of the Boc-carbamate (~24), pyrrole (<23), and trifluoroacetoamide (~17), according to their order of basicity from comparison of the p*K*a values of the protonated NH functional groups. Thus, an exact 1.0 equiv of acetic acid would protonate only the Boc-carbamate anion, and trianion **30C** would be converted into dianion **30E**. While the mixture was warming to room temperature, the remaining pyrrole anion induced condensation with the methyl ester, simultaneously generating methoxide as in the case with trianion. However, because the condensation product **30F** possesses an active NH proton, unlike the case with the trianion, methoxide does not attack the pyrrole amide but rather is quenched in abstracting the active Boc-NH proton to give the same dianion **30D**. Among various acids tried, including phenol and imide derivatives, acetic acid was found to be the best protonating reagent. In addition, precise amounts of LHMDS and acetic acid are very important for this "protonation-state switching by p*K*a game" reaction. As stated above, we succeeded in establishing an efficient method for constructing the basic core structure of palau'amine.

In the field of total synthesis, although reaction setting based on kinetics control is commonly used for selectivity control such as stereoselectivity, regioselectivity, and so on, we introduced the useful method for the control of reaction equilibrium as an example of kinetic control in this section. That is, as it is impossible to predominantly lead to **30D** that generates more unstable methoxide from the equilibrium state of **30C** and **30D** under the thermodynamic condition, the complete conversion into **30D** was achieved by switching this equilibrium state to the irreversible kinetic control condition via proton relay.

The next issue of this cascade reaction was how the amide anion of **30B** overcame the steric strain of the trans-bicyclo[3.3.0]octane skeleton to approach the C10 carbon center, although the Boc-imine moiety of **30B** was highly electron-deficient. In addition, the cyclization of **30B** was needed to occur rapidly before the rotation of the C10-11 bond and epimerization at the C10 carbon center. To solve the above issues, the coordination effect of lithium ion was focused. When the lithium countercation of the amide anion forms a coordination bond with the carbonyl group of methyl ester, the amide nitrogen (N14) and the C10 carbon center would be located at transannular positions that are close to each other due to the strain of the 8-membered ring. To investigate this coordination effect, theoretical calculations were carried out by the DFT method. The optimized structure of **30B** including lithium ions is shown in Fig. 6.6A. Interestingly, the calculation actually indicated not only the expected coordination of lithium amide to the carbonyl group of methyl ester but also the unpredictable coordination of lithium salt of pyrrole anion to the carbonyl oxygen of the Boc group. Therefore, the two trans-oriented side chains were quite close to each other by the chelation to two lithium ions, and the distance between N14 and C10 at the transannular positions was calculated to be only 2.94 Å. Because of the short distance between the two

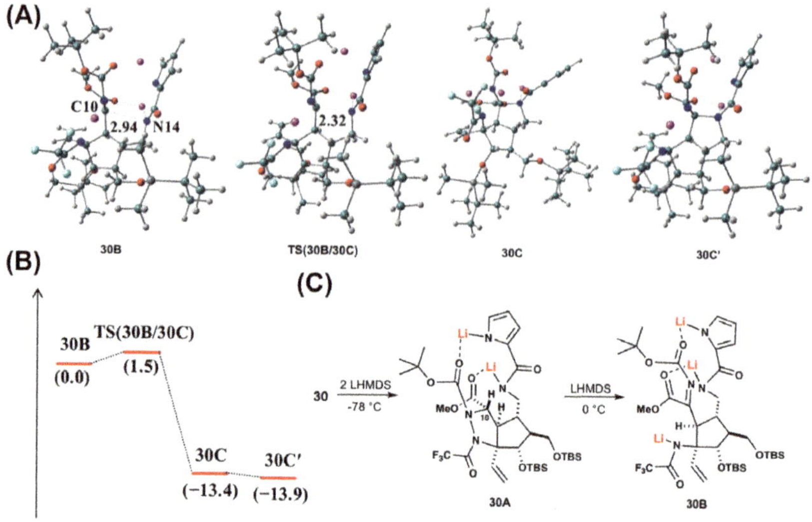

Figure 6.6 Optimized structures and energy diagram of **30**.

reaction points, the energy barrier of the cyclization reaction (**30B** → **30C**) was estimated to be only 1.5 kcal/mol, allowing the reaction to proceed smoothly (see Fig. 6.6B for the potential energy profile of the cyclization reaction). Thus, the nucleophilic addition of the amide anion (N14) occurred before the rotation of the single bond of C10—C11, which was also restricted by the coordination effect of lithium ion, to afford **31** as a single diastereomer involving the β-configuration of the NHBoc group at the C10 position. From the calculated result, it was concluded that the chelation effect of lithium ion played a significant role in the formation of the trans-bicyclo[3.3.0]octane skeleton. Actually, this cascade reaction in the presence of 3.6 equiv of hexamethylphosphoric triamide (HMPA) as a lithium ion scavenger afforded a complex mixture, and the desired cyclized (D/E ring) products and related intermediates were not detected. Furthermore, the use of other bases, such as NaHMDS and KHMDS, also did not give any desired cyclization products. In addition, the yield of tetracyclic **31** was dramatically decreased to 10%–20%, when the initial treatment of 3.0 equiv of LHMDS was directly conducted at 0°C. This result clearly suggested that the coordination of lithium ions to the two carbonyl groups must be formed first at −78°C before the abstraction of the C10 proton occurs at 0°C as shown in Fig. 6.6C. The coordination of lithium ion promotes the kinetic control by fixing to the preferred conformation for cyclization, and it induced the rapid cyclization leading to the highly strained *trans*-bicyclo[3.3.0]octane skeleton **31** before the epimerization at the C11 carbon center and the rotation of the C10–11 bond, in which the stable *cis*-bicyclo[3.3.0]octane skeleton **6** and undesired diastereomer **7** will be obtained, respectively.

5. TOTAL SYNTHESIS OF PALAU'AMINE

Having established an efficient method for constructing an ABDE ring system that overcame the most difficult barrier to the total synthesis of palau'amine, we attempted to complete the total synthesis from **31** (Fig. 6.7). We first tried to construct a C-ring by using an amino group (N9) on C10 and a carbonyl group at C6 as footholds. The amino group (N9) of **31** was converted into an isothiourea by a sequential operation of removal of the Boc group according to Ohfune's method *(14)*, formation of thiourea using *N*-benzyloxycarbonyl isothiocyanate (CbzNCS), selective reduction of the pyrrole amide at the C6 position, and conversion of the thiourea into the isothiourea to give **32** in 73% overall yield over four steps from **31**. Removal of the Boc group did not proceed if the pyrrole amide was reduced first, indicating that the planar configuration of the sp^2 carbon center reduced steric hindrance around the N9 amino group. Furthermore,

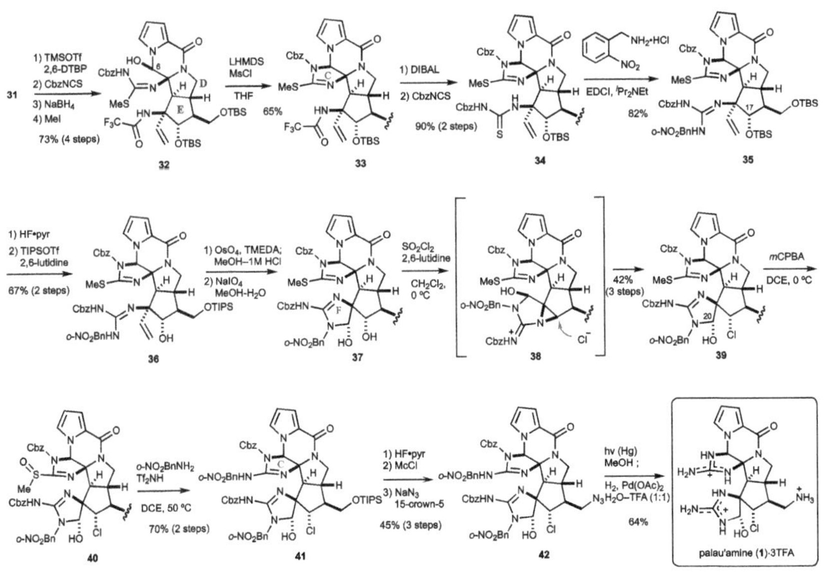

Figure 6.7 Total synthesis of palau'amine (**1**).

the use of carbodiimide derivatives for direct formation of the guanidine did not proceed, and only CbzNCS acting as a small and reactive reagent was able to react with the sterically hindered N9 amine intermediate. Because the thiourea moiety could not be directly converted into a guanidine, the thiourea moiety was converted into isothiourea **33**. At this stage, the structure of the synthetic intermediate including the *trans*-bicyclo[3.3.0]octane skeleton was unambiguously confirmed by an X-ray diffraction study. Treatment of **33** with LHMDS and methanesulfonyl chloride at −78 to −40°C induced the sequential reactions of mesylation, elimination of mesylate, and addition of the isothiourea to give pentacycle **34** in 65% yield in a manner similar to that seen in the synthesis of (+)-dibromophakellstatin reported by Nagasawa *(15)*. Having constructed the C-ring, we next tried to form the F-ring. Only reductive conditions using diisobutylaluminum hydride were successful for the removal of the trifluoroacetyl group, and subsequent treatment of the resulting amine with CbzNCS afforded thiourea **35** in good yield. The thiourea moiety was directly converted into guanidine **36** by condensation using EDCI *(16)* with *o*-nitrobenzylamine in 82% yield. However, oxidative cleavage of the vinyl group without oxidizing the pyrrole was difficult, probably due to the steric bulk of the TBS group at the C17 position. Thus, the two TBS groups were removed by HF·pyridine, and only the primary alcohol was protected again by the triisopropylsilyl (TIPS) group to give **37** in a 67% two-step yield. With the secondary hydroxyl group unprotected, dihydroxylation using OsO_4 and tetramethylethylenediamine in dichloromethane *(17)* proceeded smoothly at −78°C to give the osmate, and subsequent hydrolysis in a mixed solution of methanol and 1M HCl afforded the desired diol. Oxidative cleavage of the diol was successful only in a mixed solution of methanol and water (4:1) to give an unstable F-ring product **37**, which gradually decomposed even under the neutral conditions, probably due to a retro-aldol reaction at the C17 position. Product **37** was therefore used directly in the next reaction without purification. With the construction of the ABCDEF ring system of palau'amine complete, the functional group conversions necessary to form **1** were finally attempted. Substitution of the secondary hydroxyl group with chloride with stereoretention proceeded by exploiting a neighboring-group effect with the guanidine-containing F-ring. Secondary alcohol activation using sulfuryl chloride induced the participation of guanidine depicted in **38** analogous to "massadine aziridine,"*(18)* and subsequent nucleophilic attack of chloride afforded **39** with the α-configuration of chloride, along with a small amount of the epimer at the C20 position in the unnatural configuration. On the other hand, the hemiaminal hydroxyl group also reacted with excess sulfuryl chloride to give a dichloride when more than 1.0 equiv of sulfuryl chloride was used, but the chloride at the C20 position was readily hydrolyzed back to **38** in THF-H_2O at 50 °C. Next we had to develop a new method for converting the methylthio group into an amino group leading to a guanidine under acidic conditions, because we found that our hexacyclic intermediates were readily decomposed under basic conditions. After much examination, we achieved the desired conversion by the reaction of a trifluoromethanesulfonyl imide salt of *o*-nitrobenzylamine with sulfoxide **40** to give **41** in a 70% two-step yield. Use of the further oxidized sulfone, or of a salt other than trifluoromethanesulfonic imide, afforded only a complex mixture. We latter demonstrated that the formation of the trifluoromethanesulfonyl imide salt accelerates the introduction of various amino groups to isothioureas leading to the corresponding guanidines *(19)*. In the final sequence of functional group transformations, the TIPS group was removed, the resulting primary alcohol was converted into its chloromethanesulfonate *(20)* as a suitable leaving group, and a subsequent azidation reaction proceeded at room temperature to afford **42** a protected palau'amine. When methanesulfonate was adopted as the leaving group, the azidation reaction required a high temperature (50°C) that also induced azidation at the C17 position via a neighboring-group effect. Fortunately, the minor epimer with the unnatural configuration at the C20 position disappeared at this stage. Finally, Hg-lump irradiation followed by direct hydrogenation led to deprotection of **42** affording **1** in 64% yield as its 3TFA salt. Synthetic **1** showed spectral data (^{1}H and ^{13}C NMR, HRMS) completely identical to those of natural palau'amine **1** *(1,2)*. Furthermore, the synthetic palau'amine·3TFA salt was confirmed to exhibit strong immunosuppressive activity *(4)*.

Through this total synthesis **1** was obtained in 0.039% overall yield after 45 steps (78% average yield at each step) from commercially available cyclopentenone **8**. As a result of this synthesis, many synthetic intermediates and derivatives of **1** possessing various ring core systems were actually obtained for the first time. In future studies, we plan to perform activity evaluations for each deprotected synthetic intermediate to further elucidate the pharmacophore.

6. CONCLUSION

In summary, we have achieved the total synthesis of palau'amine **1** via the single-step construction of an ABDE tetracyclic ring system. In this single-step construction, the chelation effect of lithium salt forming the 8-membered ring was crucial for the construction of *trans*-bicyclo[3.3.0] octane skeleton (D/E ring junction) by bringing two reaction points close to each other at the transannular position, and it induced rapid cyclization before the epimerization at the C11 carbon center and the rotation of the C10-11 bond. In addition, the equilibrium between **30C** and **30D** in favor of **30C** under thermodynamic condition was switched to the kinetic condition by the addition of acetic acid that leads to the irreversible proton relay reaction, and the resulting kinetic condition predominantly afforded **30D**. They were examples using kinetic control effectively in total synthesis. In contrast, the thermodynamic condition was also efficiently used in the construction of tetrasubstituted carbon center at the C16 position. Therefore, in the study for total synthesis, it is significant to properly use the thermodynamic and the kinetic conditions depending on the situation. That is, proper application of the kinetic and thermodynamic controls to each reaction is the key to achieve the efficient total synthesis.

With the establishment of an efficient method using kinetic control for constructing the basic tetracyclic structure of palau'amine, various synthetic analogues for SAR study and chemical probes of palau'amine have become accessible. Therefore, the chemistry described here offers not only a solution to a formidable synthetic challenge but also an alternative synthetic route to elucidate the pharmacophore and the mechanistic details of the bioactivities of palau'amine. Because our synthetic route at its current stage of realization is actually too long to develop **1** as a practical immunosuppressive agent, further improvement of the key cascade cyclization reaction and the development of a shorter route to the cyclization precursors in a second-generation synthesis are also currently underway in our laboratory.

REFERENCES

1. (a) Kinnel, R. B.; Gehrken, H.-P.; Scheuer, P. J. Palau'amine: A Cytotoxic and Immunosuppressive Hexacyclic Bisguanidine Antibiotic from the Sponge Stylotella Agminata,. *J. Am. Chem. Soc.* **1993,** *115,* 3376–3377.
(b) Kinnel, R. B.; Gehrken, H.-P.; Swali, R.; Skoropowski, G.; Scheuer, P. J. Palau'amine and its Congeners: A Family of Bioactive Bisguanidines from Marine Sponge Sylotella Uranium. *J. Org. Chem.* **1998,** *63,* 3281–3286.

2. (a) Grube, A.; Köck, M. Structural assignment of tetrabromostyloguanidine: Does the relative configuration of the palau'amines need revision? *Angew. Chem. Int. Ed.* **2007,** *46,* 2320–2324.
   (b) Buchanan, M. S.; Carroll, A. R.; Addelpalli, R. Natural products, stylissadines A and B, specific antagonists of the P2X7receptor, an important inflammatory target. *J. Org. Chem.* **2007,** *72,* 2309–2317.
   (c) Buchanan, M. S.; Carroll, A. R.; Quinn, R. J. Revised structure of palau'amine. *Tetrahedron Lett.* **2007,** *48,* 4573–4574.
3. For recent reviews of synthetic studies towards palau'amine, see:
   (a) Wang, X.; Chen, C. An approach for the synthesis of nakamuric acid. *Tetrahedron* **2015,** *71,* 3690–3693.
   (b) Wang, X.; Ma, Z.; Wang, X.; De, S.; Ma, Y.; Chen, C. Dimeric pyrrole–imidazole alkaloids: synthetic approaches and biosynthetic hypotheses. *Chem. Commun.* **2014,** *50,* 8628–8639.
   (c) Moris, J. C.; Phillips, A. J. Marine natural products: Synthetic aspects. *Nat. Prod. Rep.* **2011,** *28,* 269–289.
   (d) Gaich, T.; Baran, P. S. Aiming for the ideal synthesis. *J. Org. Chem.* **2010,** *75,* 4657–4673.
4. Namba, K.; Takechi, K.; Kaihara, Y.; Oda, M.; Nakayama, A.; Nakayama, A.; Yoshida, M.; Tanino, K. Total synthesis of palau'amine. *Nat. Commun.* **2015,** *6,* 8731.
5. Seiple, I. B.; Su, S.; Young, I. S.; Lewis, C. A.; Yamaguchi, J.; Baran, P. S. Total synthesis of palau'amine. *Angew. Chem. Int. Ed.* **2010,** *49,* 1095–1098.
6. Seiple, I. B.; Su, S.; Young, I. S.; Nakamura, A.; Yamaguchi, J.; JØrgensen, L.; Rodriguez, R. A.; O'Malley, D. P.; Gaich, T.; Köck, M.; Baran, P. S. Enantioselective total syntheses of (–)-palau'amine, (–)-axinellamines, and (–)-massadines. *J. Am. Chem. Soc.* **2011,** *133,* 14710–14726.
7. Overman, L. E.; Rogers, B. N.; Tellew, J. E.; Trenkle, W. C. Stereocontrolled synthesis of the tetracyclic core of the bisguanidine alkaloids palau'amine and styloguanidine. *J. Am. Chem. Soc.* **1997,** *119,* 7159–7160.
8. (a) Ding, H.; Roberts, A. G.; Harran, P. G. Synthetic (±)-axinellamines deficient in halogen. *Angew. Chem. Int. Ed.* **2012,** *51,* 4340–4343.
   (b) Magnus, P.; Garlzt, N.; Selbert, K. A.; Ornholt, A. Synthesis of carbamates from diethoxycarbonyl hydrazine derivatives by E1cB eliminative cleavage of the *N–N'*-bond rather than reduction. *Org. Lett.* **2009,** *11,* 5646–5648.
9. Namba, K.; Nakagawa, Y.; Yamamoto, H.; Imagawa, H.; Nishizawa, M. Hg(OTf)2-catalyzed cyclization of N-tosylanilinoallylic alcohols to 2-vinylindolines. *Synlett* **2008,** 1719–1723.
10. Namba, K.; Shoji, I.; Nishizawa, M.; Tanino, K. 4-Aminopyridine catalyzed direct and regioselective acylation of N-tosylhydrazide. *Org. Lett.* **2009,** *11,* 4970–4973.
11. Namba, K.; Kaihara, Y.; Yamamoto, H.; Imagawa, H.; Tanino, K.; Williams, R. N.; Nishizawa, M. Toward palau'amine: Hg(OTf)2-catalyzed synthesis of the cyclopentane core. *Chem. Eur J.* **2009,** *15,* 6560–6563.
12. Namba, K.; Shobo, Y.; Fujimoto, K.; Shoji, I.; Yoshida, M.; Tanino, K. Synthesis of functionalized chiral carbocyclic cleft molecules complementary to Tröger's base derivatives. *Eur. J. Org. Chem.* **2014,** 5196–5203.
13. Yamamoto, H.; Yamasaki, N.; Yoshidome, S.; Sasaki, I.; Namba, K.; Imagawa, H. Carbon–nitrogen bond formation between allyl silyl ether and hydrazide promoted by mercuric triflate catalyst. *Synlett* **2012,** *23,* 1069–1073.
14. Sakaitani, M.; Ohfune, Y. Selective transformation of N-t-butoxycarbonyl group into n-alkoxy-carbonyl group via n-carboxylate ion equivalent. *Tetrahedron Lett.* **1985,** *26,* 5543–5546.

15. Imaoka, T.; Akimoto, T.; Iwamoto, O.; Nagasawa, K. Total synthesis of (+)-dibromophakellin and (+)-dibromophakellstatin. *Chem. Asian J.* **2010,** *5,* 1810–1816.
16. Shinada, T.; Umezawa, T.; Ando, T.; Kozuma, H.; Ohfune, Y. A new entry for the synthesis of *N*-acyl-*N′*-substituted guanidines. *Tetrahedron Lett.* **2006,** *47,* 1945–1947.
17. Donohoe, T. J.; Blades, K.; Moore, P. R.; Waring, M. J.; Winter, J. J. G.; Helliwell, M.; Newcombe, N. J.; Stemp, G. Directed dihydroxylation of cyclic allylic alcohols and trichloroacetamides using OsO_4/TMEDA. *J. Org. Chem.* **2002,** *67,* 7946–7956.
18. Grube, A.; Immel, S.; Baran, P. S.; Köck, M. Massadine chloride: A biosynthetic precursor of massadine and stylissadine. *Angew. Chem. Int. Ed.* **2007,** *46,* 6721–6724.
19. Takeuchi, K.; Nakayama, A.; Tanino, K.; Namba, K. Facile guanidine formation under mild acidic conditions. *Synlett* **2016,** *27,* 2591–2596.
20. Shimizu, T.; Hiranuma, S.; Nakata, T. Efficient method for inversion of secondary alcohols by reaction of chloromethanesulfonates with cesium acetate. *Tetrahedron Lett.* **1996,** *37,* 6145–6148.

CHAPTER 7

Influence of Glycosylation Pattern on Protein Biodistribution and Kinetics In Vivo Within Mice

Kenward Vong[1], Katsunori Tanaka[1,2,3,4,5]
[1]Biofunctional Synthetic Chemistry Laboratory, RIKEN, Saitama, Japan; [2]GlycoTargeting Research Team, RIKEN, Saitama, Japan; [3]Max Planck-RIKEN Joint Center for Chemical Biology Research, Saitama, Japan; [4]Biofunctional Chemistry Laboratory, A. Butlerov Institute of Chemistry, Kazan Federal University, Kazan, Russia; [5]JST PRESTO, Saitama, Japan

Contents

Kinetic Control in Synthesis and Self-Assembly
ISBN 978-0-12-812126-9
https://doi.org/10.1016/B978-0-12-812126-9.00007-9

1. INTRODUCTION

Proteins present in the cardiovascular system, whether dietary or endogenous, circulate in the bloodstream before ultimately accumulating in specific organs for elimination. Although the preferred elimination/catabolic pathway for every protein is largely dependent on size, other factors such as charge, lipophilicity, functional groups, and glycosylation patterns also play influential roles *(1)*, as summarized in Table 7.1. For the ~66 kDa human serum albumin (HSA) composed of 585 amino acids, its elimination is mainly observed in the kidney and liver via receptor-mediated endocytosis *(2)*.

Glycan (or polysaccharide) is a broad term used to describe molecular complexes of glycosidically bound monosaccharides. In humans, some of the most abundant monosaccharide units include *N*-acetylneuraminic acid (Neu5Ac), *N*-acetylglucosamine (GlcNAc), galactose (Gal), mannose (Man), and fucose (Fuc). For reference, monosaccharide unit structures are shown in Fig. 7.1, and their appearance in this chapter will either be by shortened name (shown in parentheses) or by corresponding colored shape.

Widely found in eukaryotes, *N*-linked glycans (also known as asparagine-linked glycans) conjugated to various biomolecules have been

Table 7.1 Elimination mechanisms of peptides and proteins based on molecular weight

Molecular weight (kDa)	Elimination site	Predominant elimination pathway	Major determinant
<0.5	Blood, liver	• Extracellular hydrolysis, Passive lipid diffusion	Structure, lipophilicity
0.5–1	Liver	• Carrier-mediated uptake, Passive lipid diffusion	Structure, lipophilicity
1–50	Kidney	• Glomerular filtration	Molecular weight
50–200	Kidney, liver	• Receptor-mediated endocytosis	Sugar, charge
200–400		• Opsonization	α_2-macroglobulin, IgG
>400		• Phagocytosis	Particle aggregation

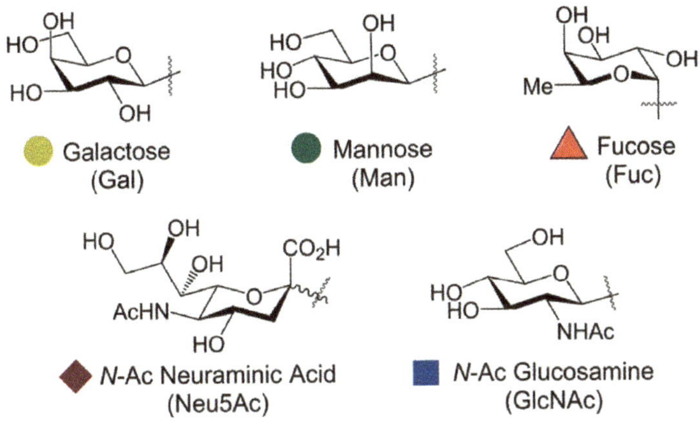

Figure 7.1 Monosaccharide *N*-glycan structures.

implicated in a wide range of important physiological processes, some of which include cell–cell recognition, adhesion, signal transduction, and quality control *(3,4)*. Common *N*-linked glycosylated biomolecules include lipids (glycolipids), peptides (glycopeptides), and proteins (glycoproteins). Of significant note is that the glycosylation patterns of glycoproteins have been observed to influence and regulate the in vivo kinetics and biodistribution of proteins. This can be best highlighted with sialic acid-containing *N*-glycans, which can enhance circulatory residency when bound to soluble proteins or peptides *(5)*. In addition, *N*-glycan modifications of bioengineered proteins, such as erythropoietin and insulin, have also shown considerably higher serum stability and prolonged bioactivity *(6,7)*.

Considering its potential for future diagnostic and therapeutic applications, one of the oft-explored areas within the glycan research field is to develop methodologies where adjusting the number, type, and composition of conjugated glycans can lead to control over in vivo protein biodistribution and kinetics. To reach this goal, numerous studies have conjugated both natural and synthetic glycans to proteins, liposomes, dendrimers, nanoparticles, cluster templates, and cell surfaces as biomimetics to study biodistribution and excretion pathways *(8–10)*. Through noninvasive imaging, a better understanding of in vivo glycan kinetics can thus be achieved.

This chapter primarily reviews recent work done on the use of *N*-glycan-linked proteins to probe and investigate biodistribution within

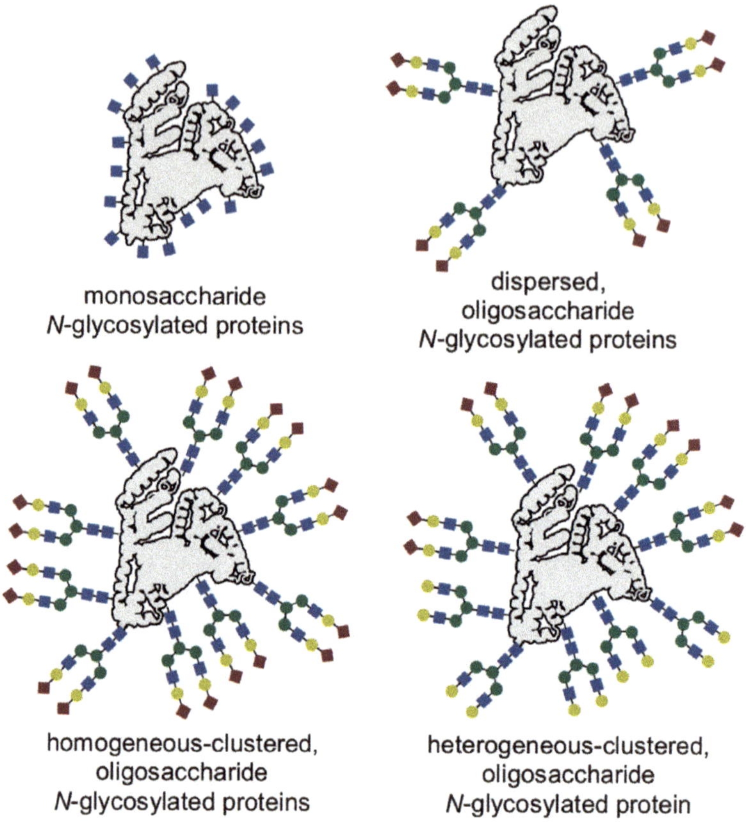

Figure 7.2 Composition types of *N*-glycosylated proteins.

living animals, such as mice. Specifically, this chapter will look at four different types of modified *N*-glycosylated proteins (Fig. 7.2). These are (1) monosaccharide *N*-glycosylated proteins, (2) dispersed, oligosaccharide *N*-glycosylated proteins, (3) homogeneous-clustered, oligosaccharide *N*-glycosylated proteins, and (4) heterogeneous-clustered, oligosaccharide *N*-glycosylated proteins. In addition, recent work that focuses on exploiting glycan-dependent pattern recognition in therapeutic applications will also be discussed. This includes some pioneering work done to develop novel strategies of glycan-based diagnostic techniques against tumor tissues/diseased organs, as well as the adaptation of targeting glycoproteins as viable in vivo catalysts to promote chemical transformations within fully developed mouse models.

2. SYNTHETIC METHODS OF *N*-GLYCOSYLATED PROTEINS

In principle, *N*-glycosylated proteins can be synthesized by adaptation of the numerous protein bioconjugation techniques that exist in literature *(11)*. This section merely aims to highlight examples where *N*-glycan protein bioconjugation has been shown to be successful, which has also been previously reviewed *(12)*. These examples have mainly employed techniques based on lysine residue conjugation, cysteine residue conjugation, bioorthogonal click-based reactions with bioengineered proteins, and some enzymatic-based approaches.

2.1 Lysine Residue Conjugation

Due largely to its chemical properties and widespread prevalence in proteins, the ε-amino group of lysine residues has been a common target for a vast majority of reports (Fig. 7.3). The earliest reports of protein–glycan conjugates generally used reductive amination protocols *(13–15)*. The main drawback of this approach, however, is that the reducing end of the sugar moiety is irreversibly transformed into a polyhydroxyamine group.

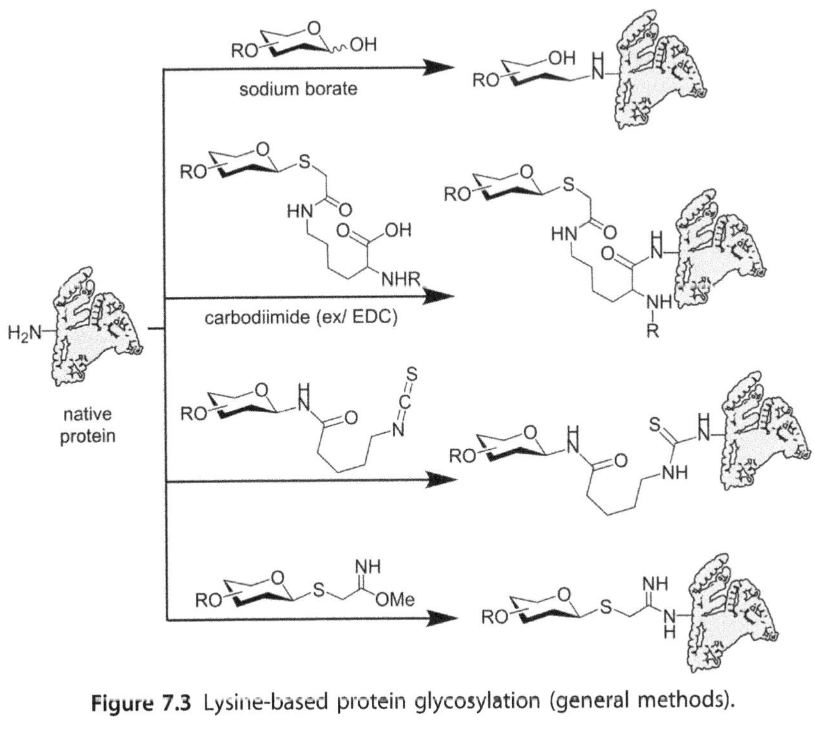

Figure 7.3 Lysine-based protein glycosylation (general methods).

This can be especially problematic for studies where the whole glycan structure is vital for functionality. Other earlier conjugation methods employed carbodiimide-based chemistry *(16)*.

To circumvent these cumbersome methods, a variety of versatile linker-mediated procedures were developed. The key to these methodologies is the intrinsic reactivity of the linkers with amines. Examples include 2-imino-2-methoxyethyl 1-thioglycoside *(17)* and isothiocyanate-based linkers *(18)*. Notably, isothiocyanate-based methods have shown utility with bulkier substrates such as *N*-glycans *(19)*, albeit with a low number of ligation sites on the protein (<4).

With the discovery and development of the RIKEN click reaction (Fig. 7.4) *(20,21)*, which operates via rapid 6π-azaelectrocyclization of unsaturated imines, many issues concerning speed, selectivity, and glycan compatibility have been overcome. Selectively acting on native lysine residues present on protein surfaces, data show reaction rates up to 1000× faster than conventional methods in almost quantitative yields under aqueous conditions at ambient temperatures. As a result, the RIKEN click reaction has allowed for the synthesis of clustered *N*-glycosylated proteins, where ligation with up to a dozen complex *N*-glycans is possible.

Using the RIKEN click reaction, three different types of clustered glycoproteins have been synthesized (Fig. 7.4), with the key to these syntheses being a linker molecule that can undergo both strain-promoted azide—alkyne cycloaddition (SPAAC) and the RIKEN click reaction. In the general protocol, the linker molecule is first mixed with azide-containing complex *N*-glycans (to promote SPAAC ligation), followed by incubation with the protein under study (to promote RIKEN click). For the synthesis of homogeneous-clustered glycoproteins, single complex *N*-glycan probes can be directly used to mix with the linker molecule/protein. The resultant homogeneous-clustered glycoprotein will generally have around 10 conjugated glycan molecules. For heterogeneous-clustered synthesis, two modified approaches can be used. The first approach is based on the sequential treatment of two different *N*-glycan probes in various ratios *(22)*. By adjusting the ratio and treatment order of the two *N*-glycan probes, a certain degree of control over the total amount of ligated glycans and their relative abundance on the protein is possible. In the second approach, one azide-containing glycan unit is instead bound with two distinct *N*-glycan moieties *(23)*. Because the location of the two glycans relative to one another is preorganized, a continuous spatial arrangement of heterogeneity all over the glycoproteins is expected.

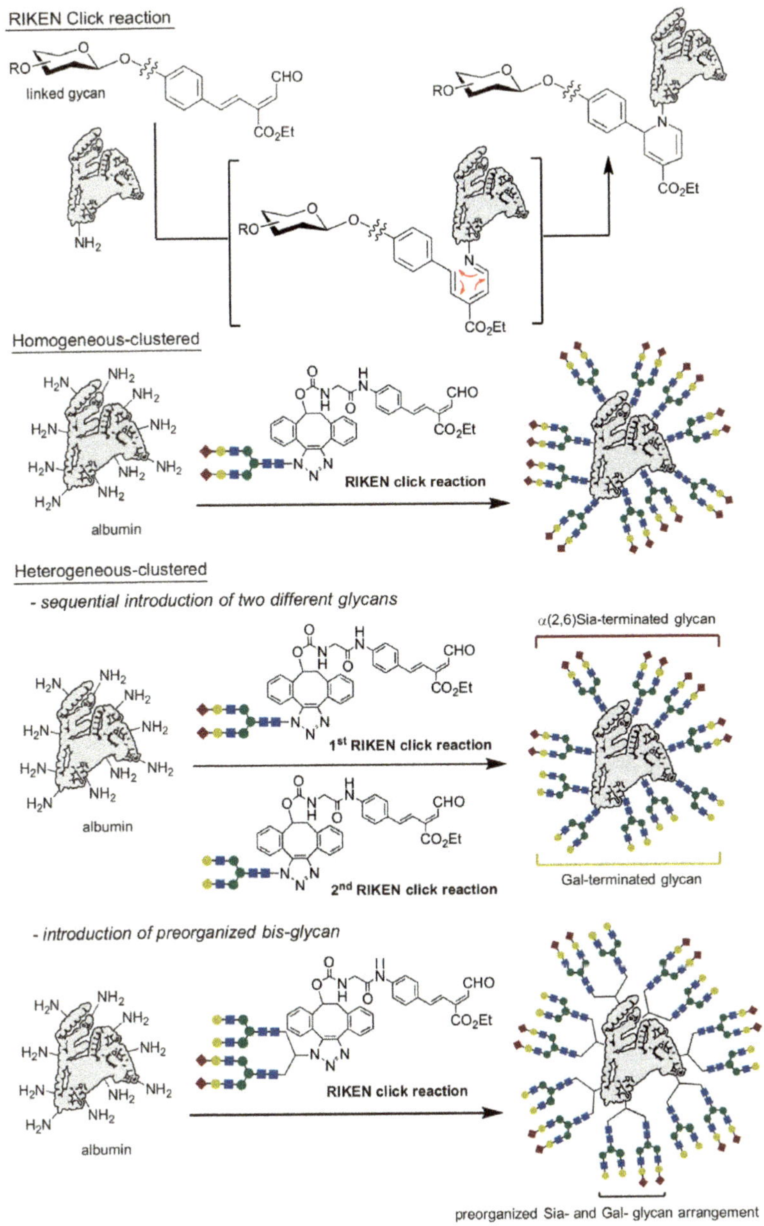

Figure 7.4 Lysine-based protein glycosylation (RIKEN click reaction). With this reaction, homogeneous- and heterogeneous-clustered oligosaccharide *N*-glycosylated proteins can be easily accessible.

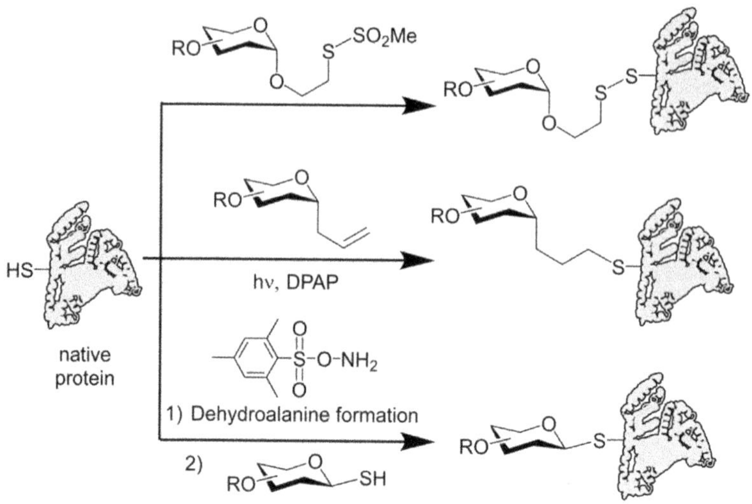

Figure 7.5 Cysteine-based protein glycosylation.

2.2 Cysteine Residue Conjugation

As an alternative, cysteine-selective approaches for protein–glycan conjugation have also found considerable utility (Fig. 7.5). As a result of the naturally low abundance of cysteine and its tendency to rapidly form intramolecular disulfide bonds, low levels of ligation are typically found with these methods. As a result, applications requiring single and site-selective glycan introduction, rather than multiple sites of ligations, have typically employed methods based on intermolecular disulfide bond formation. Furthermore, greater selectivity and control of site-specific glycosylation can be obtained through site-directed mutagenesis *(24)*.

Outside disulfide bonds, cysteine residues have also been found to form stable ligations with other chemical moieties. One thioether-based approach operates by initial oxidation of a cysteine residue to its dehydroalanine form (via sulfonylhydroxylamine), followed by reaction with thiolated glycans in an electrophilic manner *(25,26)*. Another reported example of thioether-based conjugation is dependent on a photoinitiated click reaction between cysteine and an alkenylated sugar in the presence of a metal-based photosensitizer *(27)*.

2.3 Bioorthogonal Reaction–Based Conjugation

Bioorthogonal reaction–based glycosylation has not only shown tremendous utility for protein bioconjugation (Fig. 7.6) but has also opened the

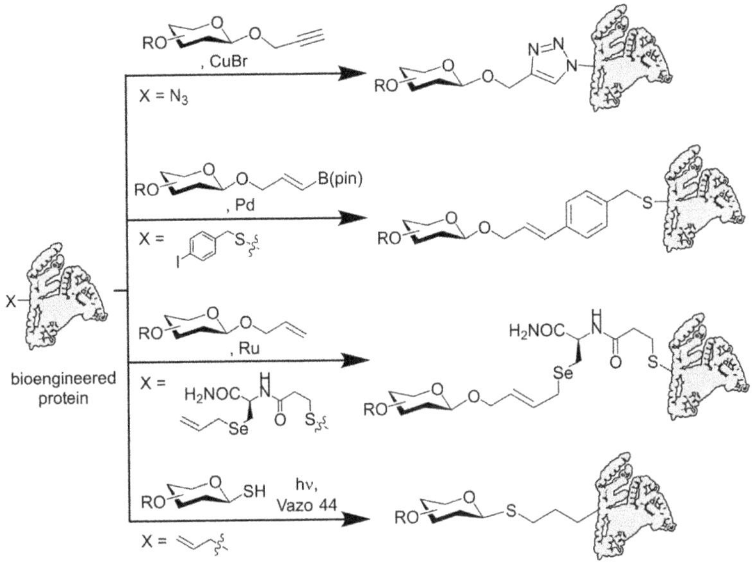

Figure 7.6 Bioorthogonal reaction-based protein glycosylation.

path toward significant therapeutic and diagnostic applications *(28)*. One method of bioorthogonal-based protein glycosylation is through the use of copper(Cu)-catalyzed cycloaddition between an azide-tagged bioengineered protein and an alkyne-bound glycan *(29)*. With this method, Davis and coworkers succeeded in monitoring rat brain inflammation with 5× increased sensitivity compared with conventional antibody-based methods. Mimicking posttranslational modifications, both Sialyl-Lewis^{X} and acidic *p*-toluenesulfonyl groups were introduced at specific sites of a P-selectin ligand, which then led to increased P-selectin binding in mice. Other methods also include ruthenium(Ru)-catalyzed olefin metathesis between allyl-modified selenopeptides and glycans *(30)* and palladium(Pd)-catalyzed Suzuki–Miyaura coupling between aryliodide-tagged proteins and vinylboronated glycans *(31)*. It is important to note that the main drawback of these approaches is the need for a metal catalyst to promote reactivity, which can largely be toxic. Other bioorthogonal-based protein glycosylation approaches include a photoinitiated click reaction between homoallylglycine and a thiolated glycan *(32)*. This technique is similar to the photoinitiated cysteine conjugation method that was previously described in Section 2.2.

Unfortunately, while the site-specific advantages of bioorthogonal-based glycosylation over other chemical-based approaches is clear, one requirement for all these methods is the need for bioengineered proteins containing unnatural amino acids. As such, application of bioorthogonal-based glycoconjugate synthesis is largely dependent on the peptides/proteins under study.

2.4 Enzymatic Glycosylation

Besides chemical-based ligation techniques, enzymatic-based protein glycosylation has also been explored in literature. Given that N-X-S/T sequence-containing peptides/proteins are known substrates for *N*-glycosyltransferases, they have been utilized by research groups to facilitate glycosylation *(33,34)*. One synthetic approach of note is to sequentially employ *N*-glycosyltransferases to produce complex and structurally diverse glycans *(35)*. Through enzymatic trimming of glycosylated end products to reinstall reactive moieties, this strategy can thus be used for multiple rounds of *N*-glycosyltransferase-based elongation *(36)*.

3. IMAGING METHODS FOR *N*-GLYCOSYLATED PROTEIN BIODISTRIBUTION

One of the challenges in developing methods of biodistribution control is to determine the best approach for monitoring and identifying accumulation in specific organs. One of the foremost conventions for in vivo glycoconjugate visualization was to dissect small animals following intravenous injection of labeled glycoconjugates and to individually analyze tissue samples. Given the vast improvements in the field of molecular imaging, many methodologies have since been developed for the "real-time" monitoring of glycan kinetics in vivo without the need to sacrifice the animal model. Although this chapter does not have a dedicated summary and explanation of each of these techniques, various examples discussed throughout use a variety of diagnostic techniques that include fluorescent imaging, positron emission tomography (PET), single-photon emission computerized tomography (SPECT), and magnetic resonance imaging (MRI) *(37,38)*.

4. BIODISTRIBUTION OF MONOSACCHARIDE *N*-GLYCOSYLATED PROTEINS

The most basic glycoprotein modification available would be the one linked only with monosaccharide sugars. In order to understand and potentially improve on therapeutic protein pharmacokinetics, numerous studies have been conducted with monosaccharide *N*-glycosylated proteins. In this section, discussion of these studies is divided between invasive (dissection-based) and noninvasive (PET-, SPECT-, and MRI-based) imaging techniques.

4.1 Invasive Imaging of Monosaccharide *N*-Glycosylated Protein Biodistribution

Starting from the earliest reports of biodistribution studies, animal dissection has always been a consistently deployed technique. The clearest benefit is that experiments can be readily performed on small animals without the need for expensive imaging equipment and complex computational tomographic analysis. The obvious drawback, however, is the need for larger animal sample sizes to carry out adequate time-dependent kinetics and distribution analyses. The following section selectively summarizes representative examples of dissection-based biodistribution studies of monosaccharide *N*-glycosylated proteins.

In one comprehensive study, Gabius and coworkers conducted a series of investigations with bovine serum albumin (BSA) conjugated with simple ^{125}I-labeled monosaccharide sugars. D-Galactose, D-glucose, D-mannose, L-fucose, GlcNAc, GalNAc, lactose, melibiose, and maltose are some of the numerous sugars utilized in this study. Overall, observations on biodistribution revealed a correlation between sugar density to serum stability; higher monosaccharide loading on proteins generally led to decreased serum stability *(39)*.

In another study, Hashida and coworkers wanted to build on known literature that galactose-modified proteins are recognized by the asialoglycoprotein receptors (ASGPR) of liver parenchymal cells, which lead to internalization by endocytosis *(40)*. To achieve this, Hashida and coworkers investigated the biodistribution of galactose-modified ^{111}In-labeled proteins in mice models *(41,42)*. Overall observations show that protein size, galactose-loading density, and injection dosage all influence the in vivo protein kinetics and organ-specific accumulation. For example, small protein glycoconjugates (ex/lysozymes) rapidly accumulated within the liver, whereas larger protein glycoconjugates (ex/γ-globulins) showed longer plasma retention times *(41)*. In terms of galactose-loading density, higher levels were able to elicit enhanced recognition by ASGPRs, which promoted preferential liver accumulation. However, this effect can be disrupted if high dosages of glycoconjugates are injected, which is mainly attributed to receptor oversaturation *(42)*.

Some studies have revealed that certain tumor cell lines overexpress the ASGPR on cell surfaces. As a consequence, unique opportunities exist for the development of galactose-modified, tumor-selective protein tracers. Recently, Kobayashi and coworkers reported the ability of galactosamine- and galactose-modified BSA to accumulate in human adenocarcinoma-bearing mice *(43–45)*. Further highlighting its sensitivity, visualization of submillimeter tumors was also possible. In additional studies, galactose-modified ^{99m}Tc-labeled HSA was employed as a viable imaging probe for intraperitoneal tumors in mice *(46,47)*.

Described by Shen and coworkers *(16)*, mannose-modified ^{125}I-labeled human placental β-glucocerebrosidase showed increased affinity with nonparenchymal cells, such as Kupffer (sinusoidal macrophage) or C-type lectin-expressing endothelial cells (mannose receptors) *(48)*. Furthermore, the groups of Arano and Jeong developed mannose-modified ^{99m}Tc-labeled proteins that showed considerable detection of sentinel lymph nodes *(49,50)*.

One possible therapeutic application of glucosamine- or mannose-modified glycoproteins is as a diagnostic tool for measuring glycan-binding capacity of salivary gland cells. For the autoimmune disease Sjogren's syndrome, this can be particularly useful *(51)*. Another interesting therapeutic application uses mannose-modified catalase to detoxify and significantly reduce the damage of reactive oxygen species like hydrogen peroxide in the liver *(52)*. This is based on the observation that mannose-modified ^{111}In-labeled catalase was found to preferentially accumulate in parenchymal cells in the liver *(53)*.

In terms of fucosylation, studies have demonstrated that fucose-modified, ^{111}In-labeled proteins tend to accumulate in the liver, especially Kupffer cells *(54)*. Das and coworkers also examined the distributions of fucose-modified HSA at the subcellular level and mainly found localization within lysosome-rich fractions *(55)*.

4.2 Noninvasive Imaging of Monosaccharide *N*-Glycosylated Protein Biodistribution

Recent trends have largely gravitated toward noninvasive imaging techniques, such as PET, SPECT, and MRI imaging. Numerous studies have used PET and SPECT imaging techniques with radioactive labeling tracers to directly monitor time-dependent kinetics and organ-specific accumulation within live animal models. Generally, PET and SPECT imaging probes are composed of biomolecules conjugated to metal chelating agents. Common biomolecules include peptides, monoclonal antibodies, oligonucleotides, and glycoconjugates, whereas the metal chelating component typically includes 1,4,7,10-tetraazacyclodecane-1,4,7,10-tetraacetic acid (DOTA), 1,4,7-triazacyclononane-1,4,7-triacetic acid (NOTA), or diethylenetriaminepentaacetic acid (DTPA). The last component to determine is the selection of a radiolabeling metal with suitable β^+ [^{68}Ga (68 min), ^{64}Cu (13 h), ^{86}Y (15 h) for PET] or γ [^{67}Ga (78 h), ^{111}In (68h), ^{99m}Tc (6h)] emission. Alternatively, near-infrared fluorescence detection can also be used for visualization.

In a study done by Jeong and coworkers, imaging of monosaccharide ^{99m}Tc-labeled glycoproteins was successfully reported in vivo with mice using gamma cameras *(50,56)*. Observations showed that mannose-modified glycoproteins were catabolized mainly in nonparenchymal cells, followed by urinary excretion. In contrast, galactose-modified glycoproteins were captured by hepatocytes through the intestinal excretion pathway *(56)*. Additionally, other monosaccharide-based glycoproteins that traffic to the lymph node, liver, and spleen have been visualized using PET imaging *(57)*.

Noninvasive fluorescence imaging techniques have also been used to investigate the biodistribution of monosaccharide-modified glycoproteins. In one study, Kobayashi and coworkers used the near-infrared fluorescence detection method, IR800, to image nude mice injected with glucose-, galactose-, mannose-, and fucose-modified HSA *(58)*. Results show that glucose- and galactose-linked albumins tended to be captured by hepatocytes, whereas mannose- and fucose-modified HSA were instead catabolized by macrophages followed by urinary excretion. These observations appear to agree well with previous observations made by Jeong *(50)*. By employing near-infrared fluorescent imaging agents activated through green light absorption, Kobayashi and coworkers have also further improved their fluorescence-based imaging methodologies *(59)*.

5. BIODISTRIBUTION OF OLIGOSACCHARIDE *N*-GLYCOSYLATED PROTEINS

In the field of glycochemical biology, the increased structural complexity of oligosaccharides is known to significantly influence the biological function and roles of natural glycoproteins. As such, numerous studies have been conducted to gain a better understanding of glycan structure complexity in hopes of demonstrating the viability and potential for pharmacological applications. In this section, three distinctive types of oligosaccharide *N*-glycosylated proteins will be summarized and discussed: dispersed, homogeneous-clustered, and heterogeneous-clustered.

5.1 Dispersed, Oligosaccharide *N*-Glycosylated Protein Biodistribution

For dispersed oligosaccharide modifications, the definition relates to protein glycosylation with only a few molecules of complex *N*-glycans. Fig. 7.7 summarizes the biodistribution and serum stability observations from several studies that employ the use of dispersed, oligosaccharide *N*-glycosylated proteins.

In addition to preliminary studies with disaccharide modifications *(39,60)*, the groups of Gabius and Unverzagt have extensively studied the structure-dependent biodistribution of biantennary complex-type *N*-glycans *(19,61–66)*. Using chemoenzymatic methods (ex/sugar elongation with glycosyltransferases), various oligosaccharide *N*-glycosylated BSA

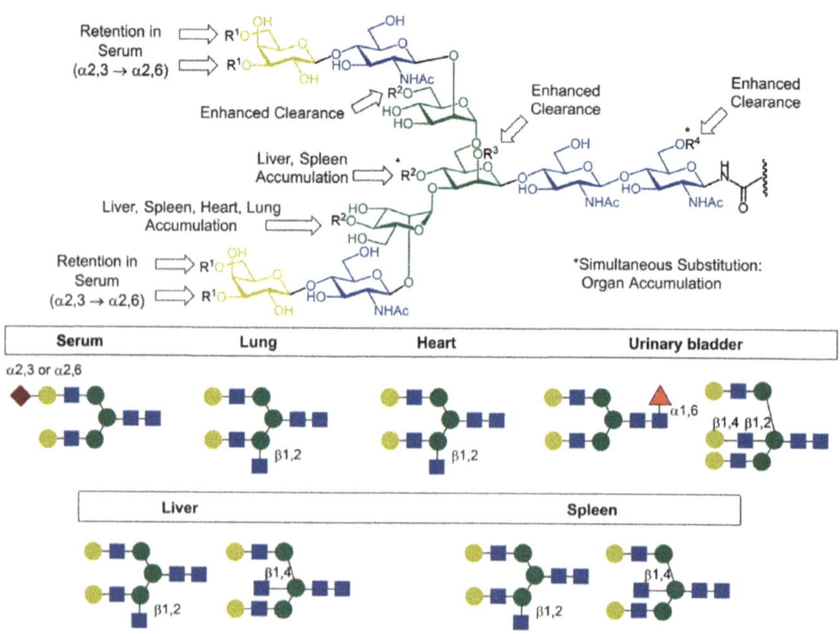

Figure 7.7 Summary of glycan-dependent kinetics and biodistribution for dispersed, oligosaccharide *N*-glycosylated proteins in mice.

proteins were prepared and labeled with ^{125}I radioisotopes. For glycoproteins with *N*Ac-neuraminic acid at the nonreducing end of galactose, observations show an increase in serum stability *(19)*. For α(2,3)- or α(2,6)-sialic acid–linked, galactose-terminated glycoproteins, radioactivity in 1 mL blood after 1 h was increased to 25.27% or 22.90% of the injected dose, respectively. This is a sizable increase from the baseline radioactivity content (3.07% of the injected dose). For core fucosylation at the reducing end of glucosamine, 1,6-arm structure conformation is calculated to be affected, which may explain the accelerated rates of serum clearance. In addition, sialoside linkage at the nonreducing end affects clearance due to the fact that α(2,6)-sialoside has been found to be cleared faster than α(2,3)-sialoside *(61)*. Studies also show that introduction of β(1,4)-bisecting glucosamine at the branching mannose could increase accumulation in the liver and spleen *(62)*. On the other hand, there are reports indicating that bisecting substitution involving β(1,2)-lactosamination of the branching mannose enhances serum clearance *(63)*.

For triantennary complex-type *N*-glycans, various studies have been performed. For example, elongation of mannose branches at the α(1,6)-arm appears to accelerate serum clearance, but branching at the α(1,3)-arm instead leads to liver, spleen, heart, and lung accumulation *(64)*. Although there is no mechanistic explanation for these observations, it can be hypothesized that glycan receptors expressed on heart and lung tissue may mediate their preferential accumulation. Similarly, simultaneous introduction of core fucose and β(1,4)-bisecting glucosamine residues led to observed liver, kidney, and spleen accumulation *(65)*.

In the Gabius and Unverzagt reports, the type of cells or glycan receptors responsible for glycan-dependent biodistribution was not examined. Having said that, the effects of sialic acid-promoted serum stability are clearly consistent with known ASGPR-mediated excretion pathways for glycoproteins in the liver *(5–7)*. Furthermore, rapid serum clearance mediated by β(1,2)-lactosamination of the branching mannose is likely due to potential interactions with C-type lectins on parenchymal cells. Overall, although a plethora of work has extensively investigated dispersed, oligosaccharide *N*-glycosylated proteins in tumor-bearing mice, there has yet to be a major breakthrough where tumor accumulation can be reached in significant levels. Despite this, these pioneering studies have offered a comprehensive look at how *N*-glycan structures can influence the biodistribution, kinetics, and serum stability of *N*-glycosylated proteins.

5.2 Homogeneous-Clustered, Oligosaccharide *N*-Glycosylated Protein Biodistribution

In this section, the focus will be on protein glycosylation with many molecules of the same complex *N*-glycans; in other words, homogeneous-clustered oligosaccharide modifications. One clear advantage of studying *N*-glycoclusters is the more profound observations of biodistribution and kinetic changes brought about by *N*-glycan trimming. This is mainly due to multivalency effects, which depend on the amplification of multiple individual interactions in a grouped cluster to reach significant levels for detection.

The first group to accomplish the synthesis of these clustered glycoproteins was Tanaka and coworkers. Using a chemical conjugation methodology discussed earlier (RIKEN click reaction), the facile preparation of glycoproteins bound with clusters of complex *N*-glycans was possible. The complex *N*-glycan structures used in these studies are shown in Fig. 7.8. With these clustered glycoproteins, targeted accumulation in specific organs/cells was then determined both by noninvasive imaging and direct analysis (via near-infrared fluorescence) of dissected tissues.

As shown in Fig. 7.9, various homogeneous-clustered *N*-glycoalbumins were injected into live mice and then noninvasively imaged for their biodistribution properties *(22)*. As expected, glycan-free native albumin was found to be spread all over the body. However, with the α(2,6)- and α(2,3)-disialoglycan-based glycoalbumins **2a**,**b**, results show increased levels of serum stability, which falls in line with previous reports by Gabius and coworkers (discussed in Section 5.1) *(19,61)*. This observation is mainly attributed to the higher stability of the α(2,3)-congener.

Another notable observation was that for asialoglycan-modified glycoalbumin **2c**. Normally, natural sialoglycoproteins in nature have sialic acid trimmed to give asialoglycoproteins, which leads to ASGPR binding and excretion *(5,40)*. With the artificial asialoglycan-modified glycoalbumin **2c**, a complete shift in the excretion pathways could be induced from the urinary bladder to the gallbladder/intestine. This is the first instance of a synthetic *N*-glycoprotein mimicking the clearance properties of natural *N*-glycoproteins. A likely explanation for these observations is that the asialoglycoalbumin **2c** is recognized and endocytosed by ASGPR in liver parenchymal cells, which subsequently transports them to the gallbladder via the polar transportation mechanism *(67)*. On the other hand, the weak

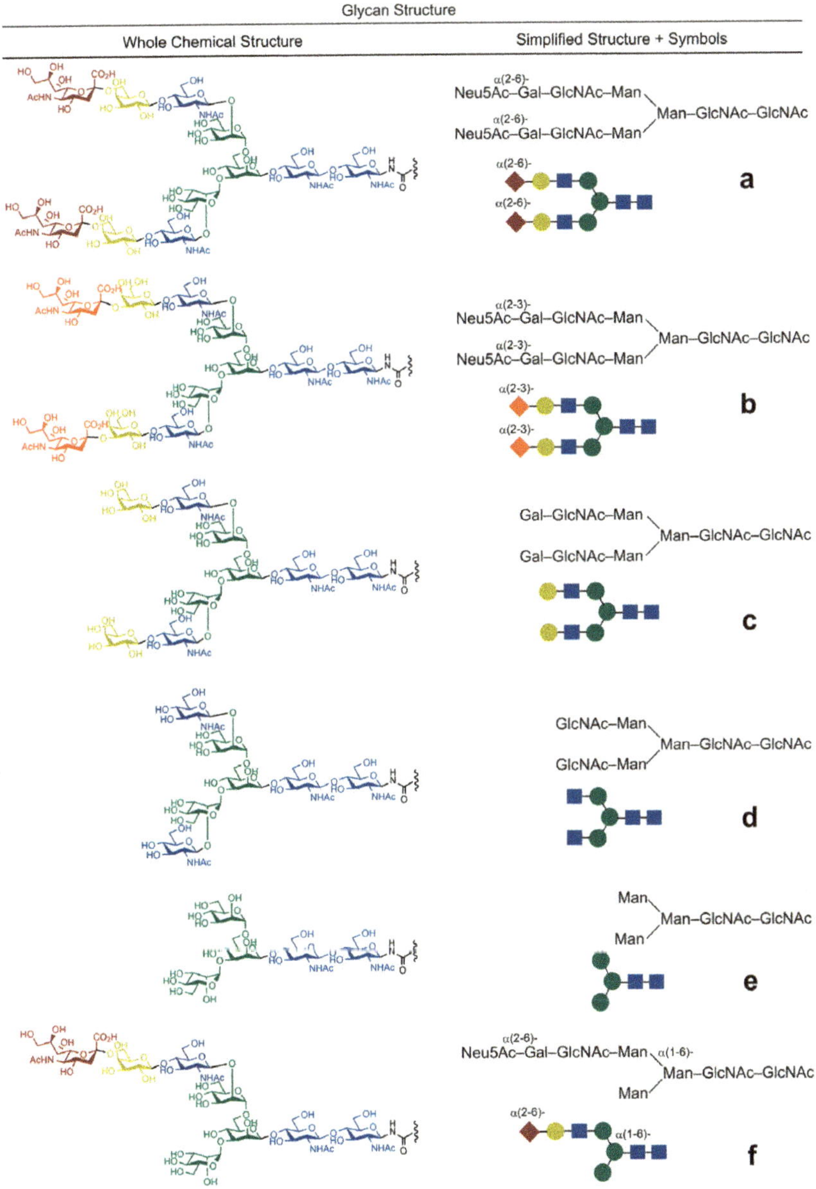

Figure 7.8 Whole and simplified chemical structures, as well as representative symbols of complex *N*-glycans used for clustered glycoprotein synthesis.

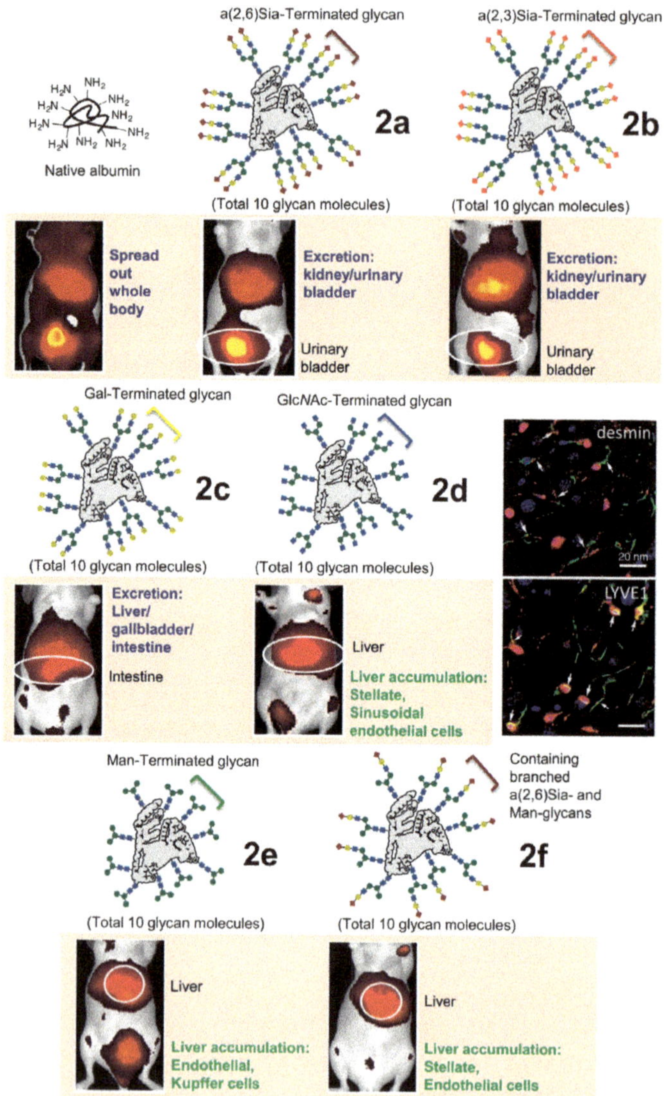

Figure 7.9 In vivo fluorescence imaging of HiLyte Fluor 750-labeled, homogeneous-clustered, *N*-glycoalbumins **2a–f** in mice (abdominal images 3 h after injection). Cell image: Co-staining of liver tissues treated with glycoalbumin **2d** with anti-desmin (to detect stellate cells) and anti-LYVE1 (to detect sinusoidal endothelial cells).

and reversible binding of sialoglycoalbumin **2a,b** to ASGPR *(68)*, likely, instead promotes excretion through biofiltration in the urinary bladder *(69)*.

For GlcNAc-terminated, mannose-terminated, and hybrid-type glycoalbumins **2d–f**, preferential excretion properties were not exhibited. Instead, these glycoconjugates are all strongly accumulated within the liver. An immunohistochemical study revealed that mannose-terminated **2e** is likely captured by Kupffer cells and/or macrophages through C-type lectin interactions *(70)*, whereas glucosamine-terminated **2d** and hybrid-type **2f** mainly interact with hepatic stellate cells, as well as sinusoidal endothelial cells. Another significant observation that was made is that GlcNAc-terminated glycoalbumin **2d** could also interact with desmin and vimentin proteins, which are expressed on the surface of hepatic stellate cells and exhibit lectin activity *(71)*. This is the first instance of these types of interactions occurring at the whole-body level. Overall, there is a clear difference in observed biodistribution properties and mode of excretion between oligosaccharide glycosylated proteins and monosaccharide glycosylated proteins *(39)*. With more research effort, possibilities may exist for the future development of *N*-glycoalbumins that have significant applicability for therapeutic treatments of liver fibrillation and cirrhosis *(72,73)*.

5.3 Heterogeneous-Clustered, Oligosaccharide *N*-Glycosylated Protein Biodistribution

In addition to homogeneous glycan clusters, the use of heterogeneous glycan clusters may elicit different in vivo kinetics and biodistribution properties. As such, Tanaka and coworkers set out to synthesize heterogeneous-clustered, oligosaccharide *N*-glycosylated protein in hopes of gaining valuable information to further the understanding of multivalency and heterogeneity effects through pattern recognition processes *(22,23)*.

As outlined in Fig. 7.10, heterogeneous-clustered glycoalbumins were focused on either combinations of α(2,6)-sialic acid and galactose (**3ac, 3ca, 4ac**) or with α(2,6)-sialic acid and mannose (**2f, 4ae, 4ae'**). As a note, although the terminal sialic acid and mannose-containing glycoalbumin **4ae** closely resembles the homogeneous glycoalbumin **2f**, the branching and spatial arrangements among neighboring *N*-glycans is largely different. Using these heterogeneous-clustered glycoalbumins, Tanaka and coworkers set out to determine their biodistribution properties and excretion pathways.

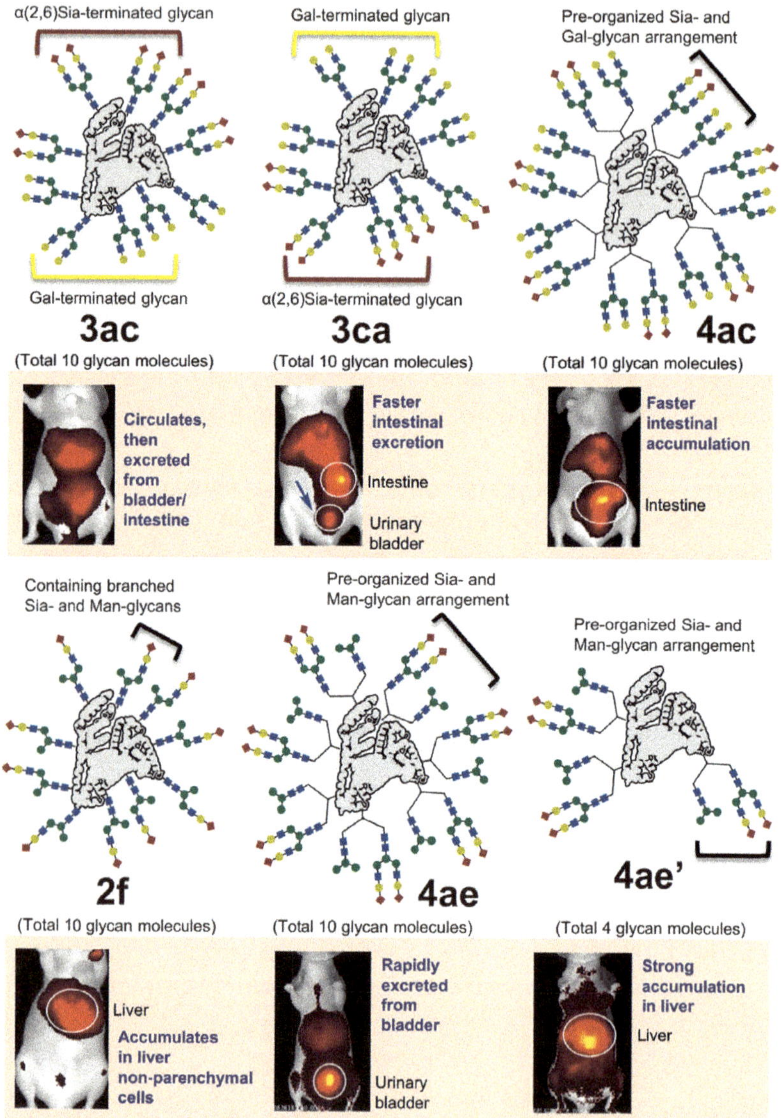

Figure 7.10 In vivo fluorescence imaging of HiLyte Fluor 750-labeled, heterogeneous-clustered, *N*-glycoalbumins **3ac**–**4ae'** in mice (abdominal images 3 h after injection).

Looking specifically at α(2,6)-sialic acid and galactose-modified glycoalbumins (**3ac, 3ca, 4ac**), the following observations were made. With glycoalbumin **3ac** (Sia/Gal glycan ratio of 1:1), the excretion properties were intermediate between those of homogeneous glycoclusters **2a** and **2c**

(Fig. 7.9). Unexpectedly, its positional isomer **3ca** (Gal/Sia glycan ratio of 1:1) showed rapid translocalization through the gallbladder to the intestine, which is found to be even faster than homogeneous Gal-terminated glycoalbumin **2c**. For Sia/Gal glycoalbumin **4ac**, clearance rates via the intestine were found to be further accelerated.

In studies with α(2,6)-sialic acid and mannose-modified glycoalbumins (**2f, 4ae, 4ae'**), results clearly show that heterogeneous glycoproteins behave differently than their homogeneous counterparts. For example, homogeneous hybrid-type glycoalbumin **2f** was previously mentioned to accumulate preferentially in nonparenchymal liver cells. Yet, for the Sia/Man heterogeneous glycoalbumins **4ae** and **4ae'**, they were instead found to be either excreted through the urinary bladder or strongly accumulated within the liver, respectively. As a note, glycoalbumins **4ae** and **4ae'** are only structurally different from each other based on the valency and number of conjugated glycans.

Although the works of Wong and Wu have previously investigated the neighboring effects of antibody—glycan interactions through microarrays to establish heterogeneous glycan environments *(74)*, these studies by Tanaka and coworkers represent the first instance of in vivo kinetic and biodistribution analysis of heterogeneous-clustered glycans on protein surfaces. With these data, it can be clearly illustrated that the dimensional arrangement of glycans are vital to influencing and enabling pattern recognition mechanisms in vivo within living biological systems. At this current state, however, a deeper exploration is still needed to increase our knowledge in this growing field of glycan molecular technologies. Through the continuing research in this area, the aim is to further discover and develop on different systems and compositions of *N*-glycosylated proteins with greater selectivity and varied biodistribution/excretion pathways.

6. THERAPEUTIC OPPORTUNITIES FOR CLUSTERED *N*-GLYCOALBUMINS

Looking toward the future, there is great potential to exploit glycan-dependent pattern recognition in therapeutic applications where precise control over organ-selective accumulation and excretion is desired. This section is focused on the pioneering work done to develop novel strategies of glycan-based diagnostic techniques against tumor tissues/diseased organs, and the adaptation of targeting glycoalbumins as viable in vivo catalysts to promote chemical reactions with therapeutic potential.

6.1 Tumor Targeting by *N*-Glycoalbumins

Previously, liposomes conjugated with Sialyl-Lewis^X have shown the ability to accumulate in tumor tissues and areas of inflammation *(75)*. Because complex *N*-glycans have much higher potential for differentiating the complicated cell surface glycan patterns between tumor and normal tissues, Tanaka and coworkers set out to use A431 tumor-bearing mice as models for in vivo kinetics and biodistribution (Fig. 7.11) *(76)*. The main

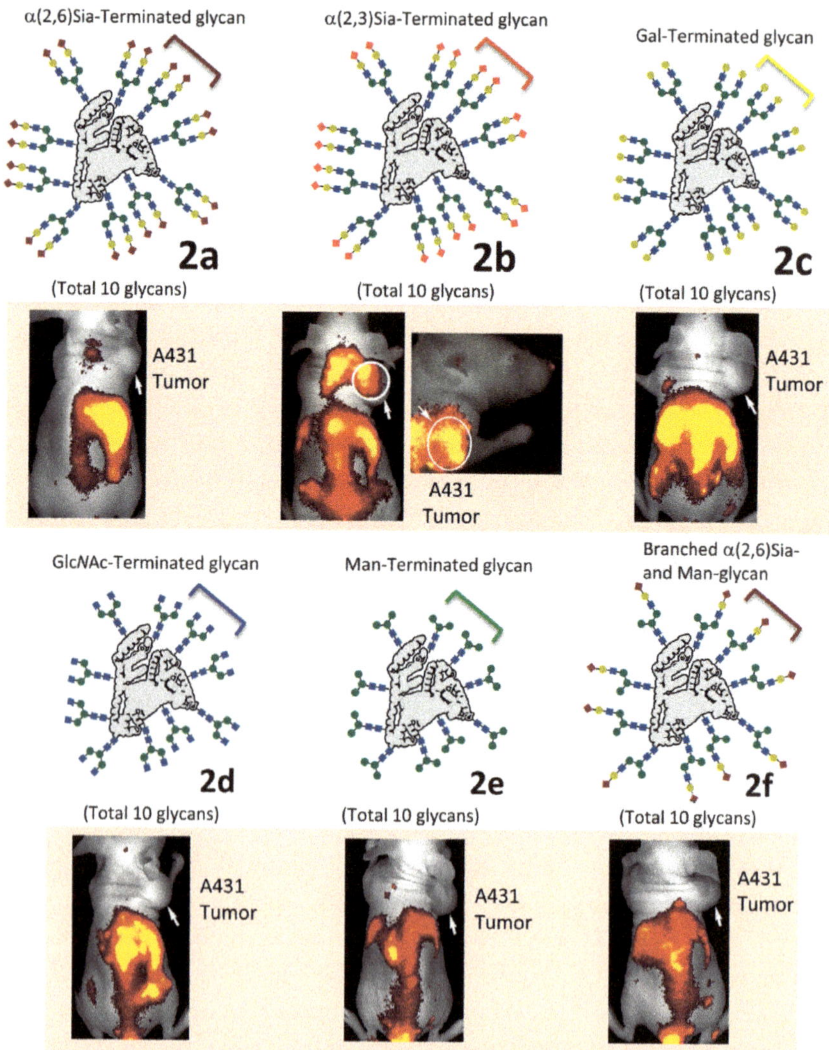

Figure 7.11 In vivo fluorescence imaging of homogeneous-clustered, *N*-glycoalbumins **2a–f** in BALB/cAJcl-nu/nu mice bearing A431 tumor tissue in the right shoulder (dorsal images 1 h after injection). Top center insert: Image of α(2,3)-disialoalbumin **2b** from right side 1 h after injection. Arrows denote the position of the tumor.

challenge in this approach is to provide sufficiently high imaging contrast for tumor detection against the whole body. Out of all the *N*-glycoalbumins examined in this research, only α(2,3)-disialoglycoalbumin **2b** could clearly visualize A431 tumors implanted in nude mice within 1 h. In terms of excretion, **2b** was found to exit the body smoothly via the urinary bladder. Because α(2,6)-disialoglycoalbumin **2a** did not display any tumor-targeting properties, it was theorized that specific receptors such as selectins (receptors for Sialyl-Lewis^X ligands) may play an influential role in the tumor-recognition process of **2b** *(75)*. Another observation from this work is the glycoalbumin accumulation times, which is significantly shorter compared with the longer times required for methods like antibodies *(38)*. As such, this work also highlights its promise as potential diagnostic probes for SPECT, PET, or MRI imaging. Overall, this area of research is still in its infancy, with much more work needed to advance the contrast and specificity of tumor-targeting *N*-glycosylated albumins.

6.2 Tumor Targeting by *N*-Glycan-Engineered Lymphocytes

Using the biocompatibility and chemoselectivity of the RIKEN click reaction, Tanaka and coworkers introduced complex *N*-glycan structures directly onto the surface of lymphocyte, via attachment to lysine and/or phosphatidylethanolamine groups commonly expressed on cell surfaces *(77–81)*. Under mild conditions, the conjugation transformations were not observed to interfere with cell viability and/or function. Lymphocytes extracted from wild-type mice were labeled with Cy5 and then modified with α(2,6)-sialo-*N*-glycan **a** (see the structure in Fig. 7.8) *(82)*. As shown in Fig. 7.12, these artificial lymphocytes were then intravenously injected into mouse models bearing DLD-1 human colon carcinoma tumors implanted into the dorsal division. Noninvasive imaging studies revealed these artificial lymphocytes were trafficked through the spleen and intestinal lymph nodes, before accumulation within the tumor region. As a control, neither freestanding *N*-glycans, corresponding glycoclusters *(20)*, nor the native lymphocytes *(77)* have the ability to target tumors in vivo. These observations suggest that the tumor-targeting mechanism likely involves the synergistic interactions of *N*-glycans and lymphocyte surfaces to selectively bind with tumor cells. With this work, the potential for novel cancer-targeting diagnostic tools based on glycan molecular recognition is highlighted. In addition, this work demonstrates that living cells can be viably modified without loss of cell function to change its biodistribution and in vivo cell-to-cell interactions.

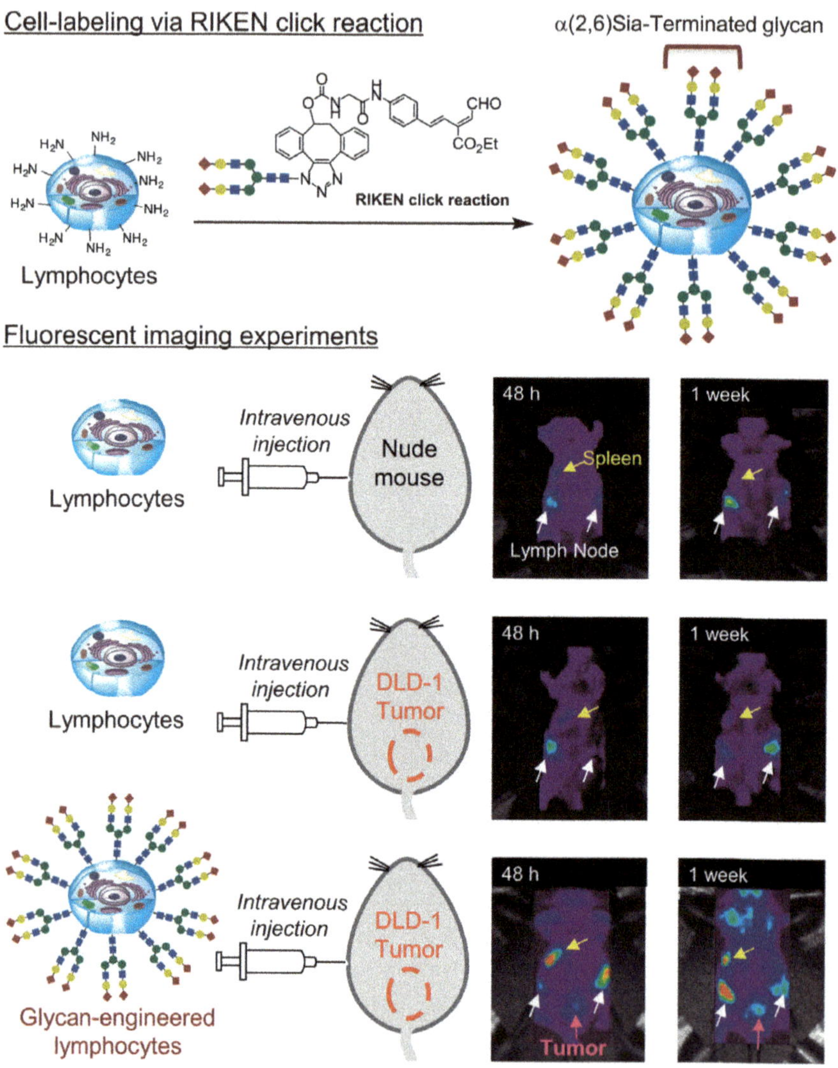

Figure 7.12 Tumor targeting by *N*-glycan-engineered lymphocytes, where glycans are introduced onto the cell surfaces via RIKEN click reaction. Noninvasive fluorescence imaging of normal and *N*-glycan-engineered Cy-5-labeled lymphocytes in DLD-1 tumor-implanted mouse models (dorsal images 48 h and 1 week after injection).

6.3 In Vivo Metal Catalyst Carriers

In literature, there is significant interest toward biocompatible metal complex catalysts *(83–87)*, with a plethora of examples shown to proceed in cells *(88–104)*, bacteria *(105–107)*, and even zebrafish *(108, 109)*.

To further explore the adaptability of glycoalbumins for other chemical biology-based applications, Tanaka and coworkers sought to modify their *N*-glycoalbumins as carriers for metal complex catalysts *(110)*. In their design (Fig. 7.13), homogeneous-clustered α(2-6)-sialoglycoalbumin **2a** and galactose-terminated glycoalbumin **2c** were used as targeting vectors to the liver and intestine, respectively. Due to the known binding affinity of coumarin to albumin *(111–113)*, a coumarin-Au(III)-linked conjugate was synthesized to anchor water-compatible cyclometallated Au(III) catalysts *(114,115)*. Following brief incubation of these coumarin–Au conjugates, glycoalbumins **Au-2a** and **Au-2c** were produced. In this study, the model chemical transformation is the Au(III)-catalyzed amidation of propargyl esters by nucleophilic amines on protein/tissue surfaces.

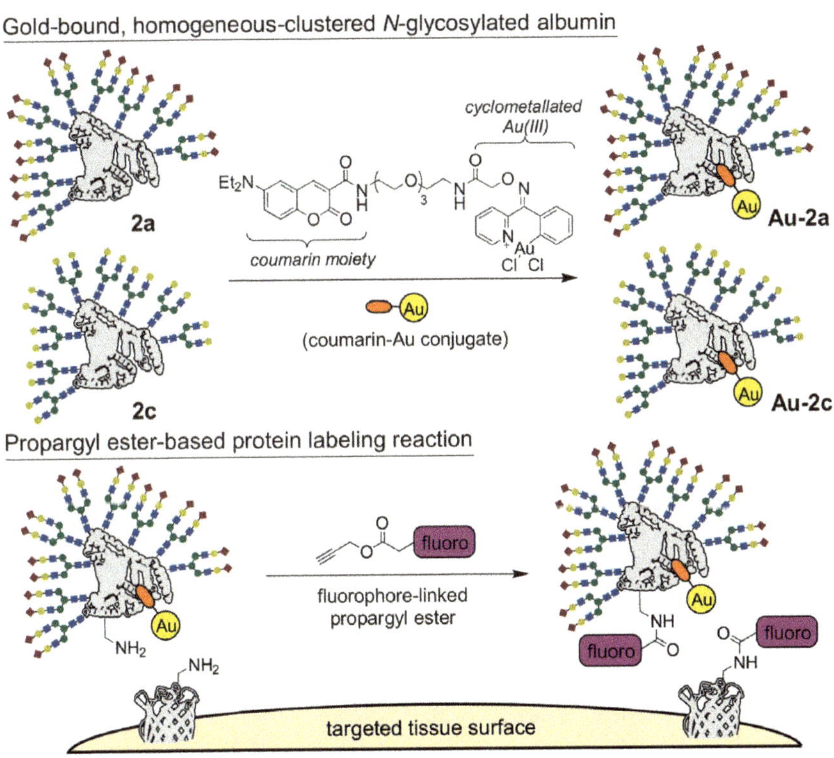

Figure 7.13 Adaptation of glycosylated proteins as metal catalyst carriers. Through the native affinity of coumarin to albumin, the anchoring of gold catalyst complexes produces **Au-2a** and **Au-2c**, which are linked with α(2,6)-sialic acid- and galactose-terminated complex *N*-glycans, respectively. In a biological environment, cell/tissue surface protein labeling via propargyl ester-based fluorophores can be facilitated by the gold complexes.

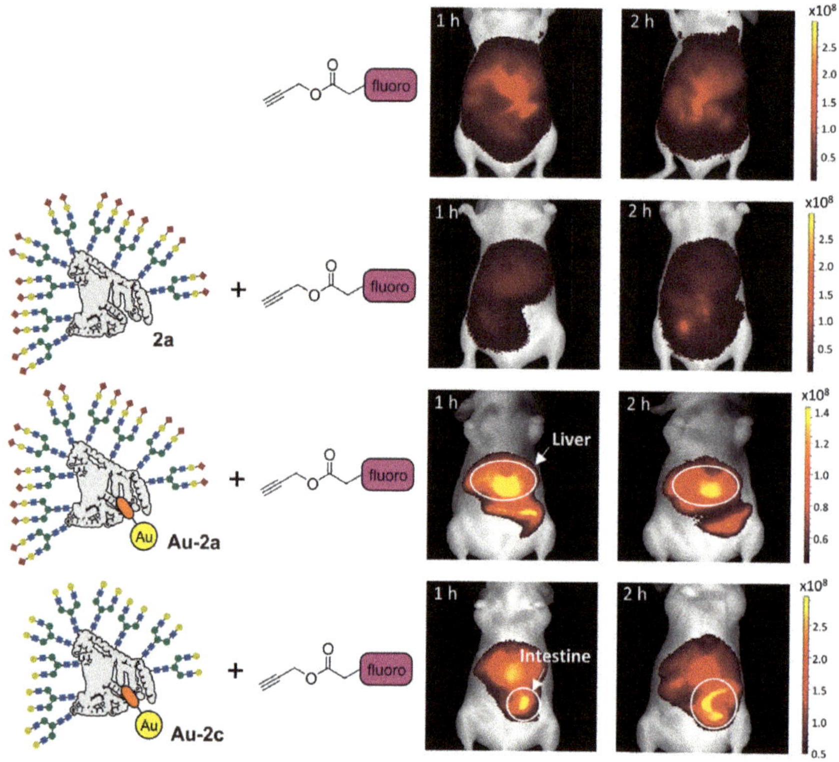

Figure 7.14 In vivo fluorescence imaging catalyzed by Au-bound, homogeneous-clustered glycosylated albumins **Au-2a** and **Au-2c** between protein surface amines and Cy7.5-linked propargyl ester probes. Following initial injection and accumulation of **Au-2a** and **Au-2c** in specific organ tissues in mice, localized reactivity of Cy7.5-linked propargyl ester probes promotes fluorescence (abdominal images 1 and 2 h after injection).

In Fig. 7.14, the imaging results of this study are shown. Following the initial injection and accumulation of **Au-2a** and **Au-2c** in mice liver and intestines, respectively, localized reactivity of the fluorescently labeled propargyl ester probes can then proceed. Under control conditions, fluorescent propargyl ester only or the use of Au-free glycoalbumin **2a**, baseline fluorescence was shown to be distributed throughout the whole body (likely via capillary vessels on the skin), before eventual excretion. With the use of **Au-2a** and **Au-2c**, however, fluorescence was clearly localization to targeted organs; **Au-2a** leads to liver localization (before eventual metabolization and urinary excretion), whereas **Au-2c** leads to intestinal localization.

Overall, using these glycoprotein–metal conjugates, the first literature example of an in vivo organ-localized, metal-catalyzed reaction was performed within a live animal. There is high potential for this work to be applied in diagnostic or therapeutic settings, especially if this system can successfully be adapted to carry other biocompatible metals (i.e., Ru, Pd, etc.) to catalyze other relevant reactions of interest.

7. CONCLUDING REMARKS

As described in this chapter, the current explorations done by numerous research groups have brought about significant knowledge concerning the in vivo biodistribution and kinetic properties that are modulated by *N*-glycan pattern recognition mechanisms. As a result, many reports of interesting organ selectivity, as well as some tumor selectivity, have been presented. In addition, the speed and selectivity of these strategies have clear benefits over conventional targeting methods such as peptide- or antibody-based approaches. Although there is much promise for the future development of glycoconjugate-based imaging tracers with favorable pharmacokinetics and selective targeting properties, current methodologies with the best outlook are limited to monosaccharide-based approaches. For example, ^{99m}Tc-galactose–modified HSA is already clinically used in hepatocyte mass determination assays.

Because of the difficulty in correlating observed glycan interactions and effects between in vivo animal and cell culture-based studies, the outlook for future studies will likely progress more and more toward systems either involving complex glycoforms (glycoproteins with several different attached saccharides) or in vivo studies with various different animal models. The reason for this train of thought is that in vivo targeting selectivity may be dependent on a variety of factors, such as synergetic effects of serum stability and pattern recognition processes of glycocluster environments on proteins. As a result of this unpredictability, synthetic glycoconjugates are best optimized directly through in vivo assays. In addition, what is also necessary for the future of this field is more protein conjugation methodologies amenable to the different functional groups found on oligosaccharides. By accomplishing this, different assemblies of glycoclusters could be rapidly prepared and structurally identified. Another aim would be to develop synthetic methods to create glycoclusters in a combinatorial fashion.

Overall, the promise of developing glycoconjugate-based protein targeting mechanisms is not only limited for use as diagnostic probes but also as potential vaccine adjuvants, virus-neutralizing pharmaceuticals, or even in vivo bioactive drug generation. Through continual improvements in the areas of glycan-compatible synthetic methods, glycocluster-based biodistribution/targeting, and biocompatible metal complex catalysts, innovative glycan molecular technologies may yet prove to make a huge breakthrough for biomedical or therapeutic applications.

REFERENCES

1. Meijer, D. K. F.; Ziegler, K. Mechanisms for the Hepatic Clearance of Oligopeptides and Proteins. In *Biological Barriers to Protein Delivery;* Audus, K. L., Raub, T. J., Eds.; Springer US: Boston, MA, 1993; pp 339–408.
2. Yoshioka, T.; Yamamoto, K.; Kobashi, H.; Tomlta, M.; Tsuji, T. Receptor-mediated Endocytosis of Chemically Modified Albumins by Sinusoidal Endothelial Cells and Kupffer Cells in Rat and Human Liver. *Liver* **1994,** *14* (3), 129–137.
3. Boons, G. J.; Lee, Y. C.; Suzuki, A.; Taniguchi, N.; Voragen, A. G. J. *Comprehensive Glycoscience: From Chemistry to Systems Biology* In: *Analysis of Glycans Polysaccharide Functional Properties,* Vol. 2; Elsevier Ltd.: United Kingdom, 2007; p 814.
4. Boons, G.-J.; Lee, Y. C.; Suzuki, A.; Taniguchi, N.; Voragen, A. G. J. *Comprehensive Glycoscience: From Chemistry to Systems Biology* In: *Biochemistry of Glycoconjugate Glycans Carbohydrate-Mediated Interactions,* Vol. 3; Elsevier Ltd.: United Kingdom, 2007; p 902.
5. Morell, A. G.; Irvine, R. A.; Sternlieb, I.; Scheinberg, I. H.; Ashwell, G. Physical and Chemical Studies on Ceruloplasmin V: Metabolic Studies on Sialic Acid-free Ceruloplasmin In Vivo. *J. Biol. Chem.* **1968,** *243* (1), 155–159.
6. Elliott, S.; Lorenzini, T.; Asher, S.; Aoki, K.; Brankow, D.; Buck, L.; Busse, L.; Chang, D.; Fuller, J.; Grant, J.; Hernday, N.; Hokum, M.; Hu, S.; Knudten, A.; Levin, N.; Komorowski, R.; Martin, F.; Navarro, R.; Osslund, T.; Rogers, G.; Rogers, N.; Trail, G.; Egrie, J. Enhancement of Therapeutic Protein In Vivo Activities through Glycoengineering. *Nat. Biotechnol.* **2003,** *21* (4), 414–421.
7. Sato, M.; Furuike, T.; Sadamoto, R.; Fujitani, N.; Nakahara, T.; Niikura, K.; Monde, K.; Kondo, H.; Nishimura, S. Glycoinsulins: Dendritic Sialyloligosaccharide-Displaying Insulins Showing a Prolonged Blood-Sugar-Lowering Activity. *J. Am. Chem. Soc.* **2004,** *126* (43), 14013–14022.
8. Vyas, S. P.; Singh, A.; Sihorkar, V. Ligand-Receptor-Mediated Drug Delivery: An Emerging Paradigm in Cellular Drug Targeting. *Crit. Rev. Ther. Drug Carrier Syst.* **2001,** *18* (1), 1–76.
9. Forssen, E.; Willis, M. Ligand-Targeted Liposomes. *Adv. Drug Deliv. Rev.* **1998,** *29* (3), 249–271.
10. Fukase, K.; Tanaka, K. Bio-imaging and Cancer Targeting with Glycoproteins and N-glycans. *Curr. Opin. Chem. Biol.* **2012,** *16* (5–6), 614–621.
11. Boutureira, O.; Bernardes, G. J. L. Advances in Chemical Protein Modification. *Chem. Rev.* **2015,** *115* (5), 2174–2195.
12. Jain, K.; Kesharwani, P.; Gupta, U.; Jain, N. K. A Review of Glycosylated Carriers for Drug Delivery. *Biomaterials* **2012,** *33* (16), 4166–4186.
13. Gray, G. R. Direct Coupling of Oligosaccharides to Proteins and Derivatized Gels. *Arch. Biochem. Biophys.* **1974,** *163* (1), 426–428.

14. Roy, R.; Katzenellenbogen, E.; Jennings, H. J. Improved Procedures for the Conjugation of Oligosaccharides to Protein by Reductive Amination. *Can. J. Biochem. Cell Biol.* **1984,** *62* (5), 270–275.
15. Gildersleeve, J. C.; Oyelaran, O.; Simpson, J. T.; Allred, B. Improved Procedure for Direct Coupling of Carbohydrates to Proteins via Reductive Amination. *Bioconjug. Chem.* **2008,** *19* (7), 1485–1490.
16. Doebber, T. W.; Wu, M. S.; Bugianesi, R. L.; Ponpipom, M. M.; Furbish, F. S.; Barranger, J. A.; Brady, R. O.; Shen, T. Y. Enhanced Macrophage Uptake of Synthetically Glycosylated Human Placental β-glucocerebrosidase. *J. Biol. Chem.* **1982,** *257* (5), 2193–2199.
17. Lee, Y. C.; Stowell, C. P.; Krantz, M. J. 2-Imino-2-methoxyethyl 1-thioglycosides: New Reagents for Attaching Sugars to Proteins. *Biochemistry* **1976,** *15* (18), 3956–3963.
18. McBroom, C. R.; Samanen, C. H.; Goldstein, I. I. Carbohydrate Antigens: Coupling of Carbohydrates to Proteins by Diazonium and Phenylisothiocyanate Reactions. *Methods Enzymol.* **1972,** *28,* 212–219.
19. Andre, S.; Unverzagt, C.; Kojima, S.; Dong, X.; Fink, C.; Kayser, K.; Gabius, H.-J. Neoglycoproteins with the Synthetic Complex Biantennary Nonasaccharide or its α2,3/α2,6-Sialylated Derivatives: Their Preparation, Assessment of Their Ligand Properties for Purified Lectins, for Tumor Cells In Vitro, and in Tissue Sections, and Their Biodistribution in Tumor-Bearing Mice. *Bioconjug. Chem.* **1997,** *8* (6), 845–855.
20. Tanaka, K.; Siwu, E. R. O.; Minami, K.; Hasegawa, K.; Nozaki, S.; Kanayama, Y.; Koyama, K.; Chen, W. C.; Paulson, J. C.; Watanabe, Y.; Fukase, K. Noninvasive Imaging of Dendrimer-Type N-glycan Clusters: In Vivo Dynamics Dependence on Oligosaccharide Structure. *Angew. Chem. Int. Ed.* **2010,** *49* (44), 8195–8200.
21. Tanaka, K.; Fukase, K.; Katsumura, S. Exploring a Unique Reactivity of 6π-azaelectrocyclization to Enzyme Inhibition, Natural Products Synthesis, and Molecular Imaging: An Approach to Chemical Biology by Synthetic Chemists. *Synlett* **2011,** *2011* (15), 2115–2139.
22. Ogura, A.; Tahara, T.; Nozaki, S.; Morimoto, K.; Kizuka, Y.; Kitazume, S.; Hara, M.; Kojima, S.; Onoe, H.; Kurbangalieva, A.; Taniguchi, N.; Watanabe, Y.; Tanaka, K. Visualizing Trimming Dependence of Biodistribution and Kinetics with homo- and Heterogeneous N-glycoclusters on Fluorescent Albumin. *Sci. Rep.* **2016,** *6,* 21797.
23. Latypova, L.; Sibgatullina, R.; Ogura, A.; Fujiki, K.; Khabibrakhmanova, A.; Tahara, T.; Nozaki, S.; Urano, S.; Tsubokura, K.; Onoe, H.; Watanabe, Y.; Kurbangalieva, A.; Tanaka, K. Sequential Double "clicks" toward Structurally Well-Defined Heterogeneous N-glycoclusters: The Importance of Cluster Heterogeneity on Pattern Recognition In Vivo. *Adv. Sci.* **2016,** 1600394.
24. Davis, B. G.; Lloyd, R. C.; Jones, J. B. Controlled Site-Selective Glycosylation of Proteins by a Combined Site-Directed Mutagenesis and Chemical Modification Approach. *J. Org. Chem.* **1998,** *63* (26), 9614–9615.
25. Bernardes, G. J. L.; Chalker, J. M.; Errey, J. C.; Davis, B. G. Facile Conversion of Cysteine and Alkyl Cysteines to Dehydroalanine on Protein Surfaces: Versatile and Switchable Access to Functionalized Proteins. *J. Am. Chem. Soc.* **2008,** *130* (15), 5052–5053.
26. Fernandez-Gonzalez, M.; Boutureira, O.; Bernardes, G. J. L.; Chalker, J. M.; Young, M. A.; Errey, J. C.; Davis, B. G. Site-Selective Chemoenzymatic Construction of Synthetic Glycoproteins Using Endoglycosidases. *Chem. Sci.* **2010,** *1* (6), 709–715.
27. Dondoni, A.; Massi, A.; Nanni, P.; Roda, A. A New Ligation Strategy for Peptide and Protein Glycosylation: Photoinduced Thiol-ene Coupling. *Chem. Eur. J.* **2009,** *15* (43), 11444–11449.

28. Chalker, J. M.; Bernardes, G. J. L.; Davis, B. G. A "tag-and-modify" Approach to Site-selective Protein Modification. *Acc. Chem. Res.* **2011,** *44* (9), 730–741.
29. van Kasteren, S. I.; Kramer, H. B.; Jensen, H. H.; Campbell, S. J.; Kirkpatrick, J.; Oldham, N. J.; Anthony, D. C.; Davis, B. G. Expanding the Diversity of Chemical Protein Modification Allows Post-translational Mimicry. *Nature* **2007,** *446* (7139), 1105–1109.
30. Lin, Y.-Y. A.; Chalker, J. M.; Davis, B. G. Olefin Cross-metathesis on Proteins: Investigation of Allylic Chalcogen Effects and Guiding Principles in Metathesis Partner Selection. *J. Am. Chem. Soc.* **2010,** *132* (47), 16805–16811.
31. Chalker, J. M.; Wood, C. S. C.; Davis, B. G. A Convenient Catalyst for Aqueous and Protein Suzuki-Miyaura Cross-Coupling. *J. Am. Chem. Soc.* **2009,** *131* (45), 16346–16347.
32. Floyd, N.; Vijayakrishnan, B.; Koeppe, J. R.; Davis, B. G. Thiyl Glycosylation of Olefinic Proteins: S-linked Glycoconjugate Synthesis. *Angew. Chem. Int. Ed.* **2009,** *48* (42), 7798–7802.
33. Schwarz, F.; Fan, Y.-Y.; Schubert, M.; Aebi, M. Cytoplasmic N-glycosyltransferase of Actinobacillus Pleuropneumoniae Is an Inverting Enzyme and Recognizes the NX(S/T) Consensus Sequence. *J. Biol. Chem.* **2011,** *286* (40), 35267–35274.
34. Wang, L.-X.; Amin, M. N. Chemical and Chemoenzymatic Synthesis of Glycoproteins for Deciphering Functions. *Chem. Biol.* **2014,** *21* (1), 51–66.
35. Lomino, J. V.; Naegeli, A.; Orwenyo, J.; Amin, M. N.; Aebi, M.; Wang, L.-X. A Two-Step Enzymatic Glycosylation of Polypeptides with Complex N-glycans. *Bioorg. Med. Chem.* **2013,** *21* (8), 2262–2270.
36. Wang, L.-X.; Lomino, J. V. Emerging Technologies for Making Glycan-Defined Glycoproteins. *ACS Chem. Biol.* **2012,** 7 (1), 110–122.
37. Tanaka, K.; Fukase, K. Recent Advances in Positron-Emission Tomography (PET) Imaging of Biomolecules: From Chemical Labeling to Cancer Diagnostics. *Mini Rev. Org. Chem.* **2008,** *5* (3), 153–162.
38. Tanaka, K.; Fukase, K. PET (Positron Emission Tomography) Imaging of Biomolecules Using Metal-DOTA Complexes: A New Collaborative Challenge by Chemists, Biologists, and Physicians for Future Diagnostics and Exploration of In Vivo Dynamics. *Org. Biomol. Chem.* **2008,** *6* (5), 815–828.
39. Kojima, S.; Ishido, M.; Kubota, K.; Kubodera, A.; Hellmann, T.; Kohnke-Godt, B.; Wosgien, B.; Gabius, H. J. Tissue Distribution of Radioiodinated Neoglycoproteins and Mammalian Lectins. *Biol. Chem. Hoppe-Seyler* **1990,** *371* (4), 331–338.
40. Ashwell, G.; Harford, J. Carbohydrate-Specific Receptors of the Liver. *Annu. Rev. Biochem.* **1982,** *51,* 531–554.
41. Nishikawa, M.; Hirabayashi, H.; Takakura, Y.; Hashida, M. Design for Cell-Specific Targeting of Proteins Utilizing Sugar-Recognition Mechanism: Effect of Molecular Weight of Proteins on Targeting Efficiency. *Pharm. Res.* **1995,** *12* (2), 209–214.
42. Nishikawa, M.; Miyazaki, C.; Yamashita, F.; Takakura, Y.; Hashida, M. Galactosylated Proteins Are Recognized by the Liver According to the Surface Density of Galactose Moieties. *Am. J. Physiol.* **1995,** *268* (5, Pt. 1), G849–G856.
43. Gunn, A. J.; Hama, Y.; Koyama, Y.; Kohn, E. C.; Choyke, P. L.; Kobayashi, H. Targeted Optical Fluorescence Imaging of Human Ovarian Adenocarcinoma Using a Galactosyl Serum Albumin-Conjugated Fluorophore. *Cancer Sci.* **2007,** *98* (11), 1727–1733.
44. Hama, Y.; Urano, Y.; Koyama, Y.; Choyke, P. L.; Kobayashi, H. D-galactose Receptor-targeted In Vivo Spectral Fluorescence Imaging of Peritoneal Metastasis Using Galactosamine-Conjugated Serum Albumin-Rhodamine Green. *J. Biomed. Opt.* **2007,** *12* (5), 051501.

45. Regino, C. A. S.; Ogawa, M.; Alford, R.; Wong, K. J.; Kosaka, N.; Williams, M.; Feild, B. J.; Takahashi, M.; Choyke, P. L.; Kobayashi, H. Two-Step Synthesis of Galactosylated Human Serum Albumin as a Targeted Optical Imaging Agent for Peritoneal Carcinomatosis. *J. Med. Chem.* **2010,** *53* (4), 1579−1586.
46. Chopra, A. *[^{99m}Tc]-Diethylenetriaminepentaacetic Acid-Galactosyl Human Serum Albumin.* https://www.ncbi.nlm.nih.gov/books/NBK23514.
47. Yao, Z.; Zhang, M.; Sakahara, H.; Saga, T.; Nakamoto, Y.; Sato, N.; Zhao, S.; Arano, Y.; Konishi, J. Imaging of Intraperitoneal Tumors with Technetium-99m GSA. *Ann. Nucl. Med.* **1998,** *12* (2), 115−118.
48. Stahl, P. D.; Wileman, T. E.; Diment, S.; Shepherd, V. L. Mannose-specific Oligosaccharide Recognition by Mononuclear Phagocytes. *Biol. Cell* **1984,** *51* (2), 215−218.
49. Takagi, K.; Uehara, T.; Kaneko, E.; Nakayama, M.; Koizumi, M.; Endo, K.; Arano, Y. ^{99m}Tc-Labeled Mannosyl-Neoglycoalbumin for Sentinel Lymph Node Identification. *Nucl. Med. Biol.* **2004,** *31* (7), 893−900.
50. Jeong, J. M.; Hong, M. K.; Kim, Y. J.; Lee, J.; Kang, J. H.; Lee, D. S.; Chung, J.-K.; Lee, M. C. Development of ^{99m}Tc-Neomannosyl Human Serum Albumin (^{99m}Tc-MSA) as a Novel Receptor Binding Agent for Sentinel Lymph Node Imaging. *Nucl. Med. Commun.* **2004,** *25* (12), 1211−1217.
51. Steinfeld, S.; Penaloza, A.; Decaestecker, C.; Rommes, S.; Andre, S.; Schuring, M.-P.; Danguy, A.; Appelboom, T.; Kiss, R.; Gabius, H.-J. Labeled Neoglycoproteins and Human Lectins as Diagnostic and Potential Functional Markers in Salivary Glands of Patients with Sjogren's Syndrome. *J. Rheumatol.* **2000,** *27* (8), 1910−1916.
52. Yabe, Y.; Kobayashi, N.; Nishikawa, M.; Mihara, K.; Yamashita, F.; Takakura, Y.; Hashida, M. Pharmacokinetics and Preventive Effects of Targeted Catalase Derivatives on Hydrogen Peroxide-Induced Injury in Perfused Rat Liver. *Pharm. Res.* **2002,** *19* (12), 1815−1821.
53. Yabe, Y.; Nishikawa, M.; Tamada, A.; Takakura, Y.; Hashida, M. Targeted Delivery and Improved Therapeutic Potential of Catalase by Chemical Modification: Combination with Superoxide Dismutase Derivatives. *J. Pharmacol. Exp. Ther.* **1999,** *289* (2), 1176−1184.
54. Opanasopit, P.; Nishikawa, M.; Yamashita, F.; Takakura, Y.; Hashida, M. Pharmacokinetic Analysis of Lectin-Dependent Biodistribution of Fucosylated Bovine Serum Albumin: A Possible Carrier for Kupffer Cells. *J. Drug Target.* **2001,** *9* (5), 341−351.
55. Sarkar, K.; Sarkar, H. S.; Kole, L.; Das, P. K. Receptor-Mediated Endocytosis of Fucosylated Neoglycoprotein by Macrophages. *Mol. Cell. Biochem.* **1996,** *156* (2), 109−116.
56. Kim, S.; Jeong, J. M.; Hong, M. K.; Jang, J.-J.; Lee, J.; Lee, D. S.; Chung, J.-K., Lee, M. C. Differential Receptor Targeting of Liver Cells Using ^{99m}Tc-Neoglycosylated Human Serum Albumins. *Arch. Pharmacal. Res.* **2008,** *31* (1), 60−66.
57. Choi, J. Y.; Jeong, J. M.; Yoo, B. C.; Kim, K.; Kim, Y.; Yang, B. Y.; Lee, Y.-S.; Lee, D. S.; Chung, J.-K.; Lee, M. C. Development of ^{68}Ga-Labeled Mannosylated Human Serum Albumin (MSA) as a Lymph Node Imaging Agent for Positron Emission Tomography. *Nucl. Med. Biol.* **2011,** *38* (3), 371−379.
58. McCann, T. E.; Kosaka, N.; Mitsunaga, M.; Choyke, P. L.; Gildersleeve, J. C.; Kobayashi, H. Biodistribution and Excretion of Monosaccharide-Albumin Conjugates Measured with In Vivo Near-Infrared Fluorescence Imaging. *Bioconjug. Chem.* **2010,** *21* (10), 1925−1932.
59. Alexander, V. M.; Sano, K.; Yu, Z.; Nakajima, T.; Choyke, P. L.; Ptaszek, M.; Kobayashi, H. Galactosyl Human Serum Albumin-NMP1 Conjugate: A Near Infrared (NIR)-Activatable Fluorescence Imaging Agent to Detect Peritoneal Ovarian Cancer Metastases. *Bioconjug. Chem.* **2012,** *23* (8), 1671−1679.

60. Lehrman, M. A.; Pizzo, S. V.; Imber, M. J.; Hill, R. L. The Binding of Fucose-Containing Glycoproteins by Hepatic Lectins: Re-Examination of the Clearance from Blood and the Binding to Membrane Receptors and Pure Lectins. *J. Biol. Chem.* **1986,** *261* (16), 7412–7418.
61. Unverzagt, C.; Andre, S.; Seifert, J.; Kojima, S.; Fink, C.; Srikrishna, G.; Freeze, H.; Kayser, K.; Gabius, H.-J. Structure-Activity Profiles of Complex Biantennary Glycans with Core Fucosylation and With/Without Additional α2,3/α2,6 Sialylation: Synthesis of Neoglycoproteins and Their Properties in Lectin Assays, Cell Binding, and Organ Uptake. *J. Med. Chem.* **2002,** *45* (2), 478–491.
62. Andre, S.; Unverzagt, C.; Kojima, S.; Frank, M.; Seifert, J.; Fink, C.; Kayser, K.; von der Lieth, C.-W.; Gabius, H.-J. Determination of Modulation of Ligand Properties of Synthetic Complex-Type Biantennary N-glycans by Introduction of Bisecting GlcNAc In Silico, In Vitro and In Vivo. *Eur. J. Biochem.* **2004,** *271* (1), 118–134.
63. Andre, S.; Kojima, S.; Prahl, I.; Lensch, M.; Unverzagt, C.; Gabius, H.-J. Introduction of Extended LEC14-Type Branching into Core-Fucosylated Biantennary N-glycan. *FEBS J.* **2005,** *272* (8), 1986–1998.
64. Andre, S.; Kojima, S.; Gundel, G.; Russwurm, R.; Schratt, X.; Unverzagt, C.; Gabius, H.-J. Branching Mode in Complex-Type Triantennary N-glycans as Regulatory Element of Their Ligand Properties. *Biochim. Biophys. Acta Gen. Subj.* **2006,** *1760* (5), 768–782.
65. Andre, S.; Kozar, T.; Schuberth, R.; Unverzagt, C.; Kojima, S.; Gabius, H.-J. Substitutions in the N-glycan Core as Regulators of Biorecognition: The Case of Core-fucose and Bisecting GlcNAc Moieties. *Biochemistry* **2007,** *46* (23), 6984–6995.
66. Gabius, H.-J.; Andre, S.; Jimenez-Barbero, J.; Romero, A.; Solis, D. From Lectin Structure to Functional Glycomics: Principles of the Sugar Code. *Trends Biochem. Sci.* **2011,** *36* (6), 298–313.
67. Hirota, M.; Sugi, K.; Inoue, M. Asialofetuin Metabolism in Animals with Obstructive Jaundice. *Int. Hepatol. Commun.* **1995,** *3* (4), 184–189.
68. Park, E. I.; Mi, Y.; Unverzagt, C.; Gabius, H.-J.; Baenziger, J. U. The Asialoglycoprotein Receptor Clears Glycoconjugates Terminating with Sialic Acid α2,6GalNAc. *Proc. Natl. Acad. Sci. U.S.A.* **2005,** *102* (47), 17125–17129.
69. Pritchard, J. B.; Miller, D. S. Renal Secretion of Organic Anions and Cations. *Kidney Int.* **1996,** *49* (6), 1649–1654. Kidney: Structure and Function in Health and Disease.
70. Martens, J. H.; Kzhyshkowska, J.; Falkowski-Hansen, M.; Schledzewski, K.; Gratchev, A.; Mansmann, U.; Schmuttermaier, C.; Dippel, E.; Koenen, W.; Riedel, F.; Sankala, M.; Tryggvason, K.; Kobzik, L.; Moldenhauer, G.; Arnold, B.; Goerdt, S. Differential Expression of a Gene Signature for Scavenger/Lectin Receptors by Endothelial Cells and Macrophages in Human Lymph Node Sinuses, the Primary Sites of Regional Metastasis. *J. Pathol.* **2006,** *208* (4), 574–589.
71. Ise, H.; Kobayashi, S.; Goto, M.; Sato, T.; Kawakubo, M.; Takahashi, M.; Ikeda, U.; Akaike, T. Vimentin and Desmin Possess GlcNAc-Binding Lectin-Like Properties on Cell Surfaces. *Glycobiology* **2010,** *20* (7), 843–864.
72. Bataller, R.; Brenner, D. A. Liver Fibrosis. *J. Clin. Investig.* **2005,** *115* (2), 209–218.
73. Zhou, W.; Inada, M.; Lee, T.-P.; Benten, D.; Lyubsky, S.; Bouhassira, E. E.; Gupta, S.; Tsai, H.-M. ADAMTS13 is Expressed in Hepatic Stellate Cells. *Lab. Investig.* **2005,** *85* (6), 780–788.
74. Liang, C.-H.; Wang, S.-K.; Lin, C.-W.; Wang, C.-C.; Wong, C.-H.; Wu, C.-Y. Effects of Neighboring Glycans on Antibody-Carbohydrate Interaction. *Angew. Chem. Int. Ed.* **2011,** *50* (7), 1608–1612.
75. Hirai, M.; Minematsu, H.; Kondo, N.; Oie, K.; Igarashi, K.; Yamazaki, N. Accumulation of Liposome with Sialyl Lewis X to Inflammation and Tumor Region: Application to In Vivo Bio-Imaging. *Biochem. Biophys. Res. Commun.* **2007,** *353* (3), 553–558.

76. Ogura, A.; Tahara, T.; Nozaki, S.; Onoe, H.; Kurbangalieva, A.; Watanabe, Y.; Tanaka, K. Glycan Multivalency Effects toward Albumin Enable N-glycan-Dependent Tumor Targeting. *Bioorg. Med. Chem. Lett* **2016,** *26* (9), 2251–2254.
77. Tanaka, K.; Minami, K.; Tahara, T.; Fujii, Y.; Siwu, E. R. O.; Nozaki, S.; Onoe, H.; Yokoi, S.; Koyama, K.; Watanabe, Y.; Fukase, K. Electrocyclization-Based Labeling Allows Efficient In Vivo Imaging of Cellular Trafficking. *ChemMedChem* **2010,** *5* (6), 841–845.
78. Tanaka, K.; Yokoi, S.; Morimoto, K.; Iwata, T.; Nakamoto, Y.; Nakayama, K.; Koyama, K.; Fujiwara, T.; Fukase, K. Cell Surface Biotinylation by Azaelectrocyclization: Easy-Handling and Versatile Approach for Living Cell Labeling. *Bioorg. Med. Chem.* **2012,** *20* (6), 1865–1868.
79. Tanaka, K.; Moriwaki, K.; Yokoi, S.; Koyama, K.; Miyoshi, E.; Fukase, K. Whole-Body Imaging of Tumor Cells by Azaelectrocyclization: Visualization of Metastasis Dependence on Glycan Structure. *Bioorg. Med. Chem.* **2013,** *21* (5), 1074–1077.
80. Tanaka, K.; Nakamoto, Y.; Siwu, E. R. O.; Pradipta, A. R.; Morimoto, K.; Fujiwara, T.; Yoshida, S.; Hosoya, T.; Tamura, Y.; Hirai, G.; Sodeoka, M.; Fukase, K. Development of bis-unsaturated ester aldehydes as amino-glue probes: Sequential double azaelectrocyclization as a promising strategy for bioconjugation. *Org. Biomol. Chem.* **2013,** *11* (42), 7326–7333.
81. Ogura, A.; Tanaka, K. Azaelectrocyclization on Cell Surface: Convenient and General Approach to Chemical Biology Research. *Tetrahedron* **2015,** *71* (26–27), 4518–4521.
82. Tanaka, K.; Minami, K.; Tahara, T.; Siwu, E. R. O.; Koyama, K.; Nozaki, S.; Onoe, H.; Watanabe, Y.; Fukase, K. A Combined 6π-azaelectrocyclization/Staudinger Approach to Protein and Cell Engineering: Noninvasive Tumor Targeting by N-glycan-Engineered Lymphocytes. *J. Carbohydr. Chem.* **2010,** *29* (3), 118–132.
83. Sasmal, P. K.; Streu, C. N.; Meggers, E. Metal Complex Catalysis in Living Biological Systems. *Chem. Commun.* **2013,** *49* (16), 1581–1587.
84. Völker, T.; Meggers, E. Transition-Metal-Mediated Uncaging in Living Human Cells — an Emerging Alternative to Photolabile Protecting Groups. *Curr. Opin. Chem. Biol.* **2015,** *25,* 48–54.
85. Li, J.; Chen, P. R. Moving Pd-Mediated Protein Cross Coupling to Living Systems. *ChemBioChem* **2012,** *13* (12), 1728–1731.
86. Yang, M.; Li, J.; Chen, P. R. Transition Metal-Mediated Bioorthogonal Protein Chemistry in Living Cells. *Chem. Soc. Rev.* **2014,** *43* (18), 6511–6526.
87. Chankeshwara, S. V.; Indrigo, E.; Bradley, M. Palladium-Mediated Chemistry in Living Cells. *Curr. Opin. Chem. Biol.* **2014,** *21,* 128–135.
88. Streu, C.; Meggers, E. Ruthenium-induced Allylcarbamate Cleavage in Living Cells. *Angew. Chem. Int. Ed.* **2006,** *45* (34), 5645–5648.
89. Völker, T.; Dempwolff, F.; Graumann, P. L.; Meggers, E. Progress towards Bioorthogonal Catalysis with Organometallic Compounds. *Angew. Chem. Int. Ed.* **2014,** *53* (39), 10536–10540.
90. Sasmal, P. K.; Carregal-Romero, S.; Parak, W. J.; Meggers, E. Light-Triggered Ruthenium-Catalyzed Allylcarbamate Cleavage in Biological Environments. *Organometallics* **2012,** *31* (16), 5968–5970.
91. Sanchez, M. I.; Penas, C.; Vazquez, M. E.; Mascarenas, J. L. Metal-Catalyzed Uncaging of DNA-Binding Agents in Living Cells. *Chem. Sci.* **2014,** *5* (5), 1901–1907.
92. Sadhu, K. K.; Lindberg, E.; Winssinger, N. In Cellulo Protein Labelling with Ru-Conjugate for Luminescence Imaging and Bioorthogonal Photocatalysis. *Chem. Commun.* **2015,** *51* (93), 16664–16666.

93. Sasmal, P. K.; Carregal-Romero, S.; Han, A. A.; Streu, C. N.; Lin, Z.; Namikawa, K.; Elliott, S. L.; Köster, R. W.; Parak, W. J.; Meggers, E. Catalytic Azide Reduction in Biological Environments. *ChemBioChem* **2012,** *13* (8), 1116–1120.
94. Wang, J.; Zheng, S.; Liu, Y.; Zhang, Z.; Lin, Z.; Li, J.; Zhang, G.; Wang, X.; Li, J.; Chen, P. R. Palladium-Triggered Chemical Rescue of Intracellular Proteins via Genetically Encoded Allene-Caged Tyrosine. *J. Am. Chem. Soc.* **2016,** *138* (46), 15118–15121.
95. Li, J.; Yu, J.; Zhao, J.; Wang, J.; Zheng, S.; Lin, S.; Chen, L.; Yang, M.; Jia, S.; Zhang, X.; Chen, P. R. Palladium-Triggered Deprotection Chemistry for Protein Activation in Living Cells. *Nat. Chem.* **2014,** *6* (4), 352–361.
96. Wang, J.; Cheng, B.; Li, J.; Zhang, Z.; Hong, W.; Chen, X.; Chen, P. R. Chemical Remodeling of Cell-Surface Sialic Acids through a Palladium-Triggered Bioorthogonal Elimination Reaction. *Angew. Chem. Int. Ed.* **2015,** *54* (18), 5364–5368.
97. Indrigo, E.; Clavadetscher, J.; Chankeshwara, S. V.; Lilienkampf, A.; Bradley, M. Palladium-Mediated In Situ Synthesis of an Anticancer Agent. *Chem. Commun.* **2016,** *52* (99), 14212–14214.
98. Weiss, J. T.; Dawson, J. C.; Fraser, C.; Rybski, W.; Torres-Sánchez, C.; Bradley, M.; Patton, E. E.; Carragher, N. O.; Unciti-Broceta, A. Development and Bioorthogonal Activation of Palladium-Labile Prodrugs of Gemcitabine. *J. Med. Chem.* **2014,** *57* (12), 5395–5404.
99. Unciti-Broceta, A.; Johansson, E. M. V.; Yusop, R. M.; Sanchez-Martin, R. M.; Bradley, M. Synthesis of Polystyrene Microspheres and Functionalization with Pd^0 Nanoparticles to Perform Bioorthogonal Organometallic Chemistry in Living Cells. *Nat. Protoc.* **2012,** 7 (6), 1207–1218.
100. Yusop, R. M.; Unciti-Broceta, A.; Johansson, E. M. V.; Sánchez-Martín, R. M.; Bradley, M. Palladium-Mediated Intracellular Chemistry. *Nat. Chem.* **2011,** *3* (3), 239–243.
101. Jung Jou, M.; Chen, X.; Swamy, K. M. K.; Na Kim, H.; Kim, H.-J.; Lee, S.-g.; Yoon, J. Highly Selective Fluorescent Probe for Au^{3+} Based on Cyclization of Propargylamide. *Chem. Commun.* **2009**, (46), 7218–7220.
102. Wang, J.-B.; Wu, Q.-Q.; Min, Y.-Z.; Liu, Y.-Z.; Song, Q.-H. A Novel Fluorescent Probe for Au(III)/Au(I) Ions Based on an Intramolecular Hydroamination of a Bodipy Derivative and its Application to Bioimaging. *Chem. Commun.* **2012,** *48* (5), 744–746.
103. Yang, Y.-K.; Lee, S.; Tae, J. A Gold(III) Ion-Selective Fluorescent Probe and its Application to Bioimagings. *Org. Lett.* **2009,** *11* (24), 5610–5613.
104. Do, J. H.; Kim, H. N.; Yoon, J.; Kim, J. S.; Kim, H.-J. A Rationally Designed Fluorescence Turn-on Probe for the Gold(III) Ion. *Org. Lett.* **2010,** *12* (5), 932–934.
105. Spicer, C. D.; Triemer, T.; Davis, B. G. Palladium-Mediated Cell-Surface Labeling. *J. Am. Chem. Soc.* **2012,** *134* (2), 800–803.
106. Li, N.; Lim, R. K. V.; Edwardraja, S.; Lin, Q. Copper-Free Sonogashira Cross-Coupling for Functionalization of Alkyne-Encoded Proteins in Aqueous Medium and in Bacterial Cells. *J. Am. Chem. Soc.* **2011,** *133* (39), 15316–15319.
107. Li, J.; Lin, S.; Wang, J.; Jia, S.; Yang, M.; Hao, Z.; Zhang, X.; Chen, P. R. Ligand-Free Palladium-Mediated Site-Specific Protein Labeling inside Gram-Negative Bacterial Pathogens. *J. Am. Chem. Soc.* **2013,** *135* (19), 7330–7338.
108. Weiss, J. T.; Dawson, J. C.; Macleod, K. G.; Rybski, W.; Fraser, C.; Torres-Sánchez, C.; Patton, E. E.; Bradley, M.; Carragher, N. O.; Unciti-Broceta, A. Extracellular Palladium-Catalysed Dealkylation of 5-fluoro-1-propargyl-uracil as a Bioorthogonally Activated Prodrug Approach. *Nat. Commun.* **2014,** *5,* 3277.
109. Clavadetscher, J.; Hoffmann, S.; Lilienkampf, A.; Mackay, L.; Yusop, R. M.; Rider, S. A.; Mullins, J. J.; Bradley, M. Copper Catalysis in Living Systems and In Situ Drug Synthesis. *Angew. Chem. Int. Ed.* **2016,** *55* (50), 15662–15666.

110. Tsubokura, K.; Vong, K. K. H.; Pradipta, A. R.; Ogura, A.; Urano, S.; Tahara, T.; Nozaki, S.; Onoe, H.; Nakao, Y.; Sibgatullina, R.; Kurbangalieva, A.; Watanabe, Y.; Tanaka, K. In Vivo Gold Complex Catalysis within Live Mice. *Angew. Chem. Int. Ed.* **2017,** *56* (13), 3579−3584.
111. Fehske, K. J.; Schläfer, U.; Wollert, U.; Müller, W. E. Characterization of an Important Drug Binding Area on Human Serum Albumin Including the High-affinity Binding Sites of Warfarin and Azapropazone. *Mol. Pharmacol.* **1982,** *21* (2), 387−393.
112. Garg, A.; Mark Manidhar, D.; Gokara, M.; Malleda, C.; Suresh Reddy, C.; Subramanyam, R. Elucidation of the Binding Mechanism of Coumarin Derivatives with Human Serum Albumin. *PLoS One* **2013,** *8* (5), e63805.
113. Zatón, A. M. L.; Ferrer, J. M.; de Gordoa, J. C. R.; Marquínez, M. A. Binding of Coumarins to Site I of Human Serum Albumin: Effect of the Fatty Acids. *Chem. Biol. Interact.* **1995,** *97* (2), 169−174.
114. Kung, K. K.-Y.; Ko, H.-M.; Cui, J.-F.; Chong, H.-C.; Leung, Y.-C.; Wong, M.-K. Cyclometalated Gold(III) Complexes for Chemoselective Cysteine Modification via Ligand Controlled C-S Bond-Forming Reductive Elimination. *Chem. Commun.* **2014,** *50* (80), 11899−11902.
115. Kung, K. K.-Y.; Lo, V. K.-Y.; Ko, H.-M.; Li, G.-L.; Chan, P.-Y.; Leung, K.-C.; Zhou, Z.; Wang, M.-Z.; Che, C.-M.; Wong, M.-K. Cyclometallated Gold(III) Complexes as Effective Catalysts for Synthesis of Propargylic Amines, Chiral Allenes and Isoxazoles. *Adv. Synth. Catal.* **2013,** *355* (10), 2055−2070.

CHAPTER 8

Covalent Bond Formation by Modular Adaptors to Locate Multiple Enzymes on a DNA Scaffold

Eiji Nakata, Shun Nakano, Arivazhagan Rajendran, Takashi Morii
Institute of Advanced Energy, Kyoto University, Uji, Kyoto, Japan

Contents

1. INTRODUCTION

Biological systems are very complex and heterogeneous, but well-organized. They use complicated nanostructural networks of macromolecules to mediate various cellular functions with extremely high efficiency and specificity. Most of these macromolecular systems have established through the spontaneous self-assembly of components into highly organized spatial structures, where the position and orientation of the macromolecules are precisely controlled to facilitate the biological reactions and functions. Various macromolecules such as polysaccharides, lipids, peptides, proteins and nucleic acids have been used as the building blocks to create the self-assembled artificial nanostructures including vesicles, nanofibers, and nanotubes *(1–3)*. Despite the exciting developments in this field, there is still a great challenge exists to definitely assemble multiple heterogeneous

Kinetic Control in Synthesis and Self-Assembly
ISBN 978-0-12-812126-9
https://doi.org/10.1016/B978-0-12-812126-9.00008-0

components into geometric patterns with nanometer precision, as seen in natural biological systems. Also, difficulties exist in the development of novel assembling methods to increase the structural complexity and heterogeneity, while maintaining the purity and quality of the assembled products. Among the various biomolecules, DNA is one of the most promising building blocks for the construction of complex biomolecular networks in nanotechnology. It is a self-assembling biopolymer that can be programmed by rigorous base pairing to form predetermined nanostructures with almost any arbitrary shape. The self-assembling and molecular recognition capabilities make DNA a unique tool in nanotechnology.

Recently, it was recognized that DNA-based nanoscale assembly could be exploited for the bottom-up fabrication of functional nanomaterials *(4,5)*. Initially, small DNA nanostructures in the size range of 10–20 nm were fabricated by using set of single-stranded DNAs. Subsequent progress led to the folding of single-stranded template into a larger DNA nanostructure, referred to as DNA origami (Fig. 8.1) *(6,7)*, has been used in the construction of a wide variety of structures, including two- and three-dimensional lattices, topologically linked arrays, nanodevices, and biosensors *(8–11)*. Moreover, as discussed in this chapter, the constructed DNA nanostructures can also be used as ideal scaffolds to locate several kinds of functional molecules, especially proteins of interest (POI), such as enzymes participating in sequential reactions, in a spatially well-defined manner *(12–17)*. Construction of a given sequential enzymatic reaction system in vitro enhances our understanding of the spatial factors and the chemistry behind the highly efficient cascade reactions and provides a principle for further application of current enzyme technology.

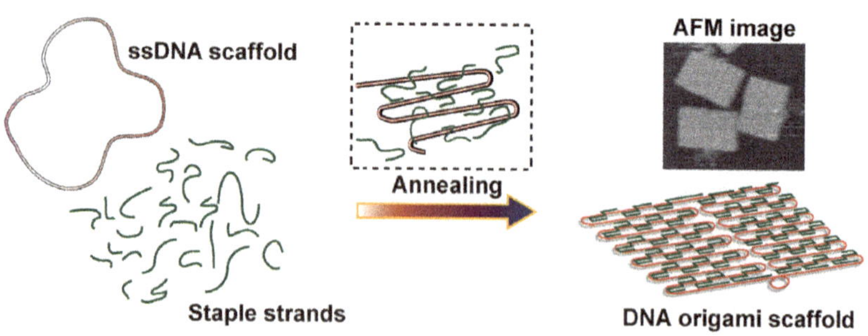

Figure 8.1 Schematic illustrations of the construction of DNA origami scaffold and representative AFM images.

2. OVERVIEW OF PROTEIN–DNA CONJUGATION METHODS

Controlled assembly of POI has been one of the main applications of DNA nanostructures as scaffolds and is an important step in creating complex nanodevices, nanoreactors, and so on. DNA scaffolds prepared by DNA origami provide reasonably large surface area that is sufficient for locating multiple POIs. Furthermore, it has the ability to incorporate additional DNA strands that can serve as the addresses for locating POIs with precise positional control. These additional DNA strands can be incorporated by a simple modification of the staple strands and can be easily functionalized by chemical modifications.

In contrast, the modification of POI requires a lot of tasks. All the modification processes of POIs have to be carried out in mild conditions to retain their activity. Until now, many methods for the conjugation of POIs with DNA have been developed *(18–21)*. These conjugation methods can be divided into two main classes, i.e., noncovalent (reversible) and covalent (irreversible) conjugations (Fig. 8.2). Each method has the various pros and cons, and thus, the choice of the conjugation method depends on the intended use.

The representative examples of noncovalent conjugation methods include the specific ligand- and protein-binding methods such as biotin–avidin interaction *(22)*, antigen–antibody interaction *(23)*, Ni-NTA–hexahistidine interaction *(24)*, aptamer-protein binding *(25)*, and specific-sequence DNA and DNA-binding protein interaction *(26–28)*. Among them, the sequence-specific DNA-binding proteins are one of the ideal candidates for locating POIs at specific positions on DNA scaffold (Fig. 8.2A) *(29)*, and thus the characteristics of DNA-binding protein as DNA-binding adaptors are described in detail in Section 8.3. These non-covalent conjugation methods in common depend on the strong interaction and high selectivity between DNA and DNA-binding adaptors. However, despite the high binding affinity of these interactions, the reversible nature of noncovalent conjugations causes difficulty in preserving the bound POIs on the DNA scaffold, meaning that they can be detached from their target positions.

Formation of a covalent linkage between protein and DNA is a promising strategy for overcoming the issues. A general method of wild-type (or commercially available) POI and DNA conjugation is achieved by a heterobifunctional cross-linkers (Fig. 8.2B) *(18–21)*. Most commonly used cross-linkers are the maleimide and *N*-hydroxysuccinimide (NHS) ester-derived cross-linkers, in which one end of the linker was covalently

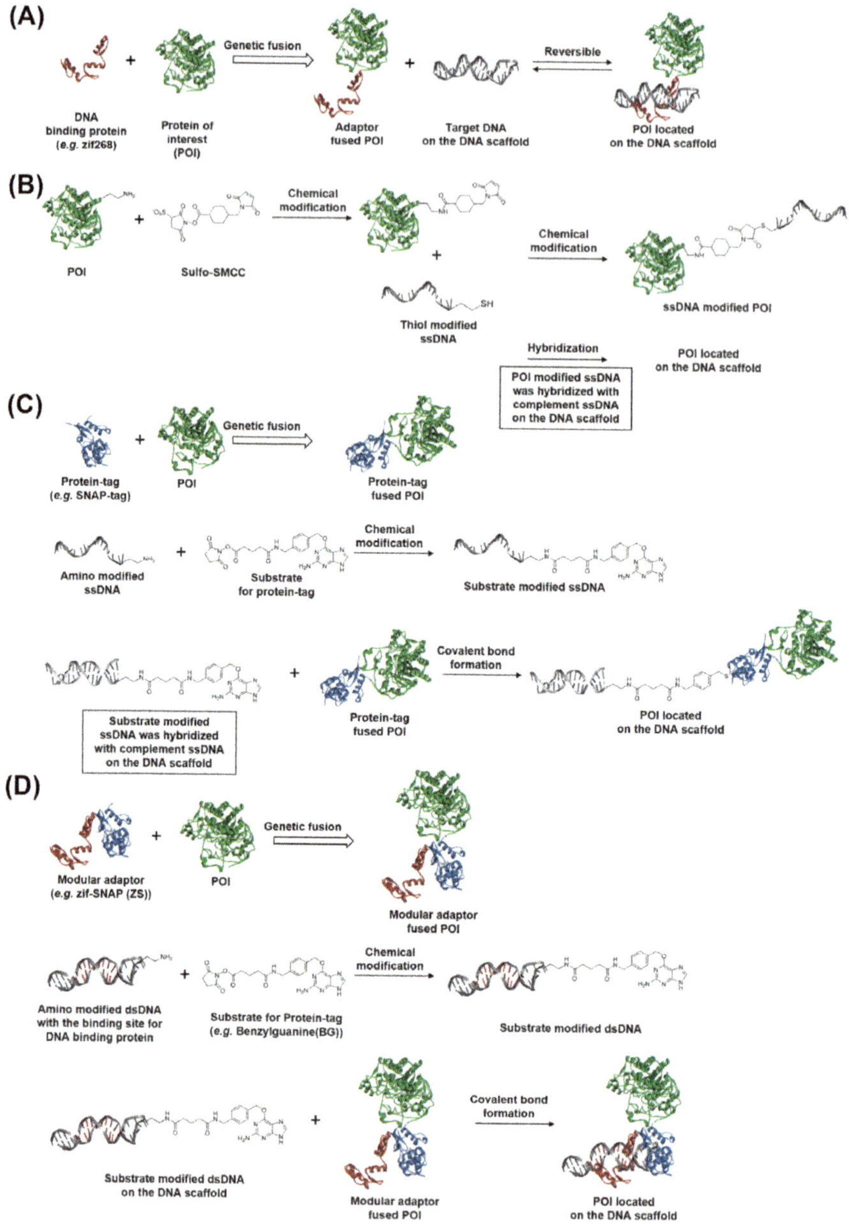

Figure 8.2 Protein–DNA conjugation methods for the loading of proteins of interests (POIs) on the DNA scaffold. (A) The binding process of DNA-binding protein (zinc finger protein) with specific double-strand DNA as an example of reversible interaction between a protein and DNA (also see Fig. 8.3). (B) Residue-specific cross-linking reaction as an example for non–site-specific covalent bond formation between a protein and DNA (sulfo-SMCC: sulfo-succinimidyl-4(*N*-maleimidomethyl)cyclohexane-1-carboxylate). (C) Cross-linking reaction between a protein and DNA by using self-ligating protein tag as an example for site-specific covalent bond formation between a protein and DNA. (D) Cross-linking reaction between a protein and DNA by using a modular adaptor as an example for site-specific covalent bond formation between a protein and DNA (also see Figs. 8.4 and 8.5).

and randomly attached to the lysine side chain or the N-terminal amino group of protein surface and the other maleimide-functionalized end was subsequently coupled to the thiol group modified on the oligo DNA. Other side chains on protein surface such as cysteine can be used for the covalent linkage by changing the functional group of cross-linkers. One problem of the protein–DNA conjugation by chemical cross-linking is that the activity of DNA tethered POI tends to decrease compared with the activity of unmodified POI *(30–33)*. Moreover, the assembly yield of the POIs on the DNA scaffold reported to be low in this method *(30–33)*. One reason for this is that, in all the cases with chemical cross-linking, the protein–DNA conjugation was carried out first and then the localization on DNA scaffold was performed by hybridization with the complemental single-strand DNA attached on the DNA scaffold (Fig. 8.2B). Thus, the assembly yields rely on the hybridization of DNA strands—one on the protein and the other on DNA scaffold. This often leads to low assembly yields, particularly when the enzymes are positioned at shorter interenzyme distances *(31)*. Another problem with the chemical cross-linking methods is the need for purified protein (or commercially available protein) prior to chemical modification and the requirement for redundant purification steps. During chemical modification with DNA, excess amount of DNA is used, and the unreacted DNAs and proteins must be removed. Such purification processes might prevent the use of unstable proteins. Furthermore, these methods have the difficulty in controlling the stoichiometry of loaded POIs. Thus, these methods are not suitable for the precise control on the orientation and spatial control of proteins on the DNA scaffold. In this view point, a genetically fused module such as the DNA-binding adaptor has the potential to control the orientation of POIs on the DNA scaffold.

As for the covalent site-specific conjugation, self-ligating protein tags such as SNAP-tag *(34)* and Halo-tag *(35,36)* were recently applied to locate POIs on the DNA scaffold (Fig. 8.2C) *(37)*. In these cases, self-ligating protein tags were genetically fused to POIs, expressed in *Escherichia coli*, and purified by the conventional methods. Separately, oligo DNAs modified with each of the respective substrates were prepared through conventional coupling reactions with the NHS functionalized substrate and the aminated oligo DNA and assembled at given positions on the DNA scaffold. Then, POI-fused self-ligating protein tags were reacted with the substrate tethered DNA on the DNA scaffold that led to the formation of covalent linkage. These self-ligating protein tags can selectively react with their respective substrates. Thus, orthogonal localization can be realized by using POIs fused to different protein tag and their respective substrate modified DNA. In some cases, the activity of POIs was influenced when self-ligating protein tag was fused. The optimization process to fuse self-ligating protein tag in the appropriate position such as N-terminal or C-terminal with suitable linkers is required. One of the serious problems in directly loading the self-ligating protein tag—fused POI on the DNA scaffold is the requirement of the long incubation and excess amounts of fusion proteins, due to the slow kinetics of the covalent bond formation of a self-ligating protein tag to its substrate on the DNA scaffold. That is not preferable to preserve the activity of POIs. When these protein tags were used to construct an assembly in the orthogonal manner, the coassembly yield was very low *(37)*, even though multiple redundant reaction sites for locating POIs were introduced on the DNA scaffold to increase the apparent loading yield. Such drawbacks cause difficulty in controlling the desired stoichiometry of located POIs.

Unlike these methods, site-specific cross-linking reactions by the modular adaptor, which consisted a sequence-specific DNA-binding zinc finger protein (ZFP) and a self-ligating protein tag, expedites the covalent linkage formation between self-ligating protein tag and its substrate modified at the target DNA sequence on DNA scaffold and leads to the reactions near saturation (Fig. 8.2D) *(38,39)*. Recently, modular adaptors with orthogonal selectivity and fast reaction kinetics to their specific DNA sequences with the respective substrate modifications were developed to enable site-specific location of different protein molecules simultaneously *(40)*. The details of the characteristics of modular adaptors are described in Section 8.4.

The covalent site-specific conjugations by using self-ligating protein tags require incorporation of their substrates on the oligo DNAs, and the currently available protein tags are limited to a few choices for orthogonal assembly of multiple POIs *(41)*. Recently, HUH endonuclease (relaxase) was used for a covalent protein—DNA conjugation *(42,43)*. HUH endonuclease reacts with its specific single-strand DNA and forms a covalent bond without the requirement for chemical modification of single-stranded DNA. Different sequence selective HUH endonucleases have been reported, thus these HUH endonucleases can be used as the orthogonal protein tags by changing DNA sequences *(42,43)*. However, HUH endonucleases—fused POIs were assembled on the DNA scaffold only in low yields, possibly due to the slow kinetics for the covalent bond formation of HUH endonucleases and target DNAs *(42)*. The kinetic aspects of HUH endonucleases need to be studied in detail. As was the case for the self-ligating protein tag, multiple redundant reaction sites were introduced on the DNA scaffold to overcome the low loading yields, which provoke difficulty in controlling the stoichiometric loading of POIs.

More comprehensive information on the classification of the protein–DNA conjugation methods can also be found in other reviews *(18–21)*.

3. DNA-BINDING PROTEINS AS THE REVERSIBLE ADAPTORS FOR SPECIFIC DNA SEQUENCES

Sequence-specific DNA-binding proteins have been utilized as the protein-based DNA-binding adaptors (Fig. 8.2A) *(29,44)*. The ZFPs are one of the best-characterized class of DNA-binding proteins, and the artificially designed ZFPs bind to a wide variety of DNA sequences *(26,27,45–49)*. Each zinc finger domain is capable of recognizing a tract of four base pairs in the major groove of a DNA duplex. A three-fingered ZFP recognizes a tract of 10 base pairs with nanomolar affinity. Basic leucine zipper protein (bZIP) are also well-studied class of DNA-binding proteins, which can recognize a specific DNA sequence around 9 or 10 base pairs with equilibrium dissociation constant (K_d) in low nanomolar range through the parallel coiled-coil formation. Thus, adaptors derived from ZFP (zif268 or AZP4) *(50,51)* and the bZIP (GCN4) *(52)* were utilized for locating POIs in the monomeric and dimeric forms, respectively. On simple construction of adaptor-fused POIs by genetic modification, the sequence-selective DNA binding of these adaptors enables orthogonal arrangements of POIs at the specific DNA sequences on the DNA scaffold with fast binding kinetics, stable complexation, and maintenance of the catalytic activities of fused enzymes (Fig. 8.3).

In particular, two ZFPs, zif268 and AZP4, targeting different DNA sequences were applied as the orthogonal zinc finger adaptors to locate monomeric auto-fluorescent proteins (Cer and YPet) as the model of POIs. The homodimeric bZIP protein GCN4 has been utilized to locate homodimeric enzyme xylitol dehydrogenase (XDH). These POI-fused DNA-binding adaptors (zif268-Cer, AZP4-YPet, and GCN4-XDH) showed high affinity toward specific sequence for the respective DNA-binding adaptors and low affinity toward the other sequences (the K_d values for the target sequences: 47 ± 14 nM for zif268-Cer, 134 ± 23 nM

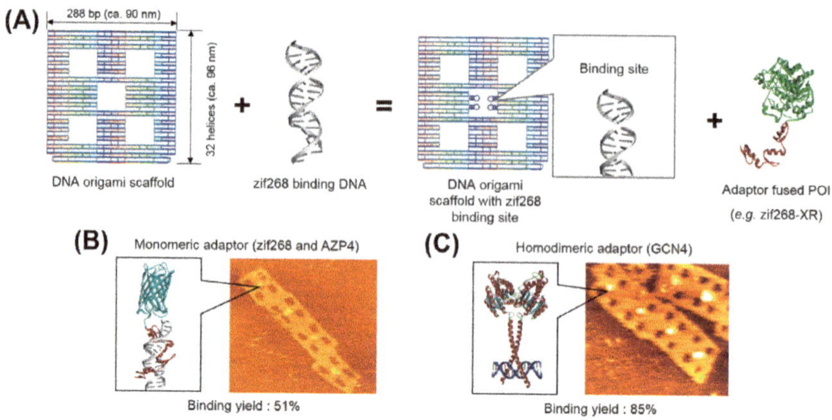

Figure 8.3 The loading of the proteins of interests (POIs) on DNA origami scaffold by specific DNA-binding adaptor. (A) Illustration of the complexation between a specific DNA sequence on DNA origami scaffold with DNA-binding adaptor-fused POI. (B) Zinc finger proteins as monomeric adaptors (e.g., zif268-fused mCer). (C) Leucine zipper proteins as dimeric adaptors (e.g., GCN4-fused XDH).

for AZP4-YPet, and 7.6 ± 2.0 nM for GCN4-XDH) *(29,44)*. It should be noted that the affinity of the adaptor-fused POI for the target sequence is consistent with its binding yield for the target-binding site on DNA scaffold. Interestingly, GCN4-XDH showed a 50% higher enzyme activity than wild-type XDH, possibly due to the higher stability of the dimer form of XDH through the homodimer formation of GCN4 *(44)*.

Adaptor-fused POIs are routinely expressed by *E. coli* and no further modification of adaptor-fused POIs was required. POIs were directly located onto the specific address on DNA scaffold even by using *E. coli* lysate *(29)*. Recently, the DNA-binding adaptors were utilized to control the stoichiometric assembly of functional membrane protein complex such as Kir3 K^+ channels at specific locations on DNA scaffold *(53)*. DNA scaffold was capable of efficiently controlling the oligomerization states and spatial arrangements of membrane proteins not only in vitro but also in living cell. On the other hands, a major drawback of this adaptor is the reversible nature of binding complex, which causes incomplete loading of POIs to the defined DNA addresses on DNA scaffold.

4. MODULAR ADAPTORS AS THE IRREVERSIBLE ADAPTORS FOR SPECIFIC DNA SEQUENCES

To overcome the foregoing drawbacks of DNA-binding adaptor systems, a chemoselective cross-linking domain was conjugated to the adaptor, which would convert the noncovalent adaptor–DNA complex into a covalently linked complex (Fig. 8.2D) *(38)*. A modular adaptor (ZF-SNAP) consists of the ZFP (zif268), and a self-ligating protein tag (SNAP-tag) expedites the covalent linkage formation between the SNAP-tag domain and its substrate present in the target DNA sequence on the DNA scaffold and leads the reaction to near saturation. In particular, the second-order rate constant for ZF-SNAP and the substrate tethered target DNA was determined to be almost 1000 times higher than that for the SNAP-tag alone (Fig. 8.2C) *(38)*. It is due to the fast binding kinetics of zif268 binding to its target DNA and the successive increase in the effective concentration (or effective molarity) (54–56) of substrate. A monomeric enzyme xylose reductase (XR) fused to ZF-SNAP (ZS-XR) was specifically loaded on the DNA scaffold with fast reaction kinetics and a high reaction yield with fully retaining the enzymatic activity (Fig. 8.4A) *(39)*. In addition, the modular adaptor is useful to strictly control the number of enzyme molecule by simply varying the number of modular adaptor–binding sites (Fig. 8.4B–D).

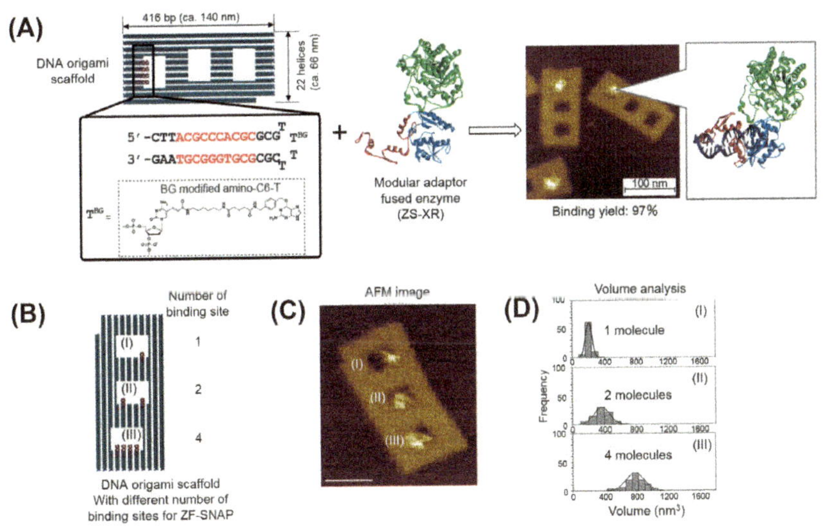

Figure 8.4 Modular adaptor for specific binding site on DNA origami scaffold. (A) Illustration of the complexation of DNA origami scaffold with the binding sites for modular adaptor (e.g., ZF-SNAP) and modular adaptor–fused enzyme (e.g., ZS-XR). (B) Illustration of DNA origami scaffold with different number of binding sites for ZF-SNAP. (C) AFM image of ZS-XR bound to the DNA origami scaffold. The scale bars represent 100 nm. (D) Frequency distributions of molecular volumes of ZS-XR for each cavity with different number of binding sites.

In order to assemble several POIs on the DNA scaffold at specific locations in functional forms, modular adaptors with orthogonality and fast reaction kinetics under mild conditions are highly desired, especially for thermally unstable proteins. A series of modular adaptors consisting of a DNA-binding domain and a self-ligating protein tag were systematically constructed to evaluate the orthogonal covalent bond formation at specific positions on the DNA scaffold (Fig. 8.5). Three DNA-binding domains

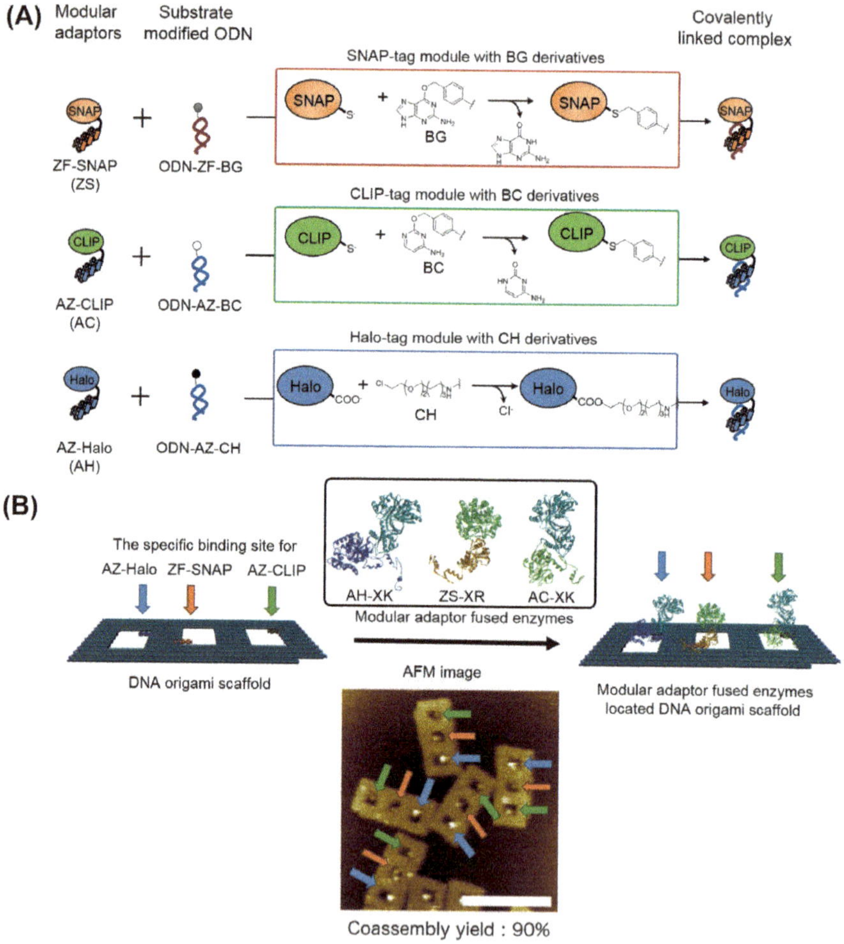

Figure 8.5 Orthogonal modular adaptors for specific binding sites on DNA origami scaffold. (A) Reaction schemes representing the cross-linking reactions between the optimized orthogonal modular adaptors and the substrates incorporated in the ODN. (B) One-pot coassembly reactions of three orthogonal modular adaptor–fused enzymes on the DNA scaffold. The scale bar represents 100 nm.

(zif268 *(50)*, AZP4 *(51)*, and GCN4 *(52)*) and three protein tags (SNAP-tag *(34)*, CLIP-tag *(57)*, and Halo-tag *(35,36)*) were adapted to construct the candidates of orthogonal modular adaptors. Among the nine possible candidates of modular adaptors, a set of three modular adaptors (ZF-SNAP, AZ-CLIP, and AZ-Halo) was successfully chosen to orthogonally react at their respective target sites with quantitative yield (Fig. 8.5A) *(40)*. It should be noted that the one-pot coassembly yield of three different modular adaptors reached to 90% at ambient temperature within 5 min or on ice for 20 min (Fig. 8.5B). The apparent rate constants for the cross-linking reaction of three different modular adaptors are in the same order of magnitude. Such characteristics of multiple cross-linking are very useful to simultaneously assemble various POIs on the DNA scaffold by a one-pot reaction. The short reaction time is very much suitable to load thermally unstable enzymes on the DNA scaffold with maintaining their activity. These orthogonal modular adaptors are highly capable for stably assembling various kinds of enzymes in their active forms on the DNA scaffold. In fact, the enzyme xylulose kinase (XK), which is reported to be thermally unstable *(58)*, was successfully loaded on the DNA scaffold with retaining the enzymatic activity. It is worth mentioning that the high loading yield achieved by the modular adaptor—fused enzyme to a single-binding site on the DNA scaffold allows us to control the orthogonality and forms covalent linkages at different DNA sequences. Application of the set of orthogonal modular adaptors enabled construction of a cascade reaction of two (XR and XDH) or three (XR, XDH, and XK) enzymes from xylose metabolic pathway on DNA scaffold *(40)*. The details of the application are described in Section 8.6.

5. KINETIC ASPECTS FOR THE DESIGN OF IDEAL MODULAR ADAPTORS

In this section, a general strategy to construct ideal modular adaptors, with complete orthogonality and retaining high reactivity, is discussed from the kinetic aspects of cross-linking reaction between modular adaptors and the substrate-modified oligodeoxyribonucleotides (ODNs).

As described above, a set of three orthogonal modular adaptors now enables successful location of three POIs at their respective positions on the DNA scaffold. To assemble more than three kinds of POIs, further expansion of the members of orthogonal modular adaptors consisting of different types of DNA-binding domains and chemoselective cross-linking domains is required. Studies on the kinetics and the selectivity of modular adaptors revealed important characteristics of the cross-linking reactions by modular adaptors that are useful for the design of orthogonal modular adaptors. To verify the kinetic aspects of cross-linking reaction between the modular adaptor (MA) and the substrate modified ODN (SUB), the reaction was considered as represented by Eq. (8.1) and Fig. 8.6, where a reversible complex of DNA-binding domain of modular adaptor and ODN was formed as a transient complex (COMP) *(59)*.

$$[\mathrm{MA}] + [\mathrm{SUB}] \underset{k_{\mathrm{off}}}{\overset{k_{\mathrm{on}}}{\rightleftarrows}} [\mathrm{COMP}] \xrightarrow{k'_{\mathrm{cov}}} [\mathrm{CROSS}] \tag{8.1}$$

The first step of Eq. (8.1) is the formation of reversible complex, COMP, by MA and SUB. The second step is an intramolecular cross-linking reaction within COMP between MA and SUB to form a cross-linking product, CROSS. The apparent rate constant of the cross-linking reaction (k_{app}) ($\mathrm{M}^{-1}\ \mathrm{s}^{-1}$) is represented as in Eq. (8.2):

$$k_{\mathrm{app}} = \frac{k'_{\mathrm{cov}} \times k_{\mathrm{on}}}{k_{\mathrm{off}} + k'_{\mathrm{cov}}} \tag{8.2}$$

where k_{on} ($\mathrm{M}^{-1}\ \mathrm{s}^{-1}$) and k_{off} (s^{-1}) are the association and dissociation rate constants of the DNA-binding domain with SUB, and k'_{cov} (s^{-1}) is the rate constant of covalent bond formation in close proximity, which is defined by the product of the rate constant of intermolecular covalent formation (k_{cov}) ($\mathrm{M}^{-1}\ \mathrm{s}^{-1}$) and the effective concentration ($\mathrm{M}_{\mathrm{eff}}$) (M). When the

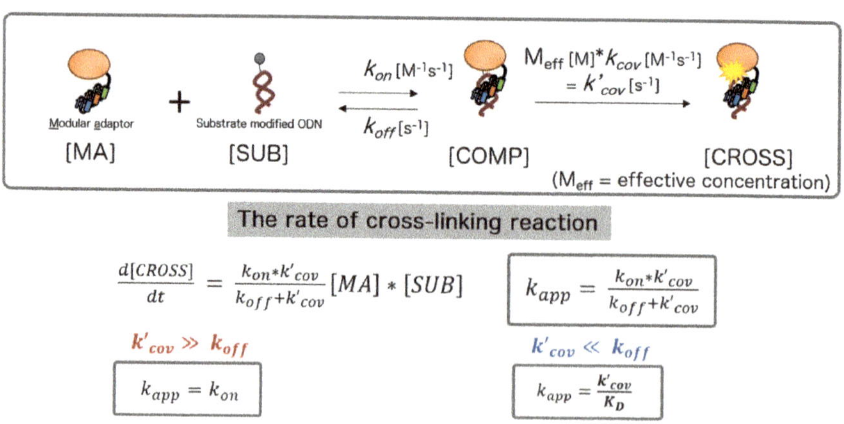

Figure 8.6 Kinetic aspects of the cross-linking reaction of modular adaptor with the substrate-modified ODN.

rate constant of covalent bond formation is much larger than the dissociation rate constant ($k'_{cov} \gg k_{off}$), Eq. (8.2) leads to Eq. (8.3):

$$k_{app} \approx k_{on} \tag{8.3}$$

On the other hand, when the rate constant of covalent bond formation is much smaller than the dissociation rate constant ($k'_{cov} \ll k_{off}$), then Eq. (8.2) leads to Eq. (8.4):

$$k_{app} \approx k'_{cov} \times k_{on}/k_{off} = k'_{cov}/K_D \tag{8.4}$$

where K_D (M) is the equilibrium dissociation constant. Firstly, the reaction of matched pairs of the modular adaptor and the substrate-modified ODN were considered. The k_{on} of DNA-binding domains including the zinc finger motif to DNA are reported to be quite large (10^6–10^8 M^{-1} s^{-1}) *(60)*. From the K_D obtained for the matched complex of the zinc finger motif (~60 nM) and the k_{on} of 10^6–10^8 M^{-1} s^{-1}, the k_{off} of COMP is expected to be 6 to 0.06 s^{-1}. The k_{cov} for the protein tag and its substrate were reported to range from 10^3–10^6 M^{-1} s^{-1} (e.g., SNAP-tag and its substrate BG ≈ 10^4 M^{-1} s^{-1}) *(34–36,57)*, which is much larger than the k_{off} of COMP (0.06 s^{-1}) by assuming that the effective concentration of the reactant within COMP is over 1 mM as previously reported for such complexes *(54–56)*. Therefore, k_{app} corresponds to the k_{on} for the complex formation between the DNA-binding domain and ODN as in Eq. (8.3). This explains the results that k_{app} obtained for the cross-linking reaction of matched complexes are around 10^5–10^6 M^{-1} s^{-1}.

Next, the reaction of unmatched pairs of the modular adaptor and the substrate-modified ODN is considered. The sequence selectivity of the zinc finger motif is governed by the k_{off}, not by the k_{on}, for the DNA complex formation *(60–64)*. Because the K_D of unmatched complex is larger than 10^{-6} M^{-1} and the k_{on} of unmatched complex is similar to that of matched complex, the k_{on} and k_{off} of unmatched complex are expected to be 10^6 M^{-1} s^{-1} and 1 s^{-1}, respectively. Even though the k_{off} for the unmatched complex (1 s^{-1}) is larger than that for the matched complex (0.06 s^{-1}), the condition of $k'_{cov} \gg k_{off}$ again stands for the reaction of the protein tag and its substrate in the unmatched complex. These considerations explain the case for the modular adaptor with the high reactivity for the protein tag and its substrate (e.g., SNAP-tag and its substrate BG ≈ 10^4 M^{-1} s^{-1}) cannot differentiate the covalent bond formation at the DNA sequence modified with its substrate solely by the sequence selectivity of DNA-binding motif. Thus, these pairs are not suitable for the orthogonal adaptors because these modular adaptors possess characteristics leading to Eq. (8.3).

Therefore, reducing the k_{cov} by the protein tag to follow the conditions of Eq. (8.4) would provide an effective strategy to prevent the cross-linking reaction at the unmatched DNA sequences by the modular adaptors bearing the same chemoselective cross-linking domains. Based on Eq. (8.4), the higher the K_D for the unmatched complex of zinc finger motif and ODN is, the lesser the undesired cross-linking reactions of unmatched complexes would proceed. A wide variation of sequence-selective DNA-binding proteins has been well characterized to date *(65)*. These DNA-binding proteins reveal a wide range of affinity and sequence selectivity. The chemoselective cross-linking domains *(41,66)* could be tuned to show different k_{cov}. Further investigation on the combination of DNA-binding domain and chemoselective cross-linking domain will provide us various sets of orthogonal modular adaptors in which the same chemoselective cross-linking domain is shared by different DNA-binding domains to complete the cross-linking reactions at different DNA sequences. Moreover, the considerations described here would stand for the site-specific chemical reactions by a modular molecules with recognition and reaction units in general, such as the cases for the covalent bond formation of sequence-specific DNA alkylating reagents *(67–70)* and the site-specific protein modification reactions *(66,71–73)*.

6. APPLICATIONS OF DNA-BINDING ADAPTORS FOR THE CONSTRUCTION OF ENZYME CASCADE REACTION SYSTEMS ON DNA SCAFFOLD

As the applications of modular adaptors to locate multiple POIs on the DNA scaffold, an artificial enzyme cascade based on the D-xylose metabolic pathway was constructed (Fig. 8.7) *(39,40)*. Within the metabolic pathway of xylose, xylose reductase (XR)–XDH pathway was focused (Fig. 8.7A–D). In the artificially designed cascade, the first enzyme XR converts xylose into xylitol by consuming the cofactor NADH. The produced xylitol and NAD^+ are both simultaneously transported to the second enzyme XDH, which converts xylitol into xylulose by consuming NAD^+ to recycle the NADH cofactor.

A DNA origami with three cavities for placing the hairpin DNAs containing the target sequences for the adaptors was utilized as a scaffold to

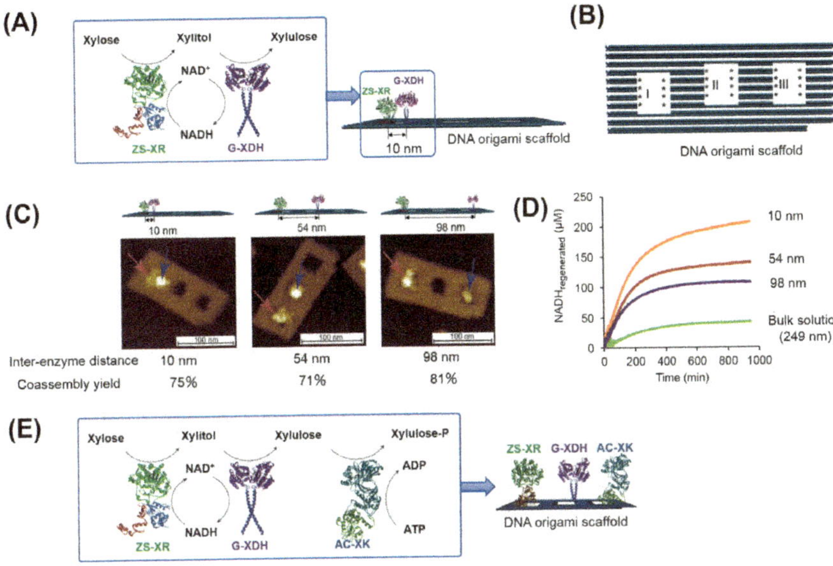

Figure 8.7 Artificial enzyme cascade reactions of xylose pathway activated on the DNA scaffold (based on the D-xylose metabolic pathway). (A–D) Two-step sequential enzymatic reactions by two enzymes ZS-XR and G-XDH on the DNA scaffold. (A) Illustration of the reactions by the coassembled enzymes. (B) Illustration of the DNA origami scaffold showing the three cavities, each of which held up to eight hairpin DNAs containing the specific binding sites for adaptor-fused enzymes at the positions indicated by *asterisks*. (C) AFM image of distance-dependent coassembled enzymes inside the cavity of the DNA origami scaffold. (D) Time-course profiles of the amount of NADH regenerated by G-XDH when the enzymes were coassembled with interenzyme distances of 10, 54, and 98 nm and for the free diffusion system (249 nm). (E) Three-step sequential enzymatic reactions by three enzymes ZS-XR, G-XDH, and AC-XK on the DNA scaffold.

coassemble the adaptor-fused enzymes, i.e., the modular adaptor (ZF-SNAP)–fused XR (ZS-XR) and GCN4 adaptor–fused XDH (G-XDH) (Fig. 8.7A). By changing the position and the number of binding sites for each adaptor, the interenzyme distance and stoichiometry of enzymes would be controlled (Fig. 8.7B). Thus, the artificial enzyme cascade system constructed on the DNA scaffold would be useful to investigate the effect of these factors for the efficiency of cascade reactions. Both the enzymes were coassembled on the DNA scaffold in a distance-dependent manner, with interenzyme distances of 10, 54, and 98 nm (Fig. 8.7C). To compare the cascade reactions with their free diffusion conditions, ZS-XR-assembled DNA scaffold was used with G-XDH existed in bulk solution (in this case, the estimated interenzyme distance is 249 nm). The sequential cascade reactions were performed by monitoring NADH absorption, and the amount of regenerated NADH in the cascade reactions was calculated and monitored against the reaction time (Fig. 8.7D). All the enzyme cascades assembled on the DNA scaffold exhibited much faster regeneration of NADH than in the case of free diffusion. Moreover, the highest enzyme activity was found for the interenzyme distance of 10 nm, and increasing the distance gradually decreased the cascade activity. Furthermore, the initial reaction NADH regeneration rate (V_{ini}), as well as the amount of NADH regenerated, and the yield of xylulose after 16 h were all found to be much higher for the enzymes on the DNA scaffold than for the free diffusion case and decreased gradually with increasing the interenzyme distance. From these results, the transport of intermediates was found to follow 3D Brownian motion. These results indicated that the interenzyme distance is one of the important parameters to regulate the efficiency of the cascade reactions by facilitating transport of intermediates.

To compare the distance dependency of the diffusion processes for bimolecular (xylitol and NAD^+) and unimolecular (NAD^+ alone) substrate, the ratio of turnover frequency at each interenzyme distance to that in the free diffusion case was calculated. The turnover frequency of the bimolecular transport system decreased sharply with increasing interenzyme distance, whereas the unimolecular transport was less sensitive. This possibly reflects faster diffusion of xylitol than NAD^+. It should be noted that the previous studies concerning the effect of interenzyme distance for the efficiency of enzyme cascade reaction on the DNA scaffold have been analyzed in the case of the transfer of a unimolecular cofactor system *(31)*. Our investigation on the actual intermediate formation and on the transfer of bimolecular substrates greatly validated the strategy of enzyme cascade reaction systems on DNA scaffold. To our knowledge, the result is the first example for validating the effect of interenzyme distance for the efficiency of enzyme cascade reactions with unimolecular and bimolecular substrates on the DNA scaffold.

By taking advantage of orthogonal modular adaptors, the XR-XDH pathway from xylose metabolic pathway was extended from two-step to three-step sequential enzymatic reactions, where a modular adaptor (AZ-CLIP)—fused xylulose kinase (AC-XK) converts xylulose into xylulose-5-phosphate by consuming ATP (Fig. 8.7E). Two types of DNA origami scaffolds that had the same number of binding sites for ZS-XR, G-XDH, and AC-XK but differed in the distance between the binding sites were prepared, and these three enzymes were loaded on these DNA scaffolds. To compare the cascade reactions, a DNA scaffold assembled with only ZS-XR and one coassembled with ZS-XR and G-XDH was also prepared. The cascade reactions by three enzymes were analyzed by HPLC to quantitate the cofactors (NADH, NAD^+, ATP, and ADP) consumed or generated in the cascade reaction at steady-state conditions after 24 h. In particular, the amount of ADP produced in the final reaction step of the cascade was used as a measure of the efficiency of the cascade reactions. When the cascade reactions of the three enzymes coassembled system was compared with that of two enzymes coassembled system and the single enzyme loaded system, the amount of ADP produced in the reaction with the three enzyme coassembled system was higher than that of the two other systems. The results indicated that the products of XR and XDH were efficiently transported within the coassembled enzymes on the DNA scaffold. Moreover, the amount of ADP produced in the three enzyme coassembled system within 10 nm was higher than that coassembled with 50 nm distance. These results were consistent with the results of two enzymes coassembled reaction system as described above. The cascade reactions consisting of three enzymes provide a useful model system to thoroughly investigate the effects of interenzyme distances, orientations, and ratios of the three enzymes.

7. SUMMARY

Rapid development in the field of structural DNA nanotechnology has recently enabled the controlled self-assembly of biomolecules and nanomaterials. Unlike most other man-made self-assembled structures, the precise and programmable base pairing of DNA provides fully addressable nanoarchitectures with various shapes and geometries as ideal scaffolds that allow assembly of functional molecules with nanometer-level precision. Proteins are particularly interesting class of molecules to assemble because of their huge functional variability, though the methods to assemble POI on the DNA scaffold have not yet been well established. Thus, further breakthrough in the method for locating POI with controlling not only the positions but also the number, stoichiometry, and the orientation is necessary to explore the spatial aspects on the function of molecular assemblies on DNA scaffolds, such as the multienzyme cascade reaction systems. As described above, the orthogonal modular adaptors are one of the promising methods to construct highly ordered protein assemblies with orientational and spatial control on the DNA scaffold in vitro. The general strategies to construct desired modular adaptors developed by the kinetic aspects of cross-linking reactions between modular adaptors and the substrate-modified DNAs would expand further the variation of useful adaptors. Moreover, the design strategy based on the kinetic aspects for the reaction of modular adaptors would stand for the site-specific chemical reactions by a modular molecule with recognition and reaction units in general. By applying the modular adaptors, three-step enzyme cascade reaction system was constructed and the cascade reactions were regulated by the interenzyme distances. More complex function of proteins assembled on the DNA scaffold will be realized by using a combination of DNA nanotechnology and a reliable chemistry for the site-specific location of proteins on DNA.

REFERENCES

1. Langer, R.; Tirrell, D. A. Designing Materials for Biology and Medicine. *Nature* **2004,** *428,* 487–492.
2. Stupp, S. I. Self-Assembly and Biomaterials. *Nano Lett.* **2010,** *10,* 4783–4786.
3. Schoffelen, S.; van Hest, J. C. M. Multi-Enzyme Systems: Bringing Enzymes Together in Vitro. *Soft Matter* **2012,** *8,* 1736–1746.
4. Seeman, N. C. DNA Nanotechnology: Novel DNA Constructions. *Annu. Rev. Biomol. Struct.* **1998,** *27,* 225–248.
5. Seeman, N. C. DNA in a Material World. *Nature* **2003,** *421,* 427–431.
6. Rothemund, P. W. K. Folding DNA to Create Nanoscale Shapes and Patterns. *Nature* **2006,** *440,* 297–302.
7. Hong, F.; Zhang, F.; Liu, Y.; Yan, H. DNA Origami: Scaffolds for Creating Higher Order Structures. *Chem. Rev.* **2017,** *117,* 12584–12640.
8. Pinheiro, A. V.; Han, D.; Shih, W. M.; Yan, H. Challenges and Opportunities for Structural DNA Nanotechnology. *Nat. Nanotechnol.* **2011,** *6,* 763–772.
9. Fu, J.; Liu, M.; Liu, Y.; Yan, H. Spatially-interactive Biomolecular Networks Organized by Nucleic Acid Nanostructures. *Acc. Chem. Res.* **2012,** *45,* 1215–1226.
10. Wang, L.; Arrabito, G. Hybrid, Multiplexed, Functional DNA Nanotechnology for Bioanalysis. *Analyst* **2015,** *140,* 5821–5848.
11. Chandrasekaran, A. R.; Anderson, N.; Kizer, M.; Halvorsen, K.; Wang, X. Beyond the Fold: Emerging Biological Applications of DNA Origami. *ChemBioChem* **2016,** *17,* 1081–1089.
12. Kazenwadel, F.; Franzreb, M.; Rapp, B. E. Synthetic Enzyme Supercomplexes: Coimmobilization of Enzyme Cascades. *Anal. Methods* **2015,** *7,* 4030–4037.
13. Linko, V.; Nummelin, S.; Aarnos, L.; Tapio, K.; Toppari, J.; Kostiainen, M. DNA-Based Enzyme Reactors and Systems. *Nanomaterials* **2016,** *6,* 139.
14. Wheeldon, I.; Minteer, S. D.; Banta, S.; Barton, S. C.; Atanassov, P.; Sigman, M. Substrate Channelling as an Approach to Cascade Reactions. *Nat. Chem.* **2016,** *8,* 299–309.
15. Küchler, A.; Yoshimoto, M.; Luginbühl, S.; Mavelli, F.; Walde, P. Enzymatic Reactions in Confined Environments. *Nat. Nanotechnol.* **2016,** *11,* 409–420.
16. Rajendran, A.; Nakata, E.; Nakano, S.; Morii, T. Nucleic-acid-templated Enzyme Cascades. *ChemBioChem* **2017,** *18,* 696–716.

17. Sanderson, K. Bioengineering: What to Make with DNA Origami. *Nature* **2010,** *464,* 158–159.
18. Niemeyer, C. M. Semisynthetic DNA–Protein Conjugates for Biosensing and Nanofabrication. *Angew. Chem. Int. Ed.* **2010,** *49,* 1200–1216.
19. Sacca, B.; Niemeyer, C. M. Functionalization of DNA Nanostructures with Proteins. *Chem. Soc. Rev.* **2011,** *40,* 5910–5921.
20. Yang, Y. R.; Liu, Y.; Yan, H. DNA Nanostructures as Programmable Biomolecular Scaffolds. *Bioconjug. Chem* **2015,** *26,* 1381–1395.
21. Trads, J. B.; Tørring, T.; Gothelf, K. V. Site-Selective Conjugation of Native Proteins with DNA. *Acc. Chem. Res.* **2017,** *50,* 1367–1374.
22. Yan, H.; Park, S. H.; Finkelstein, G.; Reif, J. H.; LaBean, T. H. DNA-Templated Self-Assembly of Protein Arrays and Highly Conductive Nanowires. *Science* **2003,** *301,* 1882–1884.
23. He, Y.; Tian, Y.; Ribbe, A. E.; Mao, C. Antibody Nanoarrays with a Pitch of ~20 Nanometers. *J. Am. Chem. Soc.* **2006,** *128,* 12664–12665.
24. Shen, W.; Zhong, H.; Neff, D.; Norton, M. L. NTA Directed Protein Nanopatterning on DNA Origami Nanoconstructs. *J. Am. Chem. Soc.* **2009,** *131,* 6660–6661.
25. Rinker, S.; Ke, Y.; Liu, Y.; Chhabra, R.; Yan, H. Self-assembled DNA Nanostructures for Distance-Dependent Multivalent Ligand–Protein Binding. *Nat. Nanotechnol.* **2008,** *3,* 418–422.
26. Klug, A.; Schwabe, J. W. Protein Motifs 5. Zinc Fingers. *FASEB J.* **1995,** *9,* 597–604.
27. Laity, J. H.; Lee, B. M.; Wright, P. E. Zinc Finger Proteins: New Insights into Structural and Functional Diversity. *Curr. Opin. Struct. Biol.* **2001,** *11,* 39–46.
28. Stains, C. I.; Porter, J. R.; Ooi, A. T.; Segal, D. J.; Ghosh, I. DNA Sequence-Enabled Reassembly of the Green Fluorescent Protein. *J. Am. Chem. Soc.* **2005,** *127,* 10782–10783.
29. Nakata, E.; Liew, F. F.; Uwatoko, C.; Kiyonaka, S.; Mori, Y.; Katsuda, Y.; Endo, M.; Sugiyama, H.; Morii, T. Zinc-finger Proteins for Site-specific Protein Positioning on DNA Origami Structures. *Angew. Chem. Int. Ed.* **2012,** *51,* 2421–2424.
30. Wilner, O. I.; Weizmann, Y.; Gill, R.; Lioubashevski, O.; Freeman, R.; Willner, I. Enzyme Cascades Activated on Topologically Programmed DNA Scaffolds. *Nat. Nanotechnol.* **2009,** *4,* 249–254.
31. Fu, J.; Liu, M.; Liu, Y.; Woodbury, N. W.; Yan, H. Interenzyme Substrate Diffusion for an Enzyme Cascade Organized on Spatially Addressable DNA Nanostructures. *J. Am. Chem. Soc.* **2012,** *134,* 5516–5519.
32. Fu, J.; Yang, Y. R.; Johnson-Buck, A.; Liu, M.; Liu, Y.; Walter, N. G.; Woodbury, N. W.; Yan, H. Multi-Enzyme Complexes on DNA Scaffolds Capable of Substrate Channelling with an Artificial Swinging Arm. *Nat. Nanotechnol.* **2014,** *9,* 531–536.
33. Rosen, C. B.; Kodal, A. L. B.; Nielsen, J. S.; Schaffert, D. H.; Scavenius, C.; Okholm, A. H.; Voigt, N. V.; Enghild, J. J.; Kjems, J.; Tørring, T.; Gothelf, K. V. Template-directed Covalent Conjugation of DNA to Native Antibodies, Transferrin and Other Metal-binding Proteins. *Nat. Chem.* **2014,** *6,* 804–809.
34. Keppler, A.; Gendreizig, S.; Gronemeyer, T.; Pick, H.; Vogel, H.; Johnsson, K. A General Method for the Covalent Labeling of Fusion Proteins with Small Molecules in Vivo. *Nat. Biotechnol.* **2003,** *21,* 86–89.
35. Los, G. V.; Encell, L. P.; McDougall, M. G.; Hartzell, D. D.; Karassina, N.; Zimprich, C.; Wood, M. G.; Learish, R.; Ohana, R. F.; Urh, M.; Simpson, D.; Mendez, J.; Zimmerman, K.; Otto, P.; Vidugiris, G.; Zhu, J.; Darzins, A.; Klaubert, D. H.; Bulleit, R. F.; Wood, K. V. H. T. A Novel Protein Labeling Technology for Cell Imaging and Protein Analysis. *ACS Chem. Biol.* **2008,** *3,* 373–382.

36. England, C. G.; Luo, H.; Cai, W. HaloTag Technology: A Versatile Platform for Biomedical Applications. *Bioconjug. Chem.* **2015,** *26,* 975–986.
37. Sacca, B.; Meyer, R.; Erkelenz, M.; Kiko, K.; Arndt, A.; Schroeder, H.; Rabe, K. S.; Niemeyer, C. M. Orthogonal Protein Decoration of DNA Origami. *Angew. Chem. Int. Ed.* **2010,** *49,* 9378–9383.
38. Nakata, E.; Dinh, H.; Ngo, T. A.; Saimura, M.; Morii, T. A Modular Zinc Finger Adaptor Accelerates the Covalent Linkage of Proteins at Specific Locations on DNA Nanoscaffolds. *Chem. Commun.* **2015,** *51,* 1016–1019.
39. Ngo, T. A.; Nakata, E.; Saimura, M.; Morii, T. Spatially Organized Enzymes Drive Cofactor-coupled Cascade Reactions. *J. Am. Chem. Soc.* **2016,** *138,* 3012–3021.
40. Nguyen, T. M.; Nakata, E.; Saimura, M.; Dinh, H.; Morii, T. Design of Modular Protein Tags for Orthogonal Covalent Bond Formation at Specific DNA Sequences. *J. Am. Chem. Soc.* **2017,** *139,* 8487–8496.
41. Hinner, M. J.; Johnsson, K. How to Obtain Labeled Proteins and what to Do with Them. *Curr. Opin. Biotechnol.* **2010,** *21,* 766–776.
42. Sagredo, S.; Pirzer, T.; Aghebat Rafat, A.; Goetzfried, M. A.; Moncalian, G.; Simmel, F. C.; dela Cruz, F. Orthogonal Protein Assembly on DNA Nanostructures Using Relaxases. *Angew. Chem. Int. Ed.* **2016,** *55,* 4348–4352.
43. Lovendahl, K. N.; Hayward, A. N.; Gordon, W. R. Sequence-Directed Covalent Protein-DNA Linkages in a Single Step Using HUH-Tags. *J. Am. Chem. Soc.* **2017,** *139,* 7030–7035.
44. Ngo, T. A.; Nakata, E.; Saimura, M.; Kodaki, T.; Morii, T. A Protein Adaptor to Locate a Functional Protein Dimer on Molecular Switchboard. *Methods* **2014,** *67,* 142–150.
45. Brayer, K. J.; Segal, D. Keep Your Fingers off My DNA: Protein–protein Interactions Mediated by C2H2 Zinc Finger Domains. *Cell Biochem. Biophys.* **2008,** *50,* 111–131.
46. Pabo, C. O.; Peisach, E.; Grant, R. A. Design and Selection of Novel Cys_2His_2 Zinc Finger Proteins. *Annu. Rev. Biochem.* **2001,** *70,* 313–340.
47. Blancafort, P.; Segal, D. J.; Barbas, C. F., III Designing Transcription Factor Architectures for Drug Discovery. *Mol. Pharmacol.* **2004,** *66,* 1361–1371.
48. Papworth, M.; Kolasinska, P.; Minczuk, M. Designer Zinc-Finger Proteins and Their Applications. *Gene* **2006,** *366,* 27–38.
49. Klug, A. The Discovery of Zinc Fingers and Their Applications in Gene Regulation and Genome Manipulation. *Annu. Rev. Biochem.* **2010,** *79,* 213–231.
50. Pavletich, N. P.; Pabo, C. O. Zinc Finger-DNA Recognition: Crystal Structure of a Zif268-dna Complex at 2.1 A. *Science* **1991,** *252,* 809–817.
51. Sera, T.; Uranga, C. Rational Design of Artificial Zinc-finger Proteins Using a Nondegenerate Recognition Code Table. *Biochemistry* **2002,** *41,* 7074–7081.
52. Ellenberger, T. E.; Brandl, C. J.; Struhl, K.; Harrison, S. C. The GCN4 Basic Region Leucine Zipper Binds DNA as a Dimer of Uninterrupted Alpha Helices: Crystal Structure of the Protein-DNA Complex. *Cell* **1992,** *71,* 1223–1237.
53. Kurokawa, T.; Kiyonaka, S.; Nakata, E.; Endo, M.; Koyama, S.; Mori, E.; Tran, N. H.; Dinh, H.; Suzuki, Y.; Hidaka, K.; Kawata, M.; Sato, C.; Sugiyama, H.; Morii, T.; Mori, Y. DNA Origami Scaffolds as Templates for Functional Tetrameric Kir3 K^+ Channels. *Angew. Chem. Int. Ed.* **2018,** *57,* 2586–2591.
54. Gargano, J. M.; Ngo, T.; Kim, J. Y.; Acheson, D. W. K.; Lees, W. J. Multivalent Inhibition of AB5 Toxins. *J. Am. Chem. Soc.* **2001,** *123,* 12909–12910.
55. Krishnamurthy, V. M.; Semetey, V.; Bracher, P. J.; Shen, N.; Whitesides, G. M. Dependence of Effective Molarity on Linker Length for an Intramolecular Protein-Ligand System. *J. Am. Chem. Soc.* **2007,** *129,* 1312–1320.

56. Hobert, E. M.; Doerner, A. M.; Walker, A. S.; Schepartz, A. Effective Molarity *Redux*: Proximity as a Guiding Force in Chemistry and Biology. *Isr. J. Chem.* **2013,** *53,* 567–576.
57. Gautier, A.; Juillerat, A.; Heinis, C.; Corrêa, I. R., Jr.; Kindermann, M.; Beaufils, F.; Johnsson, K. An Engineered Protein Tag for Multiprotein Labeling in Living Cells. *Chem. Biol.* **2008,** *15,* 128–136.
58. Pival, S. L.; Birner-Gruenberger, R.; Krump, C.; Nidetzky, B. D-Xylulose Kinase from *Saccharomyces cerevisiae*: Isolation and Characterization of the Highly Unstable Enzyme, Recombinantly Produced in *Escherichia coli*. *Protein Expr. Purif.* **2011,** *79,* 223–230.
59. Schreiber, G.; Haran, G.; Zhou, H.-X. Fundamental Aspects of Protein-Protein Association Kinetics. *Chem. Rev.* **2009,** *109,* 839–860.
60. Geertz, M.; Shore, D.; Maerkl, S. J. Massively Parallel Measurements of Molecular Interaction Kinetics on a Microfluidic Platform. *Proc. Natl. Acad. Sci. U.S.A* **2012,** *109,* 16540–16545.
61. Berg, O. G.; Winter, R. B.; von Hippel, P. H. Diffusion-Driven Mechanisms of Protein Translocation on Nucleic Acids. 1. Models and Theory. *Biochemistry* **1981,** *20,* 6929–6948.
62. Kim, J. S.; Pabo, C. O. Getting a Handhold on DNA: Design of Poly-Zinc Finger Proteins with Femtomolar Dissociation Constants. *Proc. Natl. Acad. Sci. U.S.A* **1998,** *95,* 2812–2817.
63. Halford, S. E.; Marko, J. F. How Do Site-specific DNA-Binding Proteins Find Their Targets? *Nucleic Acids Res.* **2004,** *32,* 3040–3052.
64. Halford, S. E. An End to 40 Years of Mistakes in DNA–Protein Association Kinetics? *Biochem. Soc. Trans.* **2009,** *37,* 343–348.
65. Gamsjaeger, R.; Liew, C. K.; Loughlin, F. E.; Crossley, M.; Mackay, J. P. Sticky Fingers: Zinc-Fingers as Protein-Recognition Motifs. *Trends Biochem. Sci.* **2007,** *32,* 63–70.
66. Lang, K.; Chin, J. W. Bioorthogonal Reactions for Labeling Proteins. *ACS Chem. Biol.* **2014,** *9,* 16–20.
67. Rajski, S. R.; Williams, R. M. DNA Cross-Linking Agents as Antitumor Drugs. *Chem. Rev.* **1998,** *98,* 2723–2796.
68. Bailly, C.; Chaires, J. B. Sequence-Specific DNA Minor Groove Binders. Design and Synthesis of Netropsin and Distamycin Analogues. *Bioconjug. Chem.* **1998,** *9,* 513–538.
69. Denny, W. A. DNA Minor Groove Alkylating Agents. *Curr. Med. Chem.* **2001,** *8,* 533–544.
70. Bando, T.; Sugiyama, H. Synthesis and Biological Properties of Sequence-Specific DNA-alkylating Pyrrole–imidazole Polyamides. *Acc. Chem. Res.* **2006,** *39,* 935–944.
71. Takaoka, Y.; Ojida, A.; Hamachi, I. Protein Organic Chemistry and Applications for Labeling and Engineering in Live-Cell Systems. *Angew. Chem. Int. Ed.* **2013,** *52,* 4088–4106.
72. Boutureira, O.; Bernardes, G. J. L. Advances in Chemical Protein Modification. *Chem. Rev.* **2015,** *115,* 2174–2195.
73. Chen, X.; Wu, Y.-W. Selective Chemical Labeling of Proteins. *Org. Biomol. Chem.* **2016,** *14,* 5417–5439.

CHAPTER 9

Kinetic Control in Synthesis of Polymers Using Nanoporous Metal-Organic Frameworks

Takashi Uemura[1,2]
[1]Department of Advanced Materials Science, Graduate School of Frontier Sciences, The University of Tokyo, Kashiwa, Japan; [2]CREST, Japan Science and Technology Agency (JST), Saitama, Japan

Contents

1. INTRODUCTION

One of the most important subjects in polymer chemistry is the development of controlled polymerizations that can allow the rational preparation of well-defined structures and topologies useful for desired purposes because regulating the primary structures of polymer chains, such as stereo- and regiostructures, copolymer compositions, monomer sequence, and molecular weight, and its distribution, is the key to enhance and improve the properties of the polymer materials. In addition, control of polymer assembly structures with ordered alignment and arrangement of polymer

Kinetic Control in Synthesis and Self-Assembly
ISBN 978-0-12-812126-9
https://doi.org/10.1016/B978-0-12-812126-9.00009-2

chains is of considerably interest for the construction of well-organized architectures, which can lead to anisotropic, directional, and combined properties of polymers. However, in conventional polymer synthesis system, polymerizations conducted in bulk or solution usually result in the production of polymers with poorly controlled structures. In contrast, polymerizations in nanosized confined spaces can allow effective spatial controls of primary structures of polymers and even two- and three-dimensional multilevel structuring *(1)*. For instance, a variety of nanoporous materials, such as organic hosts, zeolites, clays, and mesoporous materials, have so far played significant roles for changing the polymerizability of monomers, showing the specific size and shape effects of the nanopores on the polymerization kinetics and selectivity by through-space interactions between the pore walls and monomers *(2)*.

Since the early 1990s, study on Metal-Organic Frameworks (MOFs) consisting of metal ions and bridging organic ligands has been the subject of intense research because of their potential applications in many areas, including molecular storage, separation, and exchange *(3)*. Much research effort has been devoted to developing characteristics of MOFs that differ from conventional porous materials (Fig. 9.1), such as organic hosts, zeolites, and activated carbons: (1) crystalline regular channel structures, (2) controllable channel size close to molecular dimensions, (3) designable surface potentials and functionality, and (4) flexible structures responsive to guest molecules. These features are of key importance for the creation of attractive nanosize reaction fields in MOFs *(4)*. Thus, molecular reaction

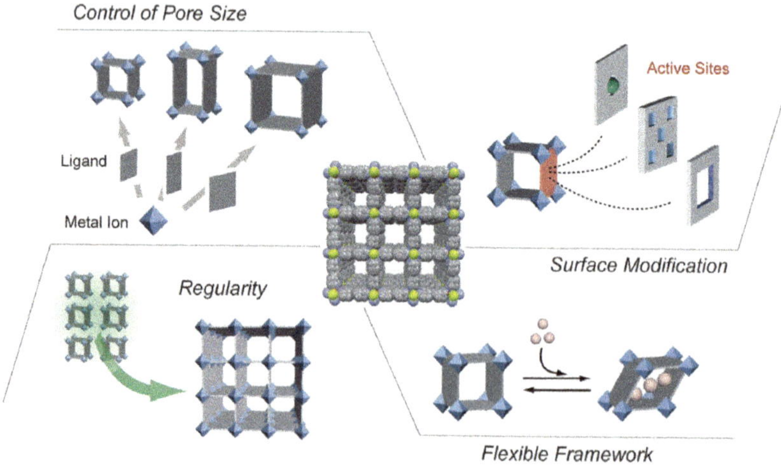

Figure 9.1 Characteristic features of metal-organic frameworks.

and transformation within the confined environments utilizing the functional channels of MOFs has recently evolved as an important topic, prompting vigorous efforts in the development of MOFs for applications in heterogeneous catalysis, size-selective reaction, and ship-in-bottle synthesis. In particular, employing the nanochannels of MOFs for controlled polymerizations is an attractive idea, which would allow multilevel controls of primary and assembly structures of resulting polymers by efficient through-space inductions and specific host–monomer interactions *(5)*. Because of the highly designable structures of MOFs, the functional MOF nanochannels can be applied to a tailor-made polymerization system to obtain preferred polymer structures.

In this chapter, I will discuss the recent progress and future perspectives of various polymerization systems utilizing the MOF nanochannels that especially form the viewpoints of kinetic controls in the synthesis and assembly of polymeric materials.

2. CHARACTERISTICS OF METAL-ORGANIC FRAMEWORKS FOR CONTROLLED POLYMER SYNTHESIS

2.1 Crystallinity and Regularity

Usually, MOFs have highly crystalline nanosized pores that are suitable to accommodate guest molecules even when the intermolecular interaction between the guest molecules and pore walls of MOFs is governed only by the dispersion force, the so-called van der Waals force (Fig. 9.1). The regular channels based on MOFs, with their narrow pore distribution, could provide many advantages in controlled polymerization because of the restricted conformation and packing of accommodated monomers. A variety of selective polymerizations (stereo- and regioregular polymerizations) can proceed in such highly regular pores. Tuning the regularity of the potential field of the channels can enable the creation of well-ordered monomer assemblies inside the MOF channels toward the target polymerization.

2.2 Pore Size and Shape

The polymerizability of guest monomers in confined nanochannels is strongly dependent on the mobility and arrangements of monomers. In general, pore sizes of MOFs range from 0.4 to 2 nm, with the combination of constituted metal ions and organic ligands, which fit for conventional monomers (Fig. 9.1). In addition, a variety of the pore shapes of MOFs, such as triangle, rectangle, and hexagonal shapes, can be prepared by the possible combinations of the components. Of considerable interest is the use of the chiral channels. In these pores, asymmetric polymerizations such as enantioselective polymerization of racemic monomers as well as asymmetric polymerization of prochiral monomers will proceed, giving chiral and helical polymers. From the viewpoint of channel dimensionality, not only one-dimensional but also layered and 3D intersecting channels can be easily prepared in the MOF materials. The designability of pore size and shape in MOFs can control the mobility, arrangement, and density of adsorbed monomers in the pores, which strongly affect the polymerization.

2.3 Pore Surface Functionality

Every MOF material has metal ions that usually act as structural directing components. Thus, preparation of MOFs containing unsaturated metal sites on the pore surfaces *(6)* can allow the utilization of MOFs for inducing a variety of polymerization reactions including oxidative polymerization, Lewis acid—catalyzed polymerization, and coordinative polymerization. In the same manner, strong interactive organic sites can be introduced in ligands of MOFs (Fig. 9.1) *(7)*, which are also responsible for catalytic polymerizations. Other interactive organic functional groups, such as carbonyl, hydroxyl, nitro, amide, etc., into the nanochannels of MOFs would affect the monomer arrangement, which may lead to stereo- and regioselective polymerizations. Introduction of polymerizable groups into organic ligands can provide copolymers and cross-linked polymers if these functional moieties can be copolymerized with other monomers in the MOF channels. The obtained networks are distinct from conventional polymers because the topology and structural information of MOFs can be finely transcribed into the polymer structure.

2.4 Flexibility

Dynamic structural transformation based on flexible frameworks is one of the most interesting and presumably characteristic property of MOFs (Fig. 9.1), which cannot be observed in conventional rigid microporous materials such as zeolites and activated carbons *(8)*. For the polymerization in nanoconfined spaces, utilizations of such flexible frameworks would be the key to highly selective recognition, alignment, and reactivity of the accommodated monomers, which is similar to the induced fit theory and allosteric effect illustrated by enzymes in biological systems.

3. KINETIC CONTROLS OF PRIMARY STRUCTURES

As distinct from conventional porous materials, one of the most advantageous features of MOFs is their modifiable pore structure that can be systematically tuned using a vast combination of organic ligands and metal ions. These properties are of key importance for the creation of unique nanospaces that can achieve precision-controlled polymerizations to regulate the primary structures of polymers, including molecular weight, stereo- and regiostructures, branching structure, and monomer composition.

3.1 Tacticity of Vinyl Polymers

Control of the stereoregularity (tacticity) in radical polymerization is very difficult because of the lack of efficient methods to provide a stereospecific environment around the propagating radical species. Tacticity in vinyl polymers is determined by the relative orientation of the terminal substituent during the addition step. Therefore, kinetic selectivity of the propagating radical for different forms of monomer addition is only influential in the tacticity of resulting polymers. Stereoregular polymers can be produced when the stereochemistry at the penultimate unit of the polymer chain induces stereospecific addition of monomers at the prochiral reactive center. Note that the type of stereocenter formed in the propagation step is dependent on both the conformation of the attacking radical and the face from which the monomer attacks the radical center, indicating that a given conformation of the propagating radical in polymers can lead to either type of stereocenter according to the face from which the monomer attacks.

An interesting nanochannel effect on the tacticity of vinyl polymers was observed in the polymerization using MOFs. In particular, the tacticity of poly(methyl methacrylate) (PMMA) was found to be affected by the pore size of $[M_2(L)_2ted]_n$ ($M = Zn^{2+}$ or Cu^{2+}, L = dicarboxylates, ted = tri ethylenediamine) (Fig. 9.2); an increase of 9% in meso diads was observed when methyl methacrylate (MMA) monomer was polymerized in the narrow nanochannels of MOF with 1,4-naphthalenedicarboxylate, compared with that obtained from the bulk polymerization system *(9)*. Because an isotactic polymer requires narrower conformational diameter than the corresponding syndiotactic one, nanochannels of MOFs could produce PMMA with less stereo-bulky isotactic units by effective through-space interaction. In this system, controlled change of the polymer stereoregularity was possible utilizing the size-tunable pores of MOF.

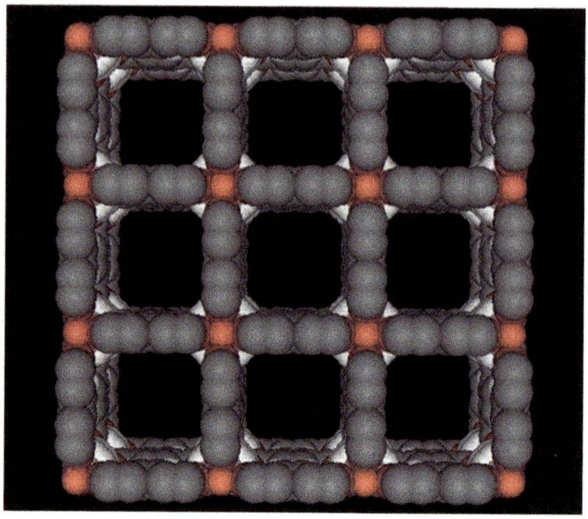

Figure 9.2 Crystal structure of $[Cu_2(terephthalate)_2(ted)]_n$.

Functionalization of $[Cu_2(L)_2(ted)]_n$ by introducing various substituents onto the organic ligand, L, could successfully regulate the tacticity of PMMA, where the tacticity of the PMMA strongly depended on the number and position of the substituent (Fig. 9.3) *(10)*. In fact, polymerization of MMA in $[Cu_2(2,5\text{-dimethoxyterephthalate})_2(ted)]_n$ gave PMMA high isotacticity (m = 55%), which is one of the most effective systems for changing the tacticity of PMMA in radical polymerization. A variety of experimental and theoretical analyses, such as IR, gas adsorption, a statics study, and molecular dynamics (MD) calculations, were performed for understanding the mechanism of this drastic stereoregularity change. Accurate MD calculations were useful to determine the most plausible structures of $[Cu_2(L)_2(ted)]_n$, and revealed that the specific channel shape (regular, narrow, and helically twisted channel) of $[Cu_2(2,5\text{-dimethoxyterephthalate})_2(ted)]_n$ could induce the large tacticity change of the resulting PMMA (Fig. 9.3). In this MOF, the effect of the polymerization temperature on the polymer tacticity was also studied. In the usual polymerization of MMA in solution or in the bulk state, syndiotacticity of PMMA continuously increases with decreasing reaction temperature. In contrast, the syndiotacticity of PMMA prepared in this MOF once increased slightly, and then decreased, as the reaction temperature decreased. Thus, PMMA obtained at lower temperatures still contained high isotactic fractions. This unique phenomenon is ascribable to the suppressed mobility (vibration and rotation) of the MOF at lower temperatures. Such quiescent channels would effectively control the

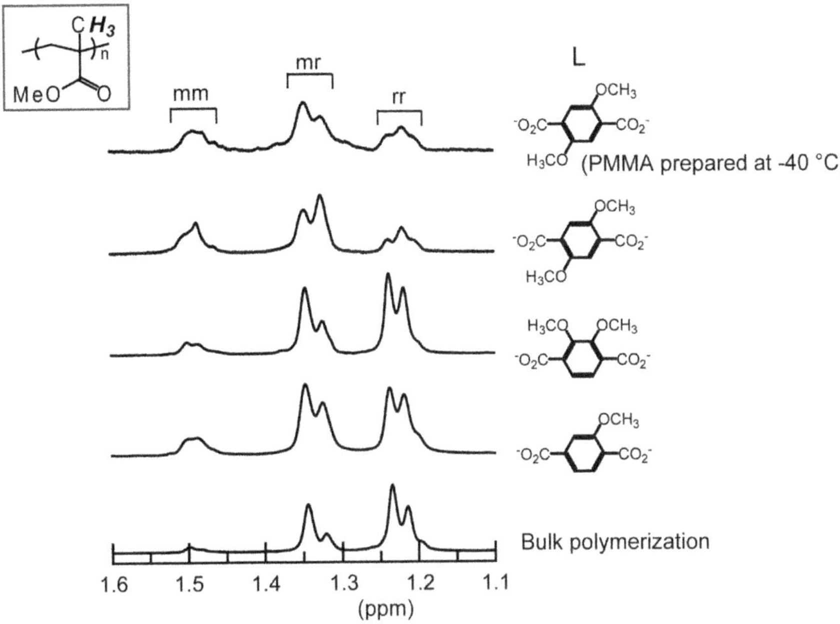

Figure 9.3 ^1H NMR spectra of PMMA in nitrobenzene-d_5 at 110°C prepared in the bulk condition and in $[Cu_2(L)_2ted]_n$ (L = methoxy-substituted ligands). The polymerizations were carried out at 70°C, otherwise noted.

arrangement of encapsulated MMA, resulting in the change of PMMA tacticity.

Similar to the cases of PMMA, the ratio of isotactic units in poly(vinyl acetate) (PVAc) increased when polymerization of vinyl acetate (VAc) was performed in the nanochannels of $[Cu_2(terephthalate)_2ted]_n$ *(9)*. In the previous reports, the only possible way to change the tacticity of PVAc was addition of fluoroalcohols to the reaction media, which could increase the syndiotactic units induced by the hydrogen-bonding interaction between the alcohols and monomers *(11)*. Despite the nonconjugated nature of VAc, polymerizations of this monomer in MOF nanochannels could inhibit the formation of branched structure, producing a linearly extended PVAc with a narrow distribution of molecular weights as well as high isotactic unit.

3.2 Tacticity of Poly(ferrocenylsilane)

Thermal ring-opening polymerization of the unsymmetrically substituted ferrocenylsilane monomer was performed in one-dimensional nanochannels of MOFs *(12)*. Interestingly, the tacticity of polyferrocenylsilane (PFS) obtained from [Al(OH) (2,6-naphthalenedicarboxylate)]$_n$ and [La(1,3,5-benzenetrisbenzoate)]$_n$ was clearly changed in comparison with that of bulk-synthesized one. According to the reported assignment of specific tacticity, PFS with higher isotactic fractions was formed in these MOF channels. Note that stereoregular PFSs are not available except for the case of solid-state ring-opening polymerization using the rather inconvenient technique of γ-ray irradiation *(13)*. Thus, utilization of MOFs is significantly important to change the tacticity of PFS. Because of the stereoregularity change, oxidation potentials of the PFS obtained from these MOFs in cyclic voltammetry were slightly different from that of the bulk-synthesized polymer.

4. KINETIC CONTROLS OF POLYMER ASSEMBLIES

Thermodynamic nonequilibrium approaches has been often used for controlling the arrangement of a polymer assembly *(14)*. This is because of the slow dynamics of polymer chains below the glass transition temperature, which can allow the kinetic manipulation of polymers to give unique polymer architectures with unusual assembly and morphology. One of the most common kinetic approaches is the utilization of nanoporous templates, including zeolites and organic cages *(15)*. In this process, accommodation of polymers inside the nanopores followed by the destruction of the nanoporous hosts result in many specific polymer conformations and assemblies. However, removal of these hosts consisting of covalent bonds requires relatively harsh conditions, often leaving the residues of host matrices and/or resulting in unfavorable rearrangements of polymers. Note that encapsulated polymers in MOFs can be easily isolated by dissolution of the supramolecular host frameworks in mild chelating solutions, which does not severely disturb the specific polymer assemblies during the isolation process.

4.1 Controlled Assemblies of Vinyl Polymers

Preparation of polymer nano- and microparticles has attracted great interest for many applications in scientific and industrial areas such as electronics, photonics, sensors, medicine, biotechnology, and environmental technology *(16)*. General methods for the formation of polymer particles, including emulsion, dispersion, suspension, and seeded polymerizations can produce spherical polymer particles with controllable sizes depending on the synthetic conditions. However, study on the preparation of anisotropic polymer particles has been rather limited despite the potential uses in industry *(17)*. In this regard, direct replica methods have produced many libraries of unique geometrical polymer particles, where many host materials have been utilized to produce anisotropic polymer particles via shape transcription from host materials to organic polymers *(15)*. MOF can be regarded as a good template for the production of anisotropic polymer particles because of their morphological crystal shapes with tunable sizes. Polymerization and subsequent dissolution of MOF molds can leave polymer particles replicated from the nonspherical pristine crystals. As described above, selective dissolution of the MOF matrices can be performed under mild conditions, which is advantageous for the rational synthesis of the replicated polymer particles.

Morphology and size of $[Zn_2(\text{terephthalate})_2(\text{ted})]_n$ can be finely controlled by simply changing the reaction conditions *(18)*. Template synthesis of polystyrene (PSt) particles using these MOF crystals was performed by the introduction of styrene (St) monomer into the pores and subsequent polymerization, followed by the removal of the host frameworks *(19)*. In this polymerization process, morphology of the host particles was found to remain completely, which showed that the polymerization proceeded only within the nanochannels. The resulting PSt was quantitatively isolated from the composite by selective dissolution of the frameworks in an aqueous sodium ethylenediaminetetraacetate (Na-EDTA) solution. Note that the polymer objects formed shape replica of the original MOF crystals despite the non—cross-linked structure. A variety of morphological particles, such as cubic, rod-like, and hexagonal polymers, were consistently obtained from the corresponding hosts with different shapes (Fig. 9.4) *(19)*. In addition to the PSt system, the same experiments using MMA could successfully give anisotropic PMMA particles that were replicated from the host MOFs *(19)*. Interestingly, the final polymer particles retain the size and morphologies of the original MOF particles, although a large mass fraction based on the frameworks was removed by dissolving away. This is because the inherent rigidity of the polymers whose glass transition temperatures (T_g) are much higher than room temperature. In fact, treatment of the obtained PSt particles above the T_g resulted in the collapse of the morphological particles. In addition, formation of porous voids in the interior of the polymer particles contributes to the maintenance of the morphology.

Poly(*N*-vinylcarbazole) (PVCz) has attracted much attention because of the relatively high hole-transporting property even for the nonconjugated

Figure 9.4 SEM images of $[Zn_2(terephthalate)_2(ted)]_n$ composites (A, C, and E) and PSt isolated from the composites (B, D, and F). Morphology of the objects were cubic (A and B), rod-like (C and D), and hexagonal (E and F).

vinyl polymer *(20)*. However, in the bulk state, the carriers do not move with uniform velocity and reside mostly at localized (trap) sites in the disordered amorphous form. To overcome this problem, alignment of carbazole groups with face-to-face conformation would be an effective method to improve the hole mobility. Our group has utilized $[La(1,3,5\text{-benzenetrisbenzoate})]_n$ for the accommodation of PVCz in 1D nanochannels *(21)*. The conformation of the confined PVCz was quite different from that of bulk PVCz and represented a single-chain conformation with linearly extended structure. Based on the fluorescent and excitation measurements, intrachain full-overlap structure of carbazole groups would be dominant in the structure, enhancing the $\pi-\pi$ stacking among the side chains. Thanks to this unusual conformation, the hole mobility of the confined PVCz increased by 10^5 times compared with that of bulk PVCz, which is comparable to those of high-performance π−conjugated polymers, such as the poly(phenylene vinylene)s and regioregular polythiophenes. Isolation of PVCz from this composite was performed by dissolving the host framework in Na-EDTA solution *(21)*. As is similar to the PSt system, morphology of the recovered PVCz objects retained microrods of the original MOF particles. We then measured the hole mobility of the isolated PVCz, which was much higher than that obtained from PVCz in the bulk state. This is probably because the 1D orientation of the PVCz chains was not severely disturbed during the isolation process, suggesting the importance of the kinetic control in chain arrangement of PVCz as a hole-transporting material.

Chain alignment can deeply influence the ultimate macroscopic properties of polymer materials; however, the synthesis of vinyl polymers usually yields amorphous materials without any ordered arrangement. To overcome this problem, several strategies have been envisaged to control chain packing in amorphous polymers, such as mechanical rubbing, microstructure-specific binding, the use of supramolecular aligner molecules, and the introduction of liquid-crystalline side chains *(22)*. However, these strategies either lack generality or are unable to produce permanently aligned assemblies of polymer chains. Indeed, such partially ordered structures can be easily perturbed by solvent or thermal treatment because they rely on weak supramolecular interactions. Ideally, a perfectly aligned polymer phase would be constituted by adjacent and parallel polymer chains that are ordered in the same direction in extended conformation. Covalently cross-linking between adjacent chains would ensure structural stability and preserve the architecture permanently from detrimental external solicitations, keeping the chains ordered in place. Ligand in MOFs can be modified, so that they can participate in the polymerization by the introduction of bifunctional vinyl moieties in the architecture. In this way, radical polymerization of guest vinyl monomers (St and MMA) can be performed with cross-linking reaction at the same time in the regular one-dimensional nanochannel structure of $[Cu_2(terephthalate)_2(ted)]_n$ *(23)*. After the polymerization is finished, the cross-linked polymers can be recovered easily by selective matrix dissolution, providing ultimately controlled uniaxial alignment of polymer chains retaining the order of the pristine compound. Notably, the newly synthetized polymers exhibited the following unconventional characteristics: (1) the cross-linking is entirely integrated in the pseudo-crystalline state and keeps the main chains in register at a defined distance; (2) the pseudo-crystallinity is realized in the presence of atactic polymer chains that are not amenable to crystallization; (3) the precise ordering of the polymer chains is stabilized because of the regular cross-linking structure.

4.2 Controlled Assemblies of Conjugated Polymers

π-Conjugated polymers, known for their unique optical and electronic properties (photoluminescence and high conductivity, respectively), have many potential applications in organic electronic devices and nonlinear optics that could be improved through the alignment of the polymer chains via the enhancement of the conductivity and anisotropic optoelectronic properties. Different from common vinyl polymers with amorphous nature, they can often form crystalline packing structures by effective π−π interactions among polymer main chains; however, we still recognize a need for improvement in the achievement of long-range ordering of the polymer chains. Kinetic control of conjugated polymers using porous MOFs as removable templates can allow the ordered polymer arrangement from one- to three-dimensional structures, providing specific functions based on the anisotropic assembling structures of polymer chains.

Unsubstituted polythiophene (PTh) is unfusible, nonmelting, and insoluble because of the strong interchain interactions; hence, the control of the packing structure is very difficult, despite its highly developed conjugation system based on the backbone planarity. Anisotropic particles of unsubstituted PTh with ordered one-dimensional chain alignment could be prepared using [La(1,3,5-benzenetrisbenzoate)]$_n$ as a host matrix (Fig. 9.5) *(24)*. For the preparation of PTh, oxidative polymerization of terthiophene using iodine vapor was carried out in the nanochannels of MOF, affording the MOF−PTh composite. Note that highly ordered alignment of polymer chains with different number of the chain assemblies could be achieved in the nanochannels, which allowed the control and enhancement of the optical and electronic properties of unsubstituted PTh, depending on the bundle thickness in the pores *(25)*. PTh was isolated from the composite by dissolution of MOF in an aqueous solution of Na-EDTA. XRPD pattern for the isolated PTh particles suggested reorganization of PTh chains into a herringbone configuration during the isolation procedure. Note that the isolated PTh exhibited narrower peak width as compared with PTh prepared by solution polymerization, suggesting that the orientation of the polymer chains in the nanochannels would contribute to enhanced

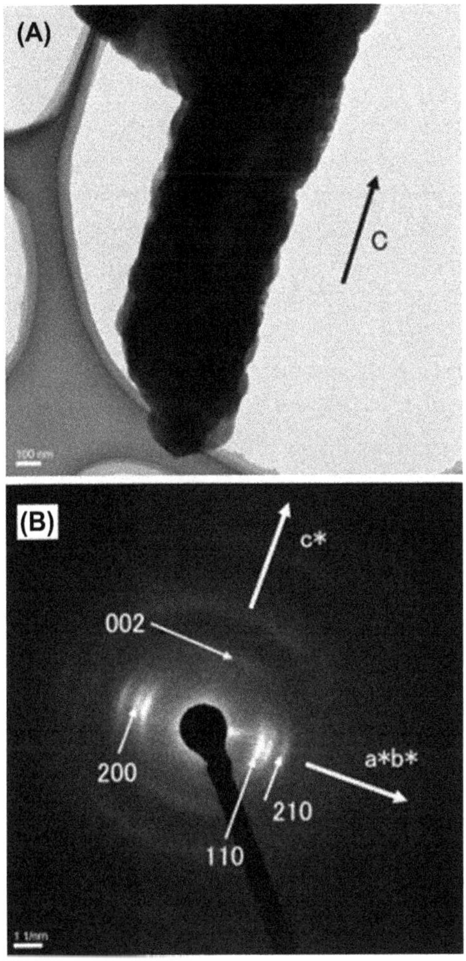

Figure 9.5 TEM image (A) and selected area electron diffraction patterns (B) of PTh from $[La(1,3,5\text{-benzenetrisbenzoate})]_n$.

crystallinity after the removal of the framework. The striking result was obtained from SEM measurement. All PTh particles retained the rod shape of the original host on micrometer-sized scale despite the removal of a large fraction of the composite and without any stabilization of the polymer chains. The selected-area electron diffraction in the TEM measurement of the isolated PTh indicates that long-range ordering of the PTh chains was preserved even after the removal of the host framework (Fig. 9.5) *(24)*. In addition, UV/vis spectra of the PTh microrods displayed an intense absorption around 600 nm corresponding to the exciton coupling among the neighboring polymer chains, allowing the long-range polaron delocalization with an efficient charge migration *(26)*. Because of the developed π-conjugated system, the isolated PTh exhibited superior electron conductivity than the bulk PTh. On doping with iodine vapor, the microrod conductivity increased from $<10^{-8}$ S/cm to 2.0×10^{-1} S/cm, three orders of magnitude higher than that of PTh prepared by solution polymerization.

MOFs can be used as templates for controlling the orientation and organization of conjugated polymer architectures at two- and three-dimensional levels. Recently, intercalative polymerization of pyrrole (Py) via oxidative polymerization was performed within a layered framework of a redox-active MOF, $[Ni(dmen)_2]_2[Fe^{III}(CN)_6](PhBSO_3)$ (dmen = 1,1-dimethylethylenediamine; $PhBSO_3^-$ = *p*-phenylbenzenesulfonate), where $[Fe^{III}(CN)_6]$ units oxidized Py monomers to form polypyrrole (PPy) within the framework *(27)*. XRPD measurements indicated that the resulting PPy was intercalated between the two-dimensional sheets. Removal of the host framework in EDTA solution allowed isolation of the intercalated PPy as an insoluble black precipitate. The controlled morphogenesis of the isolated polymer objects is evident from the SEM images. In contrast to the granular morphology of the bulk-prepared PPy, the morphology of the isolated PPy was finely platy. XRPD measurement indicated that the bulk PPy was found to be an amorphous compound; in contrast, the XRPD profile of the isolated PPy showed the accumulation of aromatic planes of PPy. The obtained PPy microsheet object showed anisotropic conductivity; the conductivity parallel to the sheet was 20 times higher than that along the direction perpendicular to the sheet. Therefore, in this system, the MOF with two-dimensional sheet structure successfully directed the two-dimensional—oriented organization of PPy assembly at the molecular level.

$[Cu_3(btc)_2]_n$ (btc = benzene-1,3,5-tricarboxylate) has three-dimensional interconnected channels and can be used for reaction catalysts because of the redox-active Cu(II) sites accessible on the pore surface *(28)*. Oxidative polymerization of Py was also performed in the pores, followed by the removal of the frameworks in various basic solutions. Porosity of the resulting polymer materials was examined by nitrogen adsorption measurements at 77K. Although PPy prepared in the bulk condition did not adsorb nitrogen gas, the PPy prepared from this MOF clearly showed an adsorption property, and the amount of adsorption gradually increased as the weaker bases were employed for the isolation of PPy. Interestingly, the pore-size distribution of the PPy isolated using ammonia solution showed a peak at 1.3 nm, where this pore size might reflect the thickness of pore walls of the host MOF (ca. 1.2 nm). Note that the porous structure of PPy was very stable and not affected by thermal and solvent treatments.

5. POLYMER BLENDING USING METAL-ORGANIC FRAMEWORKS AS TEMPLATES

Solid blending of polymers is one of the most important research fields in materials science and technology because the polymer blends with specifically tailored or synergistic properties that could find many novel uses in engineering plastics, high strength materials, electro-optical devices, and so on *(29)*. However, mixing of most polymer pairs results in macroscopic phase separation. This fatal problem poses a significant barrier to produce many commercially and scientifically relevant homogeneous polymer blends. Thus, numerous efforts have been made to reduce the propensity for phase separation of the polymers, including the use of compatibilizers, chemical modification of polymers, and rapid solvent evaporation *(30)*. However, these previous methods either lack for generality, mass-productivity, and/or compatibility at the molecular level.

Kinetic approach involves the utilization of MOFs, resulting in many specific polymer conformations and assemblies *(15)*. Recently, we have reported a novel and versatile strategy to afford ultimately mixed blends of immiscible polymers at the molecular level via MOF templating *(31)*. Use of MOFs as facilely removable templates can produce polymer blends with maintaining the well-mixed state. In this process, typical immiscible polymers, PSt and PMMA, were independently synthesized by radical polymerization of corresponding monomers in the nanochannels of $[Zn_2(\text{terephthalate})_2(\text{ted})]_n$ to randomly distribute these homopolymers at the molecular level inside the MOF particles. Treatment of the MOF–polymer composite with a mild chelating solution (EDTA solution) resulted in selective removal of the MOF matrix to give a polymer blend with homogeneous compatibilization of PSt and PMMA in the single nanometer scale, as evidenced by XRPD, NMR, and high-resolution TEM measurements (Fig. 9.6).

The molecularly mixed state of PSt and PMMA through the medium of pore walls of MOF could be substantially preserved during the isolation process due to the inability of the polymers to be self-organized at room temperature. Note that this kinetically trapped structure of polymer blend was sufficiently stable to maintain the well-mixed state for more than

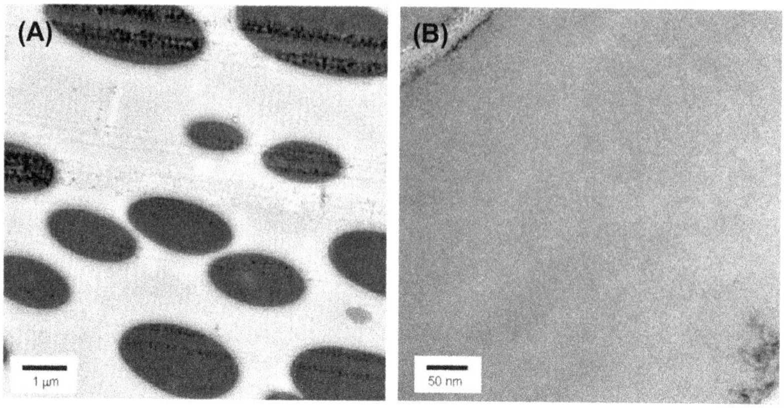

Figure 9.6 TEM images of PSt/PMMA blend obtained by casting from $CHCl_3$ solution (A) and isolated from $[Zn_2(\text{terephthalate})_2(\text{ted})]_n$ containing these polymers (B). RuO_4 was used for staining the PSt domains in both samples.

8 months at room temperature. Thanks to the unusual homogeneity at the molecular level, thermal stability of the MOF-processed PSt/PMMA blend was drastically improved in comparison with those obtained from conventional solvent-casting methods. Moreover, polyacrylonitrile could be also homogeneously compatibilized with PSt at the molecular level, despite their extremely different solubility parameters, showing high generality of the proposed methodology for enabling intimate compatibilization of highly conflicting immiscible polymers *(31)*.

The structural and chemical diversities in MOFs allow for the rational design of polymer blends with well-defined morphology, domain size, and compatibility. In addition, MOFs are able to accommodate not only polymeric materials but also other organic molecules, biomaterials, and inorganic nanoparticles, enabling the formation of functional nanohybrids for a wide range of scientific and industrial applications. This new protocol for kinetic-controlled compatibilization would contribute to the creation of functional alloys and advanced nanocomposites.

6. SUMMARY

This chapter described the methodology for controlling the primary and assembly structures of polymers using MOFs as templates. Compared to the conventional methods, these kinetic approaches could allow more precise and rational design of polymer architectures with regulated structures and controlled arrangements, where stereoregularity, chain alignment, and compatibility of polymers can be finely controlled to influence the intrinsic properties of the polymers. In addition, these approaches are useful enough to be applied to a variety of polymeric materials, including vinyl polymers, conducting polymers, biopolymers, and inorganic polymers. The methodology described here would provide access to diverse structures of polymer structures and assemblies with fascinating properties that will find industrial and medical applications.

REFERENCES

1. (a) Paleos, C. M. *Polymerization in Organized Media;* Gordon & Breach: New York, **1992**.
   (b) Sada, K.; Takeuchi, M.; Fujita, N.; Numata, M.; Shinkai, S. Post-polymerization of Preorganized Assemblies for Creating Shapecontrolled Functional Materials. *Chem. Soc. Rev.* **2007,** *36,* 415–435.
   (c) Tajima, K.; Aida, T. Controlled Polymerization with Constrained Geometries. *Chem. Commun.* **2000,** 2399–2412.
2. (a) Miyata, M. *Inclusion Polymerization* In: *Comprehensive Supramolecular Chemistry,* Vol. 10; Oxford: Pergamon, **1996**; pp 557–582.
   (b) Farina, M.; Di Silvestro, G.; Sozzani, P. *Perhydrotriphenylene: A D3 Symmetric Host* In: *Comprehensive Supramolecular Chemistry,* Vol. 10; Oxford: Pergamon, **1996**; pp 371–398.
3. (a) Kitagawa, S.; Kitaura, R.; Noro, S.-i. Functional Porous Coordination Polymers. *Angew. Chem. Int. Ed.* **2004,** *43,* 2334–2375.
   (b) Ockwig, N. W.; Delgado-Friedrichs, O.; O'Keefe, M.; Yaghi, O. M. Reticular Chemistry: Occurrence and Taxonomy of Nets and Grammar for the Design of Frameworks. *Acc. Chem. Res.* **2005,** *38,* 176–182.
   (c) Férey, G.; Mellot-Draznieks, C.; Serre, C.; Millange, F. Crystallized Frameworks with Giant Pores: Are There Limits to the Possible? *Acc. Chem. Res.* **2005,** *38,* 217–225.
4. (a) Lee, J. Y.; Farha, O. K.; Roberts, J.; Scheidt, K. A.; Nguyen, S.-B. T.; Hupp, J. T. Metal-Organic Framework Materials as Catalysts. *Chem. Soc. Rev.* **2009,** *38,* 1450–1459.
   (c) Wang, Z.; Chen, G.; Ding, K. Self-Supported Catalysts. *Chem. Rev.* **2009,** *109,* 322–359.
5. (a) Uemura, T.; Horike, S.; Kitagawa, S. Polymerization in Coordination Nanospaces. *Chem. Asian J.* **2006,** *1,* 36–44.
   (b) Uemura, T.; Yanai, N.; Kitagawa, S. Polymerization Reactions in Porous Coordination Polymers. *Chem. Soc. Rev.* **2009,** *38,* 1228–1236.
   (c) Kitao, T.; Zhang, Y.; Kitagawa, S.; Wang, B.; Uemura, T. Hybridization of MOFs and Polymers. *Chem. Soc. Rev.* **2017,** *46,* 3108–3133.
6. Higuchi, M.; Horike, S.; Kitagawa, S. Spatial and Surface Design of Porous Coordinaiton Polymers. *Supramol. Chem.* **2007,** *19,* 75–78.
7. (a) Seo, J. S.; Whang, D.; Lee, H.; Jun, S. I.; Oh, J.; Jeon, Y. J.; Kim, K. A Homochiral Metal–Organic Porous Material for Enantioselectiveseparation and Catalysis. *Nature* **2000,** *404,* 982–986.
   (b) Hasegawa, S.; Horike, S.; Matsuda, R.; Furukawa, S.; Mochizuki, K.; Kinoshita, Y.; Kitagawa, S. Three-Dimensional Porous Coordination Polymer Functionalized with

Amide Groups Based on Tridentate Ligand: Selective Sorption and Catalysis. *J. Am. Chem. Soc.* **2007,** *129,* 2607–2614.

8. (a) Férey, G.; Serre, C. Large Breathing Effects in Three-Dimensional Porous Hybrid Matter: Facts, Analyses, Rules and Consequences. *Chem. Soc. Rev.* **2009,** *38,* 1380–1399.
(b) Horike, S.; Shimomura, S.; Kitagawa, S. Soft Porous Crystals. *Nat. Chem.* **2009,** *1,* 695–704.
9. Uemura, T.; Ono, Y.; Kitagawa, K.; Kitagawa, S. Radical Polymerization of Vinyl Monomers in Porous Coordination Polymers: Nanochannel Size Effects on Reactivity, Molecular Weight, and Stereostructure. *Macromolecules* **2008,** *41,* 87–94.
10. Uemura, T.; Ono, Y.; Hijikata, Y.; Kitagawa, S. Functionalization of Coordination Nanochannels for Controlling Tacticity in Radical Vinyl Polymerization. *J. Am. Chem. Soc.* **2010,** *132,* 4917–4924.
11. (a) Yamada, K.; Nakano, T.; Okamoto, Y. Stereospecific Free Radical Polymerization of Vinyl Esters Using Fluoroalcohols as Solvents. *Macromolecules* **1998,** *31,* 7598–7605.
(b) Koumura, K.; Satoh, K.; Kamigaito, M.; Okamoto, Y. Iodine Transfer Radical Polymerization of Vinyl Acetate in Fluoroalcohols for Simultaneous Control of Molecular Weight, Stereospecificity, and Regiospecificity. *Macromolecules* **2006,** *39,* 4054–4061.
12. Kobayashi, Y.; Honjo, K.; Kitagawa, S.; Gwyther, J.; Manners, I.; Uemura, T. Thermal Ring-Opening Polymerization of an Unsymmetrical Silicon-bridged [1]ferrocenophane in Coordination Nanochannels. *Chem. Commun.* **2017,** *53,* 6945–6948.
13. Rasburn, J.; Foucher, D. A.; Reynolds, W. F.; Vancso, G. J. Solid State Polymerization of the Unsymmetrical [1]ferrocenophane Fe(η-c5 H4)2_{S}iMePh; Synthesis of the First Stereoregular Organometallic Polymer. *Chem. Commun.* **1998,** 843–844.
14. (a) Cui, H.; Chen, Z.; Zhong, S.; Wooley, K. L.; Pochan, D. J. Block Copolymer Assembly via Kinetic Control. *Science* **2007,** *317,* 647–650.
(b) Douglas, J. F.; Schneider, H. M.; Frantz, P.; Lipman, R.; Granick, S. The Origin and Characterization of Conformational Heterogeneity in Adsorbed Polymer Layers. *J. Phys. Condens. Matter.* **1997,** *9,* 7699–7718.
15. (a) Sozzani, P.; et al. Complete Shape Retention in the Transformation of Silica to Polymer Micro-objects. *Nat. Mater.* **2006,** *5,* 545–551.
(b) Bracco, S.; Comotti, A.; Ferretti, L.; Sozzani, P. Supramolecular Aggregation of Block Copolymers in the Solid State as Assisted by the Selective Formation of Inclusion Crystals. *J. Am. Chem. Soc.* **2011,** *133,* 8982–8994.
(c) Thomas, A.; Goettmann, F.; Antonietti, M. Hard Templates for Soft Materials: Creating Nanostructured Organic Materials. *Chem. Mater.* **2008,** *20,* 738–755.
(d) Tonelli, A. E. Restructuring Polymers via Nanoconfinement and Subsequent Release. *Beilstein J. Org. Chem.* **2012,** *8,* 1318–1332; (2012). Singh, V. *et al.* High thermal conductivity of chain oriented amorphous polythiophene. Nat. Nanotech. **2014**, *9*, 384-390.
16. (a) Mittal, V. *Miniemulsion Polymerization Technology;* Wiley-Scrivener, **2011**.
(b) Arshady, R. Suspension, Emulsion, and Dispersion Polymerization: A Methodological Survey. *Colloid Polym. Sci.* **1992,** *270,* 717–732.
(c) Rao, J. P.; Geckeler, K. E. Polymer Nanoparticles: Preparation Techniques and Size-Control Parameters. *Prog. Polym. Sci.* **2011,** *36,* 887–913.
17. (a) Nie, Z.; Xu, S.; Lewis, P. C.; Kumacheva, E. Polymer Particles with Various Shapes and Morphologies Produced in Continuous Microfluidic Reactors. *J. Am. Chem. Soc.* **2005,** *127,* 8058–8063.
(b) Herlihy, K. P.; Nunes, J.; DeSimone, J. M. Electrically Driven Alignment and Crystallization of Unique Anisotropic Polymer Particles. *Langmuir* **2008,** *24,* 8421–8426.

(c) Park, J.-G.; Forster, J. D.; Dufresne, E. R. High-Yield Synthesis of Monodisperse Dumbbell-Shaped Polymer Nanoparticles. *J. Am. Chem. Soc.* **2010,** *132,* 5960–5961.

18. (a) Dybtsev, D. N.; Chun, H.; Kim, K. Rigid and Flexible: A Highly Porous Metal–Organic Framework with Unusual Guest-Dependent Dynamic Behavior. *Angew. Chem. Int. Ed.* **2004,** *43,* 5033–5036.
(b) Chun, H.; Moon, J. Discovery, Synthesis, and Characterization of an Isomeric Coordination Polymer with Pillared Kagome Net Topology. *Inorg. Chem.* **2007,** *46,* 4371–4373.
19. Uemura, T.; Kaseda, T.; Kitagawa, S. Controlled Synthesis of Anisotropic Polymer Particles Templated by Porous Coordination Polymers. *Chem. Mater.* **2013,** *25,* 3772–3776.
20. Peason, G.; Stolka, M. *Poly(N-vinylcarbazole);* Gordon and Breach Science**, 1981**.
21. Uemura, T.; Uchida, N.; Asano, A.; Saeki, A.; Seki, S.; Tsujimoto, M.; Isoda, S.; Kitagawa, S. Highly Photoconducting π-Stacked Polymer Accommodated in Coordination Nanochannels. *J. Am. Chem. Soc.* **2012,** *134,* 8360–8363.
22. (a) Toney, M. F.; et al. Near-Surface Aligment of Polymers in Rubbed Films. *Nature* **1995,** *347,* 709–711.
(b) Rothemund, P. W. K. Folding DNA to Create Nanoscale Shapes and Patterns. *Nature* **2006,** *440,* 297–302.
(c) Kubo, Y.; et al. A Supramolecular Bundling Approach toward the Alignment of Conjugated Polymers. *Angew. Chem. Int. Ed.* **2006,** *45,* 1548–1553.
(d) Sozzani, P.; Bovey, F. A.; Schilling, F. C. Characterization of Isolated Polyethylene Chains in the Solid State. *Macromolecules* **1991,** *24,* 6764–6768.
(e) Lauher, J. W.; Fowler, F. W.; Goroff, N. S. Single-Crystal-to-Single-Crystal Topochemical Polymerizations by Design. *Acc. Chem. Res.* **2008,** *41,* 1215–1229.
(f) Matsumoto, A. Polymer Structure Control Based on Crystal Engineering for Material Design. *Polym. J.* **2003,** *35,* 93–121.
(g) Kissel, P.; et al. A Two-Dimensional Polymer Prepared by Organic Synthesis. *Nat. Chem.* **2012,** *4,* 287–291.
(h) Grell, M.; Bradley, D. D. C.; Inbasekaran, M.; Woo, E. P. A Glass-forming Conjugated Main-Chain Liquid Crystal Polymer for Polarized Electroluminescence Applications. *Adv. Mater.* **1997,** *9,* 798–802.
23. Distefano, G.; et al. Highly Ordered Alignment of a Vinyl Polymer by Host-Guest Cross-Polymerization. *Nat. Chem.* **2013,** *5,* 335–341.
24. Kitao, T.; et al. Preparation of Polythiophene Microrods with Ordered Chain Alignment Using Nanoporous Coordination Template. *Polym. Chem.* **2017,** *8,* 5077–5081.
25. MacLean, M. W. A.; et al. Unraveling Inter- and Intra-Chain Electronics in Polythiophene Assemblies Mediated by Coordination Nanospaces. *Angew. Chem. Int. Ed.* **2016,** *55,* 708–713.
26. Brown, P. J.; et al. Effect of Interchain Interactions on the Absorption and Emission of Poly(3-hexylthiophene). *Phys. Rev. B* **2003,** *67,* 064203.
27. Yanai, N.; et al. Fabrication of Two-Dimensional Polymer Arrays: Template Synthesis of Polypyrrole between Redox-Active Coordination Nanoslits. *Angew. Chem. Int. Ed.* **2008,** *47,* 9883–9886.
28. Uemura, T.; Kadowaki, Y.; Yanai, N.; Kitagawa, S. Template Synthesis of Porous Polypyrrole in 3D Coordination Nanochannels. *Chem. Mater.* **2009,** *21,* 4096–4098.
29. (a) Paul, D. R.; Bucknall, C. B. *Polymer Blends;* Wiley: New York**, 1999**.
(b) Freed, K. F.; Dudowicz, J. Influence of Monomer Molecular Structure on the Miscibility of Polymer Blends. *Adv. Polym. Sci.* **2005,** *183,* 63–126.
30. (a) Xanthos, M.; Dagli, S. Compatibilization of Polymer Blends by Reactive Processing. *Polym. Eng. Sci.* **1991,** *31,* 929–935.
(b) Yang, J.; Winnik, M. A. Polyurethane-Polyacrylate Interpenetrating Networks. 1

Preparation and Morphology. *Macromolecules* **1996,** *29,* 7047–7054.
(c) Rusette, A.-V.; Leibler, L. Block Copolymers in Tomorrow's Plastics. *Nat. Mater.* **2005,** *4,* 19–31.
(d) Kim, J.; Gray, M. K.; Zhou, H.; Nguyen, S. B. T.; Torkelson, J. M. Polymer Blend Compatibilization by Gradient Copolymer Addition during Melt Processing: Stabilization of Dispersed Phase to Static Coarsening. *Macromolecules* **2005,** *38,* 1037–1040.
(e) Khatua, B. B.; Lee, D. L.; Kim, H. Y.; Kim, J. K. Effect of Organoclay Plateles on Morphology of Nylon-6 and Poly(ethylene-ran-propylene) Rubber Blends. *Macromolecules* **2004,** *37,* 2454–2459.

31. Uemura, T.; et al. Mixing of Immiscible Polymers Using Nanoporous Coordination Templates. *Nat. Commun.* **2015,** *6,* 7473.

CHAPTER 10

Molecular Self-Assembly Under Kinetic Control

Sung Ho Jung, Masayuki Takeuchi, Kazunori Sugiyasu
National Institute for Materials Science (NIMS), Tsukuba, Japan

Contents

1. INTRODUCTION

Molecular self-assembly offers a powerful approach to create functional molecular systems that are generally inaccessible via conventional covalent bond—based organic synthesis. Intricate three-dimensional structures such as molecular capsules and interlocked molecules can be efficiently synthesized by careful design of angles and distances among noncovalent bonds *(1)*. Such an efficient synthesis is achieved via the self-assembly process *under thermodynamic control*; the reversible nature of noncovalent bonds enables the correction of structural defects, thereby exclusively producing the most stable structure in terms of free energy. Apart from the "closed" discrete structures such as capsules, when properly designed, noncovalent bonds can also link molecules together to produce an extended large object such as a linear polymeric chain, or so-called supramolecular polymer. In this regard, polycaps, prepared by Rebek and coworkers *(2,3)* represent an extension of a closed structure (i.e., a capsule) to a polymeric structure (Fig. 10.1A). A calix[4]arene bearing hydrogen-bonding sites on its larger rim dimerizes to form capsules (**Cap1**). However, when two such molecular entities are linked covalently at their lower rims in such a way that the hydrogen-bonding sites are held apart (**Cap2**), the two cups cannot close intramolecularly and instead polymerize intermolecularly, resulting in a polymeric capsule (i.e., polycaps).

Kinetic Control in Synthesis and Self-Assembly
ISBN 978-0-12-812126-9
https://doi.org/10.1016/B978-0-12-812126-9.00010-9

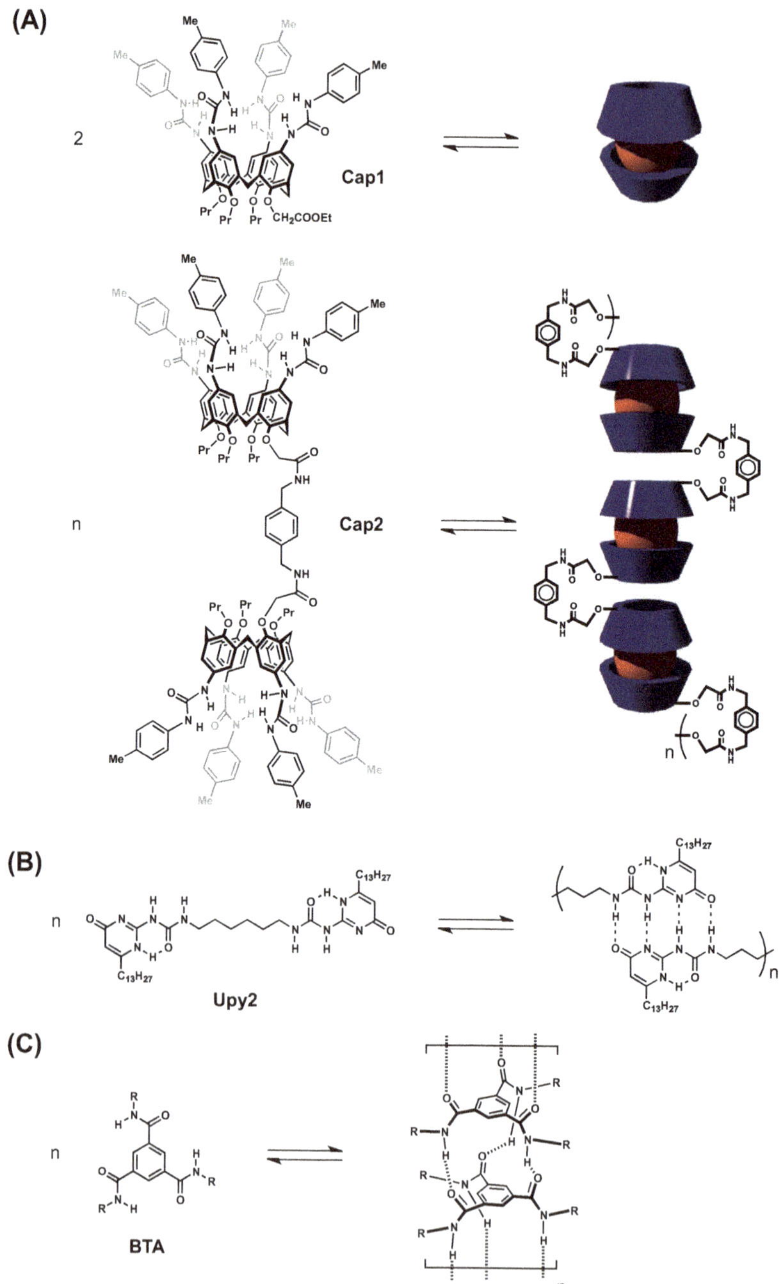

Figure 10.1 (A) Capsule formation of a calix[4]arene bearing hydrogen-bonding sites on its larger rim (**Cap1**) and supramolecular polymerization of a dimeric calix[4]arene (**Cap2**). (B) Supramolecular polymerization of **UPy2**. (C) Supramolecular polymerization of **BTA**.

A more generalized monomer design was reported by Meijer and coworkers (Fig. 10.1B) *(4)*. A bifunctional monomer (**UPy2**) consisting of two 2-ureido-pyrimidone units yielded a noncovalent polymeric chain via dimerization involving highly directional quadruple strong hydrogen bonding ($K_{dim} > 10^6\ M^{-1}$ in $CHCl_3$). The viscoelastic properties of the solution of **UPy2** were highly dependent on the concentration and temperature because of the reversible nature of hydrogen bonds. This finding encouraged supramolecular chemists to design polymeric materials featuring stimuli responsiveness, recyclability, processability, and self-healing capacity *(5)*.

Another versatile monomer is the discotic benzene-1,3,5-tricarboxamide derivative, **BTA** *(6)*. The aromatic disc tends to stack one-dimensionally and peripheral threefold hydrogen bonding cooperatively stabilizes the stack (Fig. 10.1C). The supramolecular polymerization mechanism of **BTA** was investigated in detail by temperature-dependent UV–Vis spectroscopy and circular dichroism (CD) measurements *(7)*. It was found that the cooling curves accompanied a critical temperature (T_e), indicating that supramolecular polymerization consists of two phases. In the first phase, the nucleus is formed (above T_e) and the second phase involves the elongation of the supramolecular polymer (below T_e). Meijer and colleagues modified a theory developed by Oosawa and Kasai *(8)* and established a model to characterize cooperative (nucleation–elongation) supramolecular polymerization *(9–11)*. Using this model, the enthalpy release for elongation (h_e) and dimensionless equilibrium constant of the nucleation step (K_a) can be determined, which has enabled quantitative evaluation of supramolecular polymerization.

Recently, further insights into the mechanism of supramolecular polymerization were obtained by using stochastic optical reconstruction microscopy (STROM) *(12)*. Two **BTA** derivatives, functionalized with green fluorescent Cy3 and red fluorescent Cy5 dyes (**BTACy3** and **BTACy5**, respectively, Fig. 10.2A), were synthesized to visualize the supramolecular polymers under a fluorescence microscope. The supramolecular polymers of **BTAOH**, which were stained with **BTACy3** and **BTACy5**, were prepared separately and then mixed to allow the monomer exchange to take place between them. The nucleation–elongation model assumes that monomer exchange occurs by polymerization–depolymerization at the termini of supramolecular polymers. However, STORM imaging showed unexpected results in that single-colored blocks were not observed and monomer exchange occurred homogeneously along the polymer backbone. This observation can be rationalized by the presence of disordered domains in the

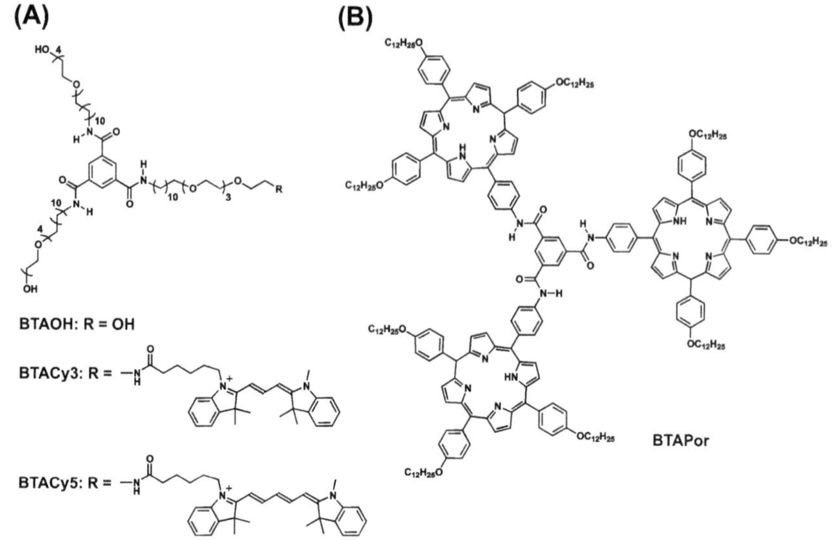

Figure 10.2 Chemical structures of monomers based on **BTA**: (A) **BTAOH, BTACy3,** and **BTACy5**; (B) **BTAPor.**

supramolecular polymers. In fact, hydrogen/deuterium exchange (HDX) mass spectrometry also revealed the structural diversity of the supramolecular polymer *(13)*. These mechanistic insights are of great significance to control the dynamic behavior of supramolecular polymers.

Meanwhile, it is worth noting that these monomers are useful building blocks for preparation of functional supramolecular materials. For example, Nolte and coworkers *(14,15)* designed a porphyrin-substituted **BTA** monomer (**BTAPor** in Fig. 10.2B), placed its chloroform solution on a mica surface, and allowed the solvent to evaporate. In due course, the authors observed that **BTAPor** had self-assembled on the surface into columnar stacks of submicrometer length. Remarkably, the supramolecular polymer of **BTAPor** spontaneously constructed highly periodic patterns at macroscopic length scales (>mm^2). These patterns were formed as a result of a delicate balance between the physical dewetting processes and highly directional supramolecular polymerization. The thus-obtained self-assembled patterns could be used to align liquid crystals in a large domain.

As represented by the **BTA** derivatives above, both fundamental mechanistic studies and applications toward materials science with regard to supramolecular polymers have evolved tremendously *(16)*. However, unlike covalent bond polymerization, control over supramolecular polymerization has remained a significant challenge. Particularly, *living supramolecular* *polymerization*, which allows for the precision synthesis of supramolecular polymers with narrow polydispersity and enhanced structural complexity (e.g., block polymers), was not realized until very recently. Note that under thermodynamic control, the polydispersity index (PDI) of supramolecular polymers, which is defined as the ratio of number average length (L_n) to the weight average length (L_w) (PDI $= L_w/L_n$), is determined to be 2.0 *(10)*. Hence, it is not surprising that molecular self-assembly under kinetic control has recently generated a lot of interest.

2. LIVING SUPRAMOLECULAR POLYMERIZATION

By a fortuitous accident, we discovered a system that was suitable for achieving living supramolecular polymerization *(17,18)*. A porphyrin-based monomer bearing two hydrogen-bonding sites (**Por**) self-assembled into nanoparticles in methylcyclohexane (MCH) (Fig. 10.3). However, the nanoparticles appeared to be a metastable species, and after several hours of

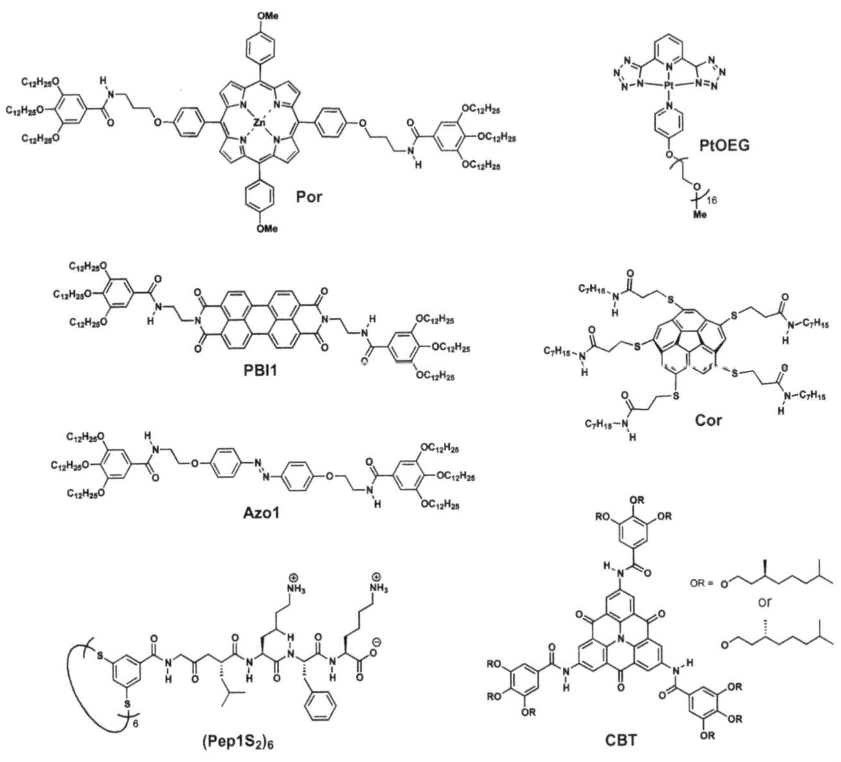

Figure 10.3 Chemical structures of monomers that can be processed in a manner of living supramolecular polymerization.

lag time, thermodynamically stable one-dimensional nanofibers (i.e., supramolecular polymers) prevailed in the system. This unique time-dependent evolution has permitted living supramolecular polymerization by which supramolecular polymer with controlled length and narrow polydispersity (PDI = 1.1) was successfully achieved.

For further details, the readers are referred to articles that highlight the recent advances in living supramolecular polymerization *(19—21)*. Herein, we have discussed the structures of monomers reported in literature that are processible in a manner of living supramolecular polymerization. It is noteworthy that all of the systems are kinetically controlled as a result of the participation of a metastable state in the self-assembly process.

Aida et al. *(22)* and Würthner et al. *(23,24)* have independently reported monomers **Cor** and **PBI1**, respectively, that form intramolecular hydrogen bonds. Although these "closed" monomers are incapable of supramolecular polymerization, the hydrogen bonds can rearrange in an intermolecular manner and lead to the formation of supramolecular polymers. These systems could be kinetically controlled owing to the coupled equilibria between the intramolecular and intermolecular hydrogen bonding.

Manners, DeCola, and coworkers and Meijer et al. have also reported kinetically controlled supramolecular polymerization of monomers **PtOEG** *(25,26)* and **CBT**, *(27)* respectively. Furthermore, we recently succeeded in achieving photo-regulated living supramolecular polymerization by using the azobenzene-based monomer, **Azo1** *(28)*.

Otto and coworkers *(29)* have established a unique dynamic combinatorial library (DCL), from which macrocyclic **(Pep1S$_2$)$_6$** was selected and underwent the nucleation—elongation process. Using this, block supramolecular copolymer was synthesized via the seeded-growth approach. A related system of the same group—mechanosensitive self-replication—will be discussed later (Fig. 10.5).

As described above, supramolecular polymerization under kinetic control provides many opportunities in creating unprecedented supramolecular materials. With this in mind, we revisited the relevant studies reported before the living supramolecular polymerization systems were realized.

3. OTHER MOLECULAR SELF-ASSEMBLIES UNDER KINETIC CONTROL

Naota and Koori *(30,31)* reported a unique molecule that self-assembled on sonication (Fig. 10.4A). This phenomenon may appear counterintuitive as sonication is usually used for the disruption of aggregates.

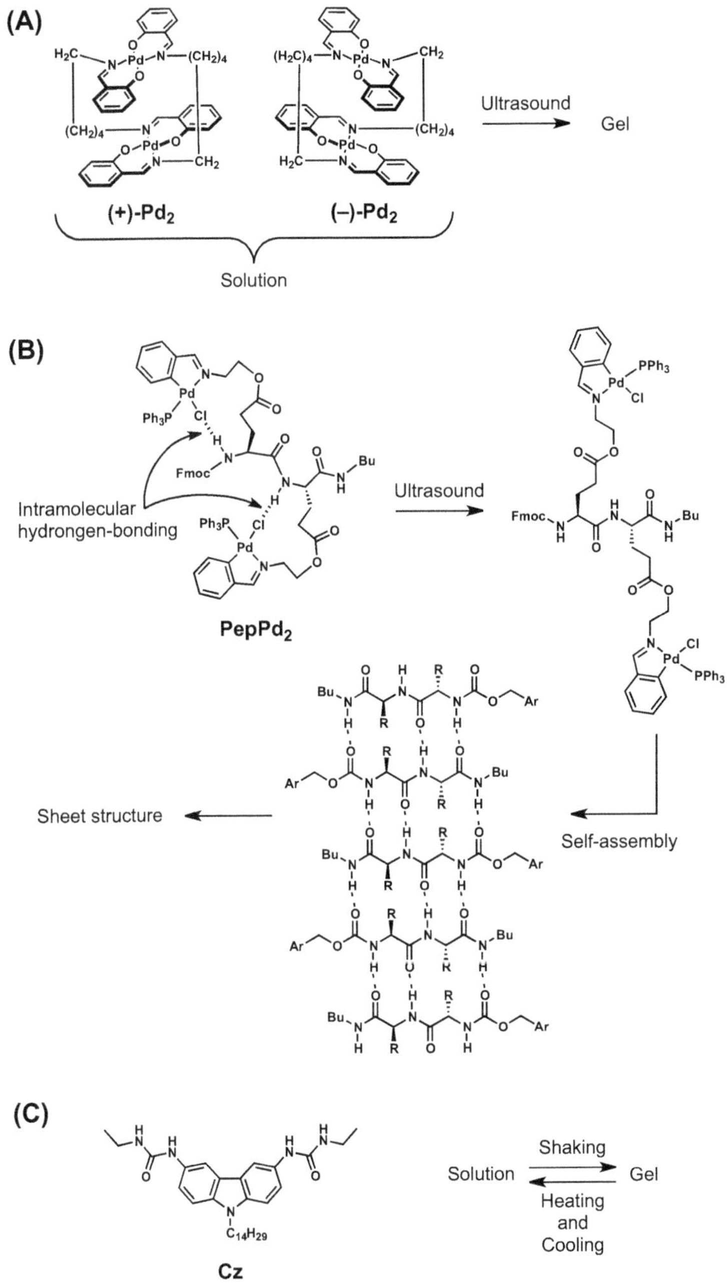

Figure 10.4 (A) Chemical structures of **(+)-Pd_2** and **(−)-Pd_2**. (B) Chemical structure of **$PepPd_2$**; intramolecular hydrogen bonding could be unlocked on sonication, which induced intermolecular hydrogen bonding (i.e., β-sheet formation) and led to gelation. (C) Chemical structure of **Cz**.

Racemic **(±)-Pd$_2$** was soluble in various organic solvents, and when the solution was left undisturbed for > 1 h, it slowly resulted in gelation. Remarkably, the application of sonication to the solution accelerated the gelation; a few seconds of applying presonication to a freshly prepared solution of **(±)-Pd$_2$** instantaneously induced the gelation (<1 min). The gel could be readily converted to the original solution on heating above the gel-to-sol transition temperature and subsequent cooling to room temperature. The sol–gel phase transition (a cycle of presonication, gelation, heating, and cooling processes) could be repeated without degradation of the gelator molecule.

Scanning electron microscopy (SEM) measurements revealed that nanoparticles with a diameter of c. 400 nm were present in solution (i.e., before sonication), and long and thin fibers were formed in the gel state (i.e., after sonication). The spontaneous gel, which was obtained by letting the solution stand at room temperature for 1 h without sonication, consisted of irregular, worm-like aggregates.

To understand how the molecules self-assemble, it is worth considering the fact that the optically pure samples, **(+)-Pd$_2$** and **(−)-Pd$_2$**, did not form a gel, irrespective of the concentration and sonication time. Single crystal X-ray analysis indicated that heterochiral molecular alignment between the linear homochiral stacks of **(+)-Pd$_2$** and **(−)-Pd$_2$** was a crucial factor responsible for close packing. In contrast, optically pure samples **(+)-Pd$_2$** and **(−)-Pd$_2$** were weakly stacked in the crystal. Therefore, the differences in the packing modes should be the origin of the gelation ability of **(±)-Pd$_2$**. It was concluded that imparting ultrasonic energy to the particles formed microcrystals consisting of interpenetrating molecular units of **(±)-Pd$_2$**.

The sol–gel phase transition of **(±)-Pd$_2$** was also intriguing in view of the phosphorescence property of the palladium complex. The particles in solution were only weakly phosphorescent because of the loose and noninterpenetrating molecular aggregation, thus generating nonradiative deactivation channels of the photoexcited state. On the other hand, conformational changes were restricted in the gel state by the tight molecular packing, which suppressed the nonradiative energy loss and gave rise to strong yellow phosphorescence. The stimuli-responsive gelation and phosphorescence enhancement is a useful attribute for sensors and light-emitting devices.

Following the above intriguing finding, the same group *(32)* designed a metalated dipeptide, **PepPd$_2$**, which also underwent ultrasound-induced gelation (Fig. 10.4B). The rate (k_{obs}) of aggregation initiated by sonication was determined by ^{1}H NMR measurements, which revealed the first-order dependence of the concentration of **PepPd$_2$**. The sonication time (t_{sonic}) also imparted an influence on the k_{obs} value, showing a linear relationship. These results suggested that the gelation process consists of a sonication-induced initiation step and a subsequent spontaneous propagation step. SEM images showed that **PepPd$_2$** was self-assembled into a belt-like structure consisting of β-sheets layers.

Interestingly, nuclear Overhauser effect spectroscopy experiments with the solution sample (i.e., before sonication) suggested the formation of *intramolecular* hydrogen bonding between the chlorine atoms and hydrogens of amide bonds (Fig. 10.4B). Therefore, it was concluded that intramolecular hydrogen bonding prevented **PepPd$_2$** from forming *inter-molecular* hydrogen bonds (i.e., β-sheet formation). On sonication, this "self-lock" was unlocked and led to the formation of a semistable initial aggregate, which further propagated into the β-sheet assembly. This self-locked monomer is similar to the molecular design concept of **Cor** and **PBI1**, whose spontaneous nucleation was also prevented as a result of intramolecular hydrogen bonding.

Browne, Feringa, and coworkers *(33)* reported an example of mechanically induced gel formation (Fig. 10.4C). These authors described a carbazole derivative, **Cz**, which was molecularly dissolved in DMSO at 80°C and formed a homogeneous solution on cooling. ^{1}H NMR spectrum of this solution at room temperature was essentially identical to that measured at 80°C, indicating that **Cz** was still fully dissolved. However, the intensity of the NMR signals of **Cz** decreased on mechanical agitation, which suggested the formation of large aggregates. Concomitantly, gelation took place as confirmed by the increase in the storage modulus *G'*. It was found that the duration of mechanical agitation and the mode of agitation (e.g., shaking, stirring, and sonication) determined the time required for gelation.

Wide-angle X-ray scattering (WAXS) and transmission electron microscopy (TEM) measurements revealed that the mechanically formed gels differed in their molecular packing from those that formed spontaneously on cooling alone. More specifically, the mechanically formed gels were relatively crystalline (from WAXS analysis) and broken, fragmented, and shorter in morphology (TEM images) compared to the spontaneously formed gels.

The authors supposed that the mechanism proposed earlier for polymer particles, proteins, etc., regarding gelation on mechanical aggregation, would also be applicable to the case of **Cz**. Thus, transient small aggregates were formed initially, and when substantial shear was applied, the induced

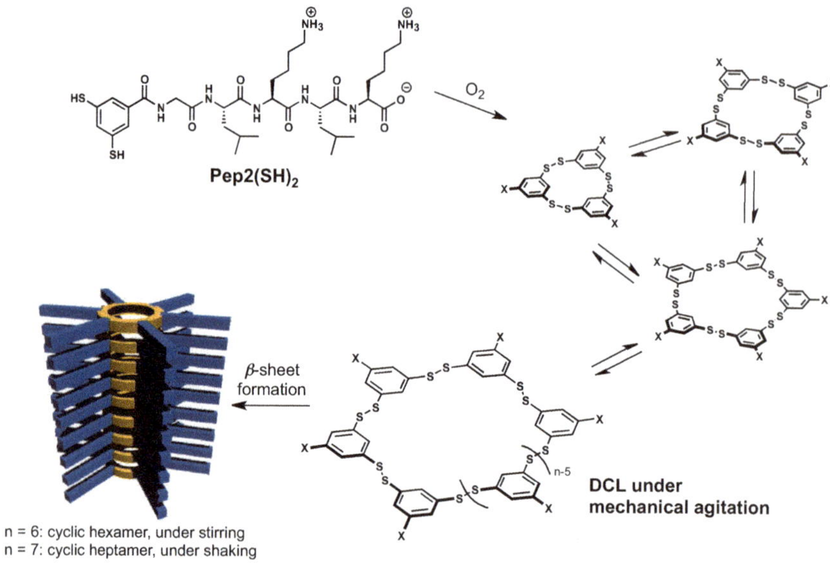

Figure 10.5 Chemical structure of **Pep2(SH)$_2$**. In the presence of oxygen, **Pep2(SH)$_2$** formed cyclic oligomers: **(Pep2S$_2$)$_n$** (n = 3, 4, 5, and 6 are shown). The mixture of **(Pep2S$_2$)$_n$** constructed a dynamic combinatorial library (DCL) from which **(Pep2S$_2$)$_6$** and **(Pep2S$_2$)$_7$** were selected under stirring and shaking conditions, respectively, through self-assembly.

shear diminished the activation barrier and accelerated the aggregation. In the meantime, mechanical agitation broke the preexisting fibrils, which increased the number of fibril termini that acted as nucleation points, thus speeding up the gelation process. However, the manner of retardation of spontaneous aggregation at the molecular level to achieve the metastable supersaturated state is not clear as yet.

Otto and coworkers *(34,35)* investigated the self-replication of macrocycles that were formed from **Pep2(SH)$_2$** in the presence of oxygen (Fig. 10.5). The reversible disulfide bond formation created so-called dynamic combinatorial libraries (DCLs), a mixture of macrocycles with different sizes (**(Pep2S$_2$)$_n$**) that were exchangeable in rapid equilibrium (Fig. 10.5). The population of these macrocycles in the DCL was determined by high-performance liquid chromatography (HPLC). In the initial stages, cyclic trimers and tetramers were formed in the solution, and when left undisturbed, these species predominated the overall composition over a period of 7.5 months. However, under stirring conditions, a sudden change in composition occurred within several days, and the cyclic heptamer became the dominant species. It was interesting that the outcome depended critically on the mode of agitation; with shaking instead of stirring, the cyclic hexamer was preferentially formed over the heptamer.

The composition of the hexamer and heptamer under mechanical agitation increased in a sigmoidal manner, suggesting that these macrocycles were able to promote their own formation (i.e., self-replication). Using cryogenic transmission electron microscopy (cryo-TEM), long thin fibers of length 1–2 μm and width 4.7–4.9 nm were visualized. These dimensions were comparable with the diameter of the cyclic oligomer comprised of peptide chains extending radially in a β-sheet conformation. CD and fluorescence (using thioflavin T) spectral measurements corroborated the β-sheet formation of the peptide side chains. The authors concluded that the fibers acted as kinetic traps of the similarly sized macrocycle, i.e., the hexamer (or heptamer) units in solution were selectively recruited to form the hexamer (or heptamer) fibers, which decreased the concentration of these units and reequilibrated the overall distribution in the DCLs. In fact, if a small amount of fibers of either the hexamer or heptamer, which had been prepared beforehand under the corresponding mechanical conditions, was added to the DCL solution, these fibers acted as "seeds" to increase the proportion of the respective macrocycles.

Mechanical switching of the resultant assembly (hexamer vs. heptamer) was rationalized by the competition between the elongation and fragmentation of the fibers. Thus, under stirring, which rather localized the shear stress and was consequently less effective to the whole system, the heptamer could dominate because of more efficient elongation via more hydrogen-bonding sites. On the other hand, under shaking, the hexamer was efficiently fragmented because of the lesser number of hydrogen-bonding sites, and thereby generated more seeds for propagation and dominated the system. These results suggested that mechanical forces could act as a selection pressure in the evolution of molecules, an important step in the origin of life.

Würthner and coworkers *(36,37)* found a unique time-dependent stereomutation of a helical supramolecular assembly. When a tetrahydrofuran (THF: a good solvent) solution of bis(merocyanine) dye, ***(R,R)*-BMC** (Fig. 10.6), was mixed with methylcyclohexane (MCH: a poor solvent), the dye instantaneously self-assembled by strong Coulomb interactions between the dipolar **BMC** units ($K_{\mathrm{dim}} = 10^6\ \mathrm{M}^{-1}$ in low polarity solvents) and formed a supramolecular oligomer/polymer (**D**). It was found that the aggregated species **D** further organized into H-aggregate species (**H1**) following first-order kinetics (within an hour), as

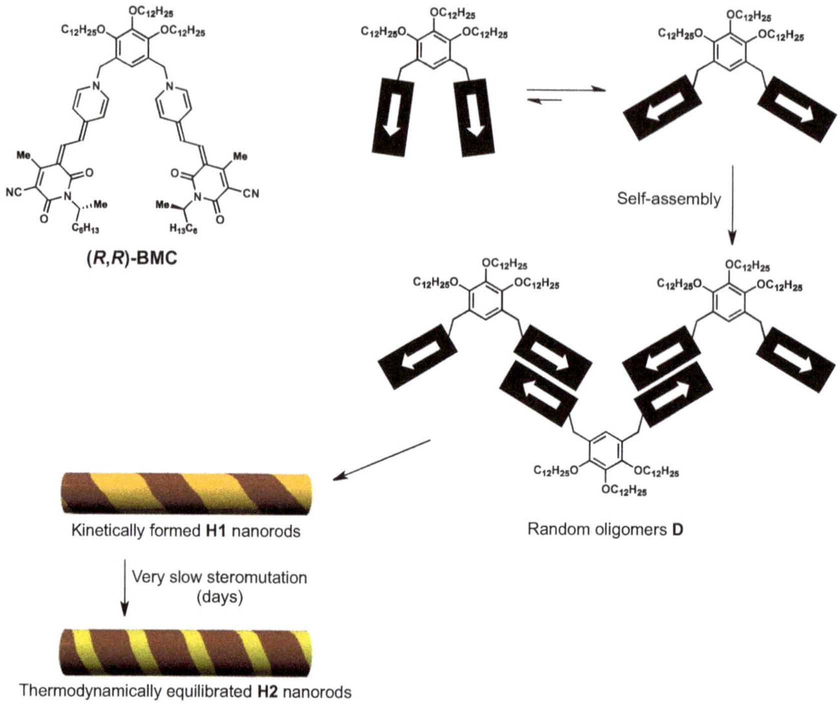

Figure 10.6 Chemical structure of ***(R,R)*-BMC** and schematic representation of its self-assembly through dipolar interaction. A random oligomer (**D**) was formed, which then organized into **H1** and further transmutated into **H2**. The majority-rule effect was observed for the kinetically formed **H1** aggregate when ***(R,R)*-BMC** and ***(S,S)*-BMC** were mixed.

characterized by a time-dependent hypsochromic shift of the absorption maximum. The **H1** species showed a strong bisignate negative CD signal in the H absorption band region.

Interestingly, although the absorption spectra remained almost unchanged, inversion of the CD signal of **H1** was observed on a much longer experimental time-scale (>10 h). Another species with the opposite exciton chirality was denoted as **H2**.

The mechanism of **D** → **H1** → **H2** transformation was investigated by tapping mode AFM, which visualized a less-structured supramolecular polymer (**D**) and right-handed nanorods, **H1** and **H2**, having different helical pitches (10.4 ± 0.6 and 4.9 ± 0.6 nm, respectively).

Amplification of chirality can occur in helical supramolecular polymers, and the collective effect by which a slight enantiomeric excess (ee) of chiral monomers determines the overall helical sense is referred to as "majority rules."

Based on this unique stereomutation phenomenon, Lohr and Würthner investigated the majority-rules effect *under kinetic control*. Thus, a THF solution of **(*R*,*R*)-BMC** and **(*S*,*S*)-BMC** at a given ratio was mixed with MCH to induce the formation of **H1** nanorods, and subsequent stereomutation into **H2** nanorods was probed by CD spectroscopy.

After a long time (8 days), the systems were equilibrated and CD spectra reflected the peaks of only the thermodynamically stable **H2** nanorods. The CD intensities showed an almost linear dependence with respect to the ee values, indicating that the majority-rules effect was not operative under thermodynamic control. However, before the equilibrium was reached, the nonlinear dependence of ee on the CD intensity was observed for the kinetically formed **H1** nanorods. A closer look at the time-dependent CD profile during the formation of **H1** nanorods revealed a two-step kinetic behavior, which could be interpreted as a nucleation elongation process. Thus, it appeared that the **D** → **H1** transformation accompanied the formation of the precursors **H1*** as nuclei. In the first step (up to 110 s: **D** → **H1***), the majority-rules effect was apparently not involved. However, chiral amplification was observed after 110 s (**H1*** → **H1**). The authors evaluated the kinetics by nonlinear curve fitting and found that an autocatalytic mechanism governed the chiral amplification in the **H1*** → **H1** process. The obtained mechanistic insights into the unique kinetic behavior are of particular importance to understand the origin of homochirality in nature.

Yagai, Würthner, and coworkers *(38)* reported a coassembly of ditopic perylene bisimide (**PBI2**) and azobenzene-functionalized melamine derivative (**Azo2**) (Fig. 10.7). A mixture of **PBI2** and **Azo2** (1:1 molar ratio) in MCH formed ill-defined morphologies of less than 50 nm in size. Interestingly, however, the addition of another equivalent of **Azo2** to the mixture (i.e., the final molar ratio of **PBI2** to **Azo2** was 1:2) induced well-defined J aggregate formation, as confirmed by the red-shifted absorption maximum of the perylene core. The transformation accompanied a substantial equilibration time (>10 h). A possible packing model is shown in Fig. 10.7: **PBI2** and **Azo2** appeared to assemble linearly in a manner of ≡**PBI2**≡**Azo2**=**Azo2**≡**PBI2**≡**Azo2**=**Azo2**≡ (where "≡" and "=" represent triple and double hydrogen bonds, respectively). The assembly was then folded in the *syn* orientation into a helically coiled structure because of the dodecyl chains pointing outward and enhanced π−π stacking. In this system, the outcome strongly depended on the sample preparation protocol of mixing the two components. Furthermore, the stoichiometry of the

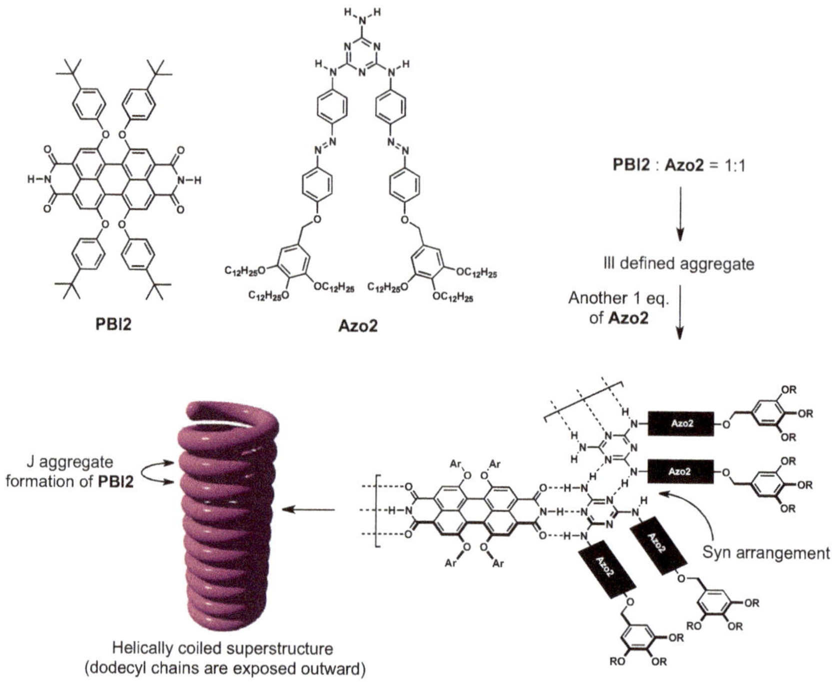

Figure 10.7 Chemical structures of **PBI2** and **Azo2**. The mixture of **PBI2** and **Azo2** (in a 1:2 molar ratio) formed a superhelical structure wherein **PBI2** adopted the J aggregation mode.

components also played an important role in dictating the final nanostructures and the kinetic behavior. As such, this study is an instructive step toward the synthesis of multicomponent supramolecular polymers.

Rybtchinski and coworkers *(39)* reported a kinetic study of the transformation of supramolecular polymers (Fig. 10.8). In water/THF-mixed solvent systems, a bolaamphiphilic perylene bisimide dimer **PBI3** self-assembled to form short fibers (with J aggregate character) and was designated as **PBI3g** because of its green color. It was found that **PBI3g** slowly transformed into long fibers, which were designated as **PBI3p** (brownish purple color with an H aggregate character). The authors observed that increasing the THF content accelerated the transformation of **PBI3g** to **PBI3p**; the organic solvent enhanced the dynamics of self-assembly by lowering the barriers arising from the hydrophobic effect.

The authors first investigated the seeding effect to gain an insight into the nucleation mechanism. A solution of **PBI3p** seed was added to a solution of **PBI3g**; however, this operation hardly affected the transformation

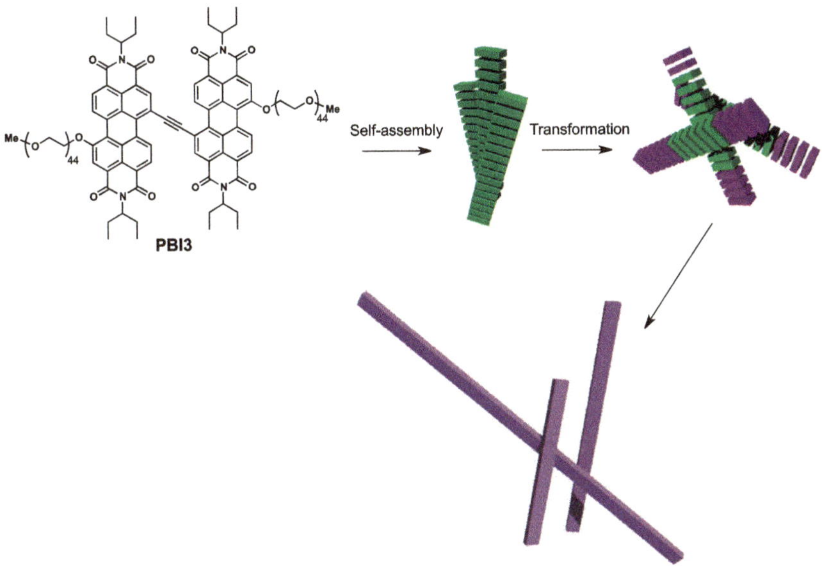

Figure 10.8 Chemical structure of **PBI3** and schematic representation of its supramolecular polymerization and transformation.

kinetics. This result indicated that a strong hydrophobic effect made the self-assembly of **PBI3g** nondynamic, and the monomeric **PBI3** was entrapped and prevented from interacting with **PBI3p**. This observation is significant and worth considering when it is required to control the self-assembly pathway starting from a metastable state.

To probe the transformation kinetics more precisely, a solution of **PBI3** (in a molecularly dissolved state) in THF was injected into water to induce the formation of **PBI3g**, on which absorption spectral changes were monitored. The kinetics of the transformation of **PBI3g** to **PBI3p** followed a sigmoidal trend, which is characteristic of nucleation–growth autocatalytic processes. The sigmoidal kinetics was analyzed by using the Kolmogorov, Johnson, Mehl, and Avrami nucleation–growth isothermal transformation model, and the rate constants for nucleation (k_{nuc}) and growth (k_{grow}) processes were obtained. The enthalpy and entropy of activation for each process were determined using the Eyring equation. It was found that negative activation entropy was a major contributor to the activation barrier, suggesting that a change in the solvation played an important role in the transformation.

Meijer and coworkers *(40)* found a pathway complexity in supramolecular polymerization (Fig. 10.9) and unveiled the mechanism of such

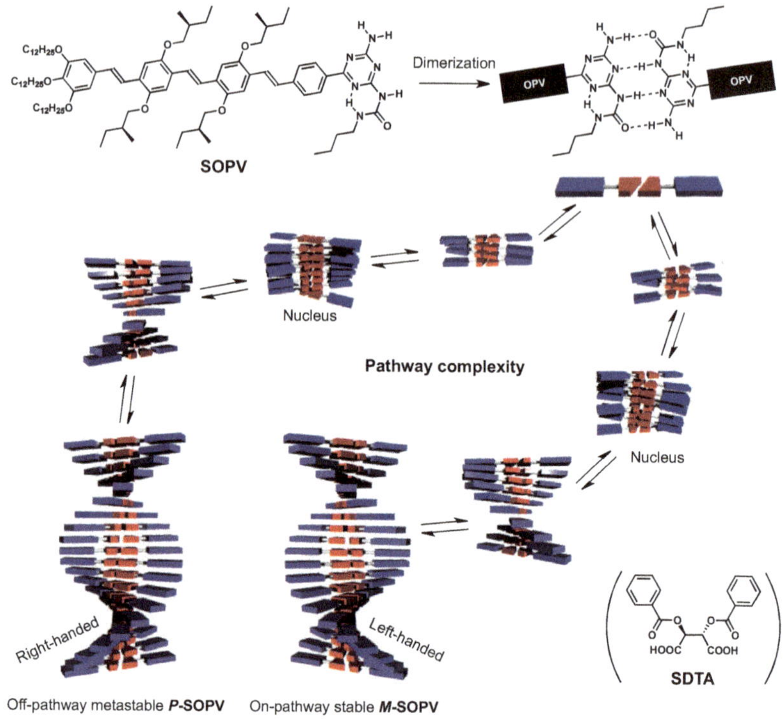

Figure 10.9 Pathway complexity of **SOPV**.

complex kinetic behavior through stopped-flow experiments. Previously, the π-conjugated oligomer *S*-chiral oligo(*p*-phenylenevinylene) (**SOPV**) was found to dimerize through hydrogen bonding (such as 2-ureido-pyrimidone [**UPy**] unit, see Fig. 10.1B), and then self-assemble into a left-handed *M*-type helical stack (***M*-SOPV**) via the nucleated growth mechanism. These authors further investigated the supramolecular polymerization process under nonequilibrium conditions in detail.

Using a stopped-flow experiment setup, a concentrated solution of molecularly dissolved **SOPV** in chloroform (good solvent) was mixed with an excess amount of MCH (poor solvent) to initiate self-assembly, and the time-dependent evolution was probed by using CD spectroscopy. Interestingly, on mixing, the right-handed *P*-type aggregate (***P*-SOPV**) was formed initially and then converted into ***M*-SOPV**. The concentration- and temperature-dependent experiments revealed that the former was an off-pathway metastable intermediate, whereas the latter was thermodynamically the more stable product. As such, two nucleated assembly pathways competed kinetically. By extending the models developed in the field of protein fibrillization, the pathway complexity was simulated, which fitted well with the experimental aggregation kinetics.

Based on these mechanistic understandings, the authors sought to synthesize metastable supramolecular assemblies. They used *S*-chiral dibenzoyl tartaric acid (**SDTA**) as a chiral auxiliary, which bonded to the **SOPV** aggregate through hydrogen bonding, thereby biasing the helicity toward ***P*-SOPV** against ***M*-SOPV**. **SDTA** was then removed by aqueous extraction at a low temperature (273 K), leaving the metastable ***P*-SOPV** in the organic phase. The obtained metastable ***P*-SOPV** underwent stereomutation at 298 K toward ***M*-SOPV**, as probed by the time-dependent CD spectral inversion. This result illustrates a noncovalent synthesis of a supramolecular structure, which was otherwise inaccessible under thermodynamic control.

Ulijn and coworkers *(41)* reported an example of enzyme-assisted self-assembly (Fig. 10.10). In a sense that an enzyme assists the equilibration, the system is under thermodynamic control; but the enzymatic reaction shows time-dependent evolution and is also intriguing from the viewpoint of kinetics.

In the presence of a protease enzyme, **Fmoc-F** (F: phenylalanine, Fig. 10.10) and $\mathbf{F_2}$ (diphenylalanine peptide) undergo a reversible condensation reaction, producing building blocks for self-assembly. The time course of changes in the peptide distribution was monitored by HPLC. The system initially appeared as a milky suspension and formed a

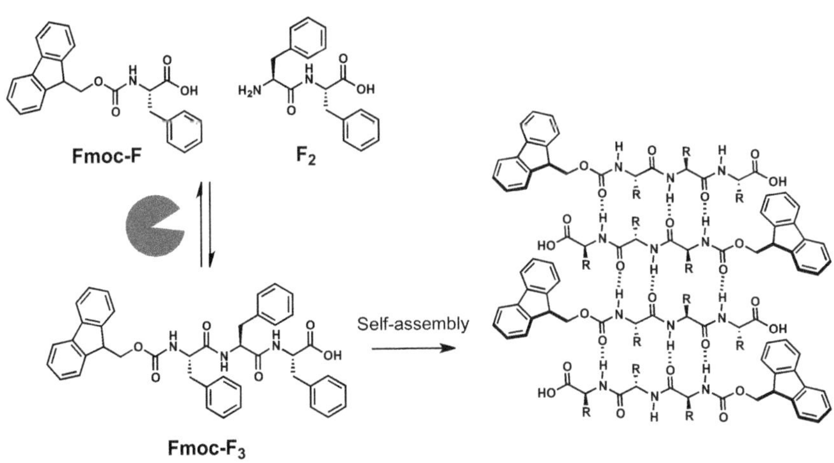

Figure 10.10 Enzyme-assisted self-assembly under thermodynamic control. In the presence of an enzyme, **Fmoc-F** and $\mathbf{F_2}$ afforded $\mathbf{Fmoc\text{-}F_3}$, which self-assembled via β-sheet formation.

transparent hydrogel over 60 min, in which **Fmoc-F$_3$** predominated. The propagation of nanofibers was visualized over time by TEM.

The self-assembly of the peptides, initiated by the enzymatic reaction, occurs via the nucleation–growth mechanism. Thus, it appears likely that nucleation occurs near the enzyme. Taking advantage of such a localizable phenomenon and reaction kinetics, spatiotemporal self-assembly was achieved by using a substrate on which enzymes were immobilized. This approach is intriguing in view of the fact that biomolecular systems function in a confined environment.

Fukushima, Aida, and coworkers *(42)* reported a supramolecular linear heterojunction composed of electronically dissimilar semiconducting molecular assemblies (Fig. 10.11). Previously, they succeeded in creating graphite-like nanotubular objects through self-assembly of a hexabenzocoronene (HBC) derivative. The newly designed **HBCBPy** molecule, which carries two bipyridine (bipy) units, was also found to be capable of forming the nanotube. Furthermore, the coordination of Cu^{2+} with bipy units stabilized the self-assembled nanotubular structure through cross-linking and dispersed the "seed" homogeneously by electrostatic

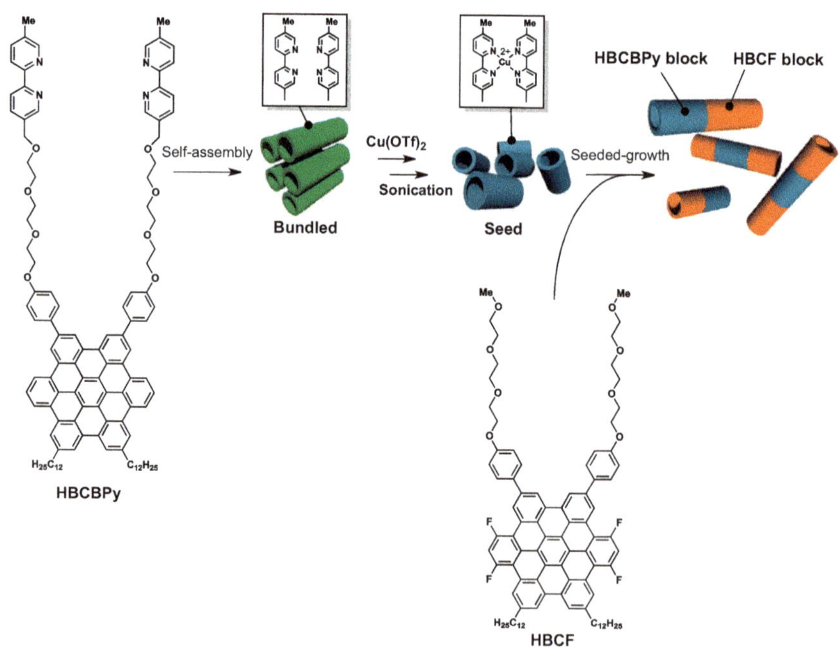

Figure 10.11 Chemical structures of **HBCBPy** and **HBCF**. Schematic representation of the synthesis of a linear heterojunction structure.

repulsion. It is important to note that the π-stacking geometry of the HBC core remained intact through this treatment.

Another component was designed by substituting four electron-withdrawing fluorine atoms (**HBCF**) such that the HBC core became electronically dissimilar. Owing to the skeletal distortion caused by the fluorine substituents, **HBCF** was unable to form the nanotube in all common organic solvents except acetone.

The seed of **HBCBPy** with Cu^{2+} was prepared in acetone, pre-sonicated for a few minutes to cut into short pieces, and then mixed at 50°C with an acetone solution of **HBCF** (Fig. 10.11). The mixture was allowed to cool and stand at 25°C, which induced the assembly of **HBCF**.

SEM and AFM images clearly showed that the resultant nanotubes consisted of two block segments, although three block segments were also observed as a minority. TEM energy-dispersive X-ray spectroscopy (TEM-EDX) revealed that copper was localized in one of the blocks, evidencing the expected linear heterojunction structure. The fact that multiblock nanotubes were not observed indicates that postconnection of the two different nanotubes was unlikely and that the nanotube of **HBCF** was elongated from the seed of **HBCBPy**.

Fluorescence spectral measurements suggested that two nanotubes are communicated, allowing excitation energy transfer over the heterojunction interface. Furthermore, flash-photolysis time-resolved microwave conductivity measurements revealed that the charge carriers generated in the linear heterojunction were long-lived; the observed lifetime ($\tau_{1/e} = 8.8 \times 10^{-6}$ s) was roughly five times longer than those of individual nanotubes of **HBCBPy** (1.4×10^{-6} s) and **HBCF** (2.5×10^{-6} s). This study demonstrates the advanced synthetic methodology of sophisticated supramolecular nanoarchitectures.

Finally, we highlight intriguing systems reported by Ito et al. *(43)* and Yagai et al. *(44)* although these systems may not be conceptually similar to the other molecular self-assemblies introduced in this chapter, one can find a mechanistic similarity in terms of the dynamic behavior under kinetic control.

When rapidly crystallized, **NCAu** formed crystals exhibiting blue photoluminescence (Fig. 10.12, polymorphic crystal **Ib**). Single crystal X-ray analysis revealed that **NCAu** units were arranged in a head-to-tail herringbone fashion with large distances (>4.65 Å) between the gold atoms of adjacent molecules, indicating the absence of Au—Au interactions. In contrast, the slow crystallization produced polymorphic crystals (**IIy**), which showed a strong yellow photoluminescence. The **NCAu** units in **IIy**

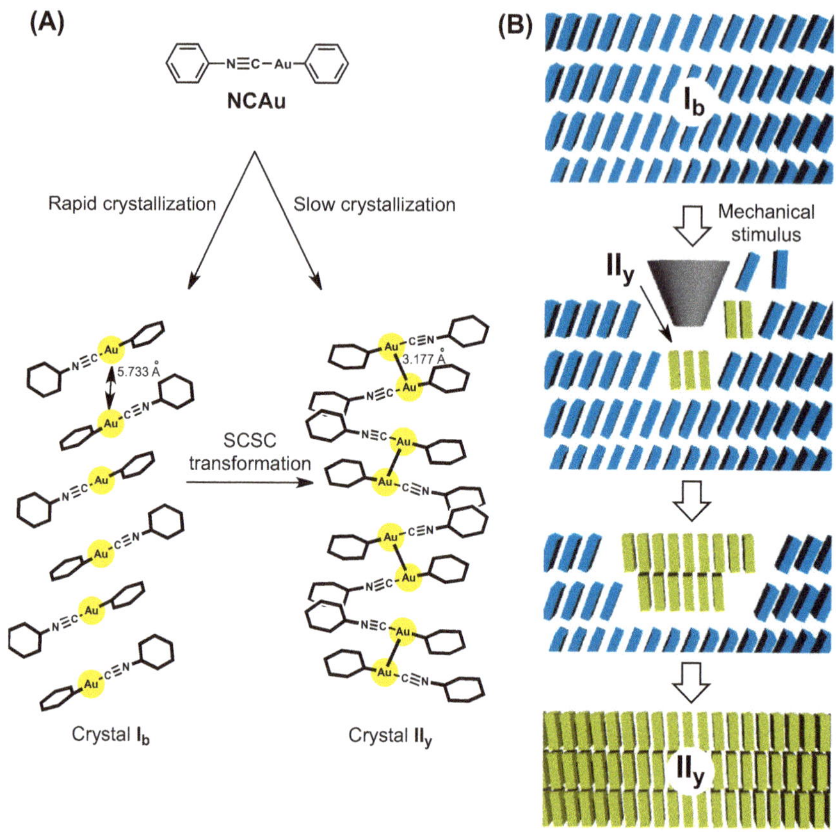

Figure 10.12 (A) Chemical structure of **NCAu** and its crystal packing in **Ib** and **IIy** polymorphs. (B) Schematic representation of the mechanical transformation from **Ib** to **IIy**.

adopted a nearly flat conformation, forming oblique head-to-tail dimers with a short Au—Au distance of 3.177 Å. This result suggests the presence of significant aurophilic interactions in **IIy**, which is consistent with the observed yellow emission.

Given the crystallization protocols, i.e., rapid versus slow crystallization, **Ib** should be a kinetically formed crystal, while **IIy** is a thermodynamically stable crystal. In fact, **Ib** was transformed into **IIy** with a mechanical stimulus. A small pit was formed on the surface of an **Ib** crystal using a needle; as a result, a yellow luminescent spot emerged at the location of the small pit. This observation suggests a single-crystal-to-single-crystal (SCSC) transformation from **Ib** to **IIy**. Interestingly, the domain of **IIy** gradually propagated into the **Ib** crystal (Fig. 10.12B).

The SCSC transformation could also be triggered by contacting crystals of **IIy** and **Ib**. From the point of contact, yellow emissive **IIy** invaded the **Ib** crystal. This solid-seeding phase change represents a self-replicating progression of phase transformation. The crystal transformation of **Ib** to **IIy** was unambiguously confirmed by single-crystal X-ray analysis. It is intriguing that such a dynamic process can occur even in the solid crystal without a solvent, and so it was concluded that the molecules were able to diffuse across the gap between the two phases and rearrange into the thermodynamically stable phase by the formation of intermolecular aurophilic interactions. Ito and coworkers referred to such behavior as "molecular dominoes."

To achieve a mechanoresponsive luminescence color change, Yagai et al. *(44)* rationally designed amphiphilic dipolar π-conjugated molecule (**OPV**, Fig. 10.13). In general, dipolar molecules tend to assemble in an antiparallel arrangement so that the dipoles can be intermolecularly canceled (see Fig. 10.6). Nevertheless, dipoles of **OPV** molecules are oriented in parallel because of its amphiphilic character. Such an orientation is energetically unfavorable due to the dipolar repulsion, yet it is achieved as a result of a delicate balance/conflict between the intermolecular interactions. As such, the molecular packing can be sensitively influenced by external mechanical stimuli.

It was found that a drop cast film of **OPV** exhibited yellow photoluminescence, but after the sample was pressed, the luminescence color changed into orange. Remarkably, application of anisotropic rubbing operation further induced a luminescence color change from orange to green. These three states, denoted as **OPVy**, **OPVo**, and **OPVg**, were characterized as aggregate, mesophase, and crystal, respectively, thoroughly by polarized optical microscopy, powder X-ray diffraction, and differential scanning calorimetry. Thus, photoluminescence color change from **OPVy** to **OPVo** was rationalized by the loss of excitonic interactions between the dipolar π-conjugated systems in the mesophase, while that from **OPVo** to **OPVg** was attributed to the increased rigidity of the molecular structure on crystallization.

Interestingly, **1O** phase could be stabilized when the film was immersed into aqueous CF_3SO_3Li solution; lithium ions chelate the ethylene glycol chains of **OPV**, thereby preventing the **OPVo** from crystallization into **OPVg** phase. Accordingly, a pattern fabricated on the **OPVo** film via inkjet printing of CF_3SO_3Li solution preserved the orange luminescence, whereas the background (i.e., the part where lithium ions were absent)

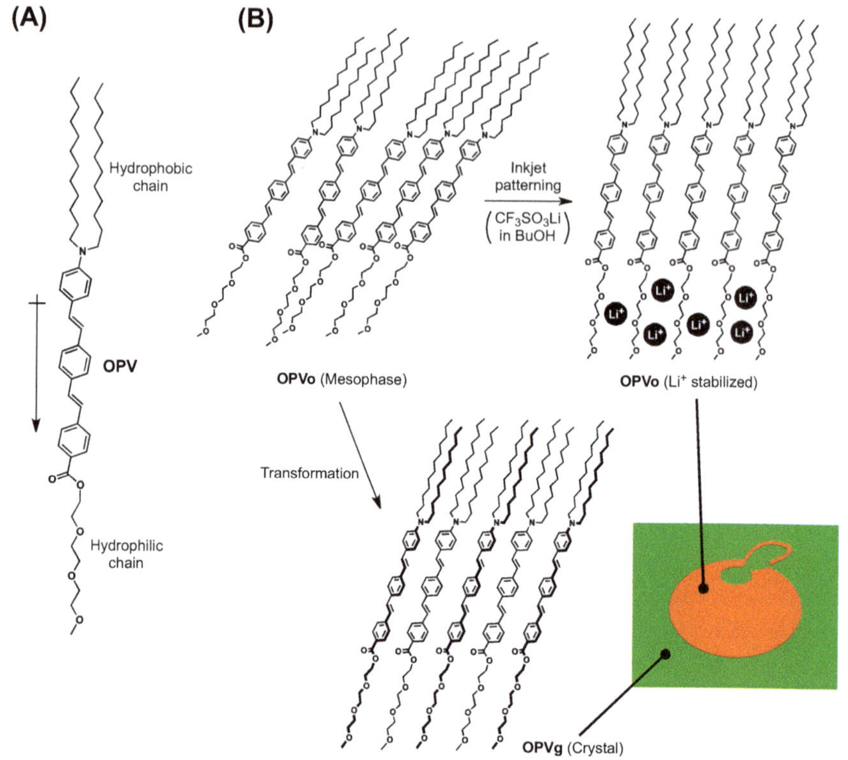

Figure 10.13 (A) Chemical structure of **OPV**. Because of the hydrophobic and hydrophilic side chains (amphiphilicity), the dipole can be arranged in parallel in the condensed phase as shown in (B). (B) Schematic representation of inkjet patterning: orange mark was patterned on a film of metastable **OPVo** phase via inkjet process. Orange photoluminescence was preserved at the area where lithium ions were spread, while the background can be transformed into more stable **OPVg** phase.

transformed into more stable **OPVg** phase exhibiting green photoluminescence (Fig. 10.13B). This new type of π-conjugated molecule featuring tunable luminescent properties can be applied in displays, sensors, and optical devices with stimuli responsiveness.

4. CONCLUSION

This chapter highlights molecular self-assemblies under kinetic control as these offer new opportunities for the creation of unprecedented supramolecular systems. Some of the systems have led to living supramolecular polymerization, and in this way, more complex supramolecular structures such as block supramolecular polymers can be obtained. Although not included in this chapter, other intriguing systems based on chemical reactions have recently attracted significant attention. For example, fuel-driven supramolecular polymerization, which is reminiscent of the formation of microtubules and actin filaments, has been recently established *(45,46)*. We believe that synthetic molecular self-assemblies showing kinetic behaviors have the potential to function as biomolecular counterparts do in nature.

REFERENCES

1. Steed, J. W.; Atwood, J. L. *Supramolecular Chemistry,* 2nd ed.; WILEY, 2009.
2. Castellano, R. K.; Rudkevich, D. M.; Rebek, J., Jr. Polycaps: Reversibly Formed Polymeric Capsules. *Proc. Natl. Acad. Sci. U.S.A.* **1997,** *94,* 7132−7137.
3. Castellano, R. K.; Clark, R.; Craig, S. L.; Nuckolls, C.; Rebek, J., Jr. Emergent Mechanical Properties of Self-assembled Polymeric Capsules. *Proc. Natl. Acad. Sci. U.S.A.* **2000,** *97,* 12418−12421.
4. Sijbesma, R. P.; Beijer, F. H.; Brunsveld, L.; Folmer, B. J. B.; Ky Hirschberg, J. H. K.; Lange, R. F. M.; Lowe, J. K. L.; Meijer, E. W. Reversible Polymers Formed from Self-complementary Monomers Using Quadruple Hydrogen Bonding. *Science* **1997,** *278,* 1601−1604.
5. de Greef, T. A. F.; Meijer, E. W. Supramolecular Polymers. *Nature* **2008,** *453,* 171−173.
6. Cantekin, S.; de Greef, T. F. A.; Palmans, A. R. A. Benzene-1,3,5-tricarboxamide: a Versatile Ordering Moiety for Supramolecular Chemistry. *Chem. Soc. Rev.* **2012,** *41,* 6125−6137.
7. Smulders, M. M. J.; Schenning, A. P. H. J.; Meijer, E. W. Insight into the Mechanisms of Cooperative Self-assembly: The "Sergeants-and-soldiers" Principle of Chiral and Achiral C3-symmetrical Discotic Triamides. *J. Am. Chem. Soc.* **2008,** *130,* 606−611.
8. Oosawa, F.; Kasai, M. A Theory of Linear and Helical Aggregation of Macromolecules. *J. Mol. Biol.* **1962,** *4,* 10−21.
9. Jonkheijm, P.; van der Schoot, P.; Schenning, A. P. H. J.; Meijer, E. W. Probing the Solvent-assisted Nucleation Pathway in Chemical Self-assembly. *Science* **2006,** *313,* 80−83.
10. de Greef, T. F. A.; Smulders, M. M. J.; Wolffs, M.; Schenning, A. P. H. J.; Sijbesma, R. P.; Meijer, E. W. Supramolecular Polymerization. *Chem. Rev.* **2009,** *109,* 5687−5754.
11. Smulders, M. M. J.; Nieuwenhuizen, M. M. L.; de Greef, T. F. A.; van der Schoot, P.; Schenning, A. P. H. J.; Meijer, E. W. How to Distinguish Isodesmic from Cooperative Supramolecular Polymerization. *Chem. Eur J.* **2010,** *16,* 362−367.
12. Albertazzi, L.; van der Zwaag, D.; Leenders, C. M. A.; Fitzner, R.; van der Hofstad, R. W.; Meijer, E. W. Probing Exchange Pathways in One-dimensional Aggregates with Super-resolution Microscopy. *Science* **2014,** *344,* 491−495.
13. Lou, X.; Lafleur, R. P. M.; Leenders, C. M. A.; Schoenmakers, S. M. C.; Matsumoto, N. M.; Baker, M. B.; van Dongen, J. L. J.; Palmans, A. R. A.; Meijer, E. W. Dynamic Diversity of Synthetic Supramolecular Polymers in Water as Revealed by Hydrogen/Deuterium Exchange. *Nat. Commun.* **2017,** *8,* 15420. https://doi.org/10.1038/ncomms15420.
14. van Hameren, R.; Schön, P.; van Buul, A. M.; Hoogboom, J.; Lazarenko, S. V.; Gerritsen, J. W.; Engelkamp, H.; Christianen, P. C. M.; Heus, H. A.; Maan, J. C.; Rasing, T.; Speller, S.; Rowan, A. E.; Elemans, J. A. A. W.; Nolte, R. J. M.

Macroscopic Hierarchical Surface Patterning of Porphyrin Trimers via Self-assembly and Dewetting. *Science* **2006,** *314,* 1433–1436.
15. van Hameren, R.; van Buul, A. M.; Castriciano, M. A.; Villari, V.; Micali, N.; Schön, P.; Speller, S.; Scolaro, L. M.; Rowan, A. J.; Elemans, A. A. W.; Nolte, R. J. M. Supramolecular Porphyrin Polymers in Solution and at Solid-liquid Interface. *Nano Lett.* **2008,** *8,* 253–259.
16. Aida, T.; Meijer, E. W.; Stupp, S. I. Functional Supramolecular Polymers. *Science* **2012,** *335,* 813–817.
17. Ogi, S.; Sugiyasu, K.; Manna, S.; Samitsu, S.; Takeuchi, M. Living Supramolecular Polymerization Realized through a Biomimetic Approach. *Nat. Chem.* **2014,** *6,* 188–195.
18. Fukui, T.; Kawai, S.; Fujinuma, S.; Matsushita, Y.; Yasuda, T.; Sakurai, T.; Seki, S.; Takeuchi, M.; Sugiyasu, K. Control over Differentiation of a Metastable Supramolecualr Assembly in One-and Two Dimensions. *Nat. Chem.* **2017,** *9,* 493–499.
19. Würthner, F. Living it up. *Nat. Chem.* **2014,** *6,* 171–173.
20. van der Zwaag, D.; de Greef, T. F. A.; Meijer, E. W. Programmable Supramolecular Polymerization. *Angew. Chem. Int. Ed.* **2015,** *54,* 8334–8336.
21. Mukhopadhyay, R. D.; Ajayaghosh, A. Living Supramolecular Polymerization. *Science* **2015,** *349,* 241–242.
22. Kang, J.; Miyajima, D.; Mori, T.; Inoue, Y.; Itho, Y.; Aida, T. Rational Strategy for the Realization of Chain-growth Supramolecular Polymerization. *Science* **2015,** *347,* 646–651.
23. Ogi, S.; Stepanenko, V.; Sugiyasu, K.; Takeuchi, M.; Würthner, F. Mechanism of Self-assembly Process and Seeded Supramolecular Polymerization of Perylene Bisimide Organogelator. *J. Am. Chem. Soc.* **2015,** *137,* 3300–3307.
24. Ogi, S.; Stepanenko, V.; Thein, J.; Würthner, F. Impact of Alkyl Spacer Length on Aggregation Pathways in Kinetically Controlled Supramolecular Polymerization. *J. Am. Chem. Soc.* **2016,** *138,* 670–678.
25. Robinson, M. E.; Lunn, D. J.; Nazemi, A.; Whittell, G. R.; De Cola, L.; Manners, I. Length Control of Supramolecular Polymeric Nanofibers Based on Stacked Planar Platinum(II) Complexes by Seeded-growth. *Chem. Commun. (J. Chem. Soc. Sect. D)* **2015,** *51,* 15921–15924.
26. Robinson, M. E.; Nazemi, A.; Lunn, D. J.; Hayward, D. W.; Boot, C. E.; Hsiao, M.-S.; Harniman, R. L.; Davis, S. A.; Whittell, G. R.; Richardson, R. M.; De Cola, L.; Manners, I. Dimensional Control and Morphological Transformations of Supramolecular Polymeric Nanofibers Based on Cofacially-stacked Planar Amphiphilic Platinum(II) Complexes. *ACS Nano* **2017,** *11,* 9162–9175.
27. Haedler, A. T.; Meskers, S. C. J.; Zha, R. H.; Kivala, M.; Schmidt, H.-W.; Meijer, E. W. Pathway Complexity in the Enantioselective Self-assembly of Functional Carbonyl-bridged Triarylamine Trisamides. *J. Am. Chem. Soc.* **2016,** *138,* 10539–10545.
28. Endo, M.; Fukui, T.; Jung, J. H.; Yagai, S.; Takeuchi, M.; Sugiyasu, K. Photoregulated Living Supramolecular Polymerization Established by Combining Energy Landscapes of Photoisomerization and Nucleation-Elongation Processes. *J. Am. Chem. Soc.* **2016,** *138,* 14347–14353.
29. Malakoutikhah, A. P. A.; Leonetti, G.; Tezcan, M.; Colomb-Delsuc, M.; Nguyen, V. D.; van der Gucht, J.; Otto, S. Controlling the Structure and Length of Self-synthesizing Supramolecular Polymers through Nucleated Growth and Disassembly. *Angew. Chem. Int. Ed.* **2015,** *54,* 7852–7856.
30. Naota, T.; Koori, H. Molecules that Assemble by Sound: An Application to the Instant Gelation of Stable Organic Fluids. *J. Am. Chem. Soc.* **2005,** *127,* 9324–9325.

31. Komiya, N.; Muraoka, T.; Iida, M.; Miyanaga, M.; Takahashi, K.; Naota, T. Ultrasound-induced Emission Enhancement Based on Structure-dependent Homo-and Heterochiral Aggregations of Chiral Binuclear Platinum Complexes. *J. Am. Chem. Soc.* **2011,** *133,* 16054−16061.
32. Isozaki, K.; Takaya, H.; Naota, T. Ultrasound-induced Gelation of Organic Fluids with Metalated Peptide. *Angew. Chem. Int. Ed.* **2007,** *46,* 2855−2857.
33. van Herpt, J. T.; Stuart, M. C. A.; Browne, W. B.; Feringa, B. L. Mechanically Induced Gel Formation. *Langmuir* **2013,** *29,* 8763−8767.
34. Carnall, J. M. A.; Waudby, C. A.; Belenguer, A. M.; Stuart, M. C. A.; Peyralans, J. J.-P.; Otto, S. Mechanosensitive Self-replication Driven by Self-organization. *Science* **2010,** *327,* 502−506.
35. Mattia, E.; Pal, A.; Leonetti, G.; Otto, S. Mechanism of Building Block Exchange in Stacks of Self-replicating Macrocycles. *Synlett* **2017,** *28,* 103−107.
36. Lohr, A.; Lysetska, M.; Würthner, F. Supramolecular Stereomutation in Kinetic and Thermodynamic Self-assembly of Helical Merocyanine Dye Nanorods. *Angew. Chem. Int. Ed.* **2005,** *44,* 5071−5074.
37. Lohr, A.; Würthner, F. Evolution of Homochiral Helical Dye Assemblies: Involvement of Autocatalysis in the "Majority-Rules" Effect. *Angew. Chem. Int. Ed.* **2008,** *47,* 1232−1236.
38. Yagai, S.; Hamamura, S.; Wang, H.; Stepanenko, V.; Seki, T.; Unoike, K.; Kikkawa, Y.; Karatsu, T.; Kitamura, A.; Würthner, F. Unconventional Hydrogen-bond-directed Hierarchical Co-assembly between Perylene Bisimide and Azobenzene-functionalized Melamine. *Org. Biomol. Chem.* **2009,** *7,* 3926−3929.
39. Baram, J.; Weissman, H.; Rybtchinski, B. Supramolecular Polymer Transformation: A Kinetic Study. *J. Phys. Chem. B* **2014,** *118,* 12068−12073.
40. Korevaar, P. A.; George, S. J.; Markvoort, A. J.; Smulders, M. M. J.; Hibers, P. A. J.; Schenning, A. P. H. J.; de Greef, T. A. F.; Meijer, E. W. Pathway Complexity in Supramolecular Polymerization. *Nature* **2012,** *481,* 492−496.
41. Williams, R. J.; Smith, A. M.; Collins, R.; Hodson, N.; Das, A. K.; Ulijn, R. V. Enzyme-assisted Self-assembly under Thermodynamic Control. *Nat. Nanotechnol.* **2009,** *4,* 19−24.
42. Zhang, W.; Jin, W.; Fukushima, T.; Saeki, T.; Seki, S.; Aida, T. Supramolecular Linear Heterojunction Composed of Graphite-like Semiconducting Nanotubular Segment. *Science* **2011,** *334,* 340−343.
43. Ito, H.; Muromoto, M.; Kurenuma, S.; Ishizaka, S.; Kitamura, N.; Sato, H.; Seki, T. Mechanical Stimulation and Solid Seeding Trigger Single-Crystal-to-single-Crystal Molecular Domino Transformation. *Nat. Commun.* **2013,** *4,* 2009. https://doi.org/10.1038/ncomms3009.
44. Yagai, S.; Okamura, S.; Nakano, Y.; Yamauchi, M.; Kishikawa, K.; Karatsu, T.; Kitamura, A.; Ueno, A.; Kuzuhara, D.; Yamada, H.; Seki, T.; Ito, H. Design Amphiphilic Dipolar Π-Systems for Stimuli-responsive Luminescent Materials Using Metastable States. *Nat. Commun.* **2014,** *5,* 4013. https://doi.org/10.1038/ncomms5013.
45. Boekhoven, J.; Hendriksen, W. E.; Koper, G. J. M.; Elkema, R.; van Esch. Transient Assembly of Active Materials Fueled by a Chemical Reaction. *Science* **2015,** *349,* 1075−1079.
46. Kumar, M.; Brocorens, P.; Tonnelé, C.; Beljonne, D.; Surin, M.; George, S. J. A Dynamic Supramolecular Polymer with Stimuli-responsive Handedness for in Situ Probing of Enzymatic ATP Hydrolysis. *Nat. Commun.* **2014,** *5,* 5793. https://doi.org/10.1038/ncomms6793.

CHAPTER 11

Kinetic Control Over the Topology of Curved Supramolecular Polymers

Shiki Yagai[1,4] Mitsuaki Yamauchi[1,3], Bimalendu Adhikari[2]
[1]Department of Applied Chemistry and Biotechnology, Graduate School of Engineering, Chiba University, Chiba, Japan; [2]Department of Chemical Sciences, Indian Institute of Science Education and Research (IISER) Mohali, Punjab, India; [3]Department of Applied Chemistry for Environment, School of Science and Technology, Kwansei Gakuin University, Sanda, Japan; [4]Institute for Global Prominent Research (IGPR), Chiba University, Chiba, Japan

Contents

1. INTRODUCTION

Every reader of this chapter know that proteins are polymer of amino acids but may not be able to answer exactly what is the origin of their amazing functions. Most of the sophisticated functions such as molecular recognition, catalysis, and information storage of proteins are stemmed from their higher-order structures. Polypeptide chains of proteins are folded by noncovalent interactions, primarily hydrogen bonds between amide units to form a specific shape composed of secondary structures, which can be defined as topology *(1)*. The close relationship between functions and topologies of proteins epitomizes the importance of topological control in the development of synthetic polymers. In fact, controlling topological features in synthetic polymers have realized new properties and functions, which is one of the current trends in polymer research *(2,3)*.

Kinetic Control in Synthesis and Self-Assembly
ISBN 978-0-12-812126-9
https://doi.org/10.1016/B978-0-12-812126-9.00011-0

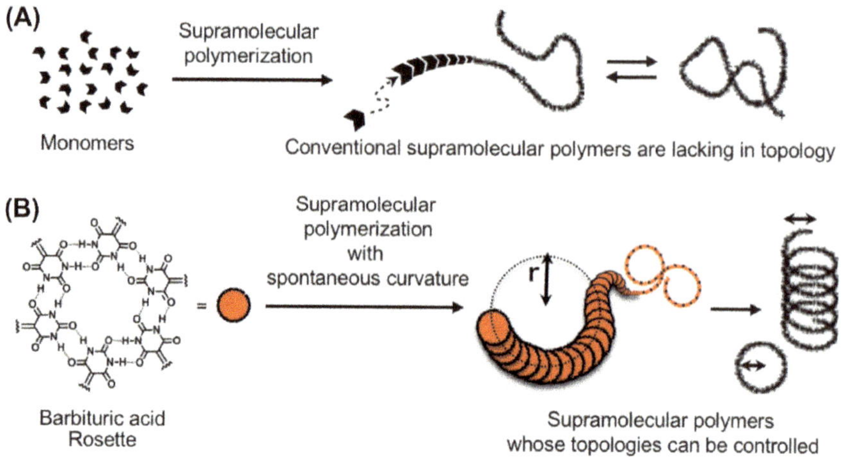

Figure 11.1 Comparison of (A) conventional supramolecular polymer and (B) novel supramolecular polymer with topological features.

In contrast to covalent and biopolymers, the main chains of supramolecular polymers are formed by noncovalent bonds *(4,5)*, and hence it is difficult to control their higher-order structures noncovalently by using the same interactions (Fig. 11.1A) *(6)*. Moreover, many of the recently studied supramolecular polymers are often formed with monomers possessing π-conjugated molecules to impart functions to the polymer backbone *(7)*. Consequently, supramolecular polymers are mostly formed by a combination of multiple interactions such as π−π stacking and hydrogen bonding *(8)*. Although degrees of internal order and functionality are considerably increased through such monomer design *(7)*, their main chains become very rigid, making a topological control extremely difficult.

The research group of Yagai and coworkers has been working toward the development of supramolecular polymers with well-defined topological features. The supramolecular polymers are formed hierarchically from hydrogen-bonded cyclic aggregates so-called macrocycles/rosettes *(9)* of barbituric acid *(10)*, and these rosettes are supramolecularly polymerized with spontaneous curvature (Fig. 11.1B). In this chapter, we reviewed how this novel type of supramolecular polymers has been developed, particularly by focusing on the kinetic aspect of their supramolecular polymerization process.

2. TOROIDAL SUPRAMOLECULAR POLYMER

When the structure of a monomer, capable of forming supramolecular polymers, is flexible, cyclic "oligomers" can be preferentially formed at low concentrations, and these cyclic species result in a kinetic trap that hinders the formation of supramolecular polymers *(11)*. As the monomer concentration is raised, eventually it reaches a critical concentration defined as effective molarity, which is the ratio of the intramolecular rate constant to the intermolecular rate constant *(12)*, and above this critical concentration, supramolecular polymerization preferentially occurs through intermolecular interactions than cyclization at oligomeric level. When cyclic oligomers consist of a few monomers, they cannot be regarded as "cyclized supramolecular polymers" in terms of molecular weight. However, the "cyclized supramolecular polymer" in this chapter is referred to as a supramolecular polymer in terms of molecular weight, and it is an large aggregate having a closed structure in which the two ends are linked together. For example, Lee and coworkers reported that supramolecular polymers composed of specifically designed amphiphilic π-conjugated molecules result in ring-shaped nanostructures (nanoring) in water *(13)*. We also accidentally found that a rosette, formed by complementary hydrogen bonding between an *N,N'*-disubstituted melamine bearing π-conjugated units and an *N*-alkylated cyanuric acid forms toroidal nanoaggregates in nonpolar solvents *(14)*. Although we did not refer to these toroidal nanoaggregates as a cyclized supramolecular polymer at that time, the result essentially suggests that a supramolecular polymer with length of approximately 67 nm (corresponding to the circumference) cyclizes into a discrete toroidal nanostructure. Although such short aggregate might be out of the definition of supramolecular polymer documented by Meijer et al. in their seminal review *(4)*, we would like to classify them as supramolecular polymers because they contain on average 150 rosettes which is equivalent to 900 monomers.

Can similar cyclized supramolecular polymers be obtained from structurally relevant other compounds? Because the rosette can also be obtained by complementary hydrogen bonding between *N,N'*-disubstituted melamine and barbituric acid whose chemical structure is similar to cyanuric acid, we designed barbituric acid-equipped several π-conjugated and aliphatic molecules. Although these barbituric acid compounds were chemically stable even in solution by themselves, it was found that the barbituric acid unit was eliminated to form aromatic aldehyde precursors through retro-Knoevenagel reaction within a day when they were mixed with melamine derivatives which are basic in nature. This disappointed us, but later we fortuitously discovered that some of the synthesized barbituric acids self-assembled in low-polar solvents such as methylcyclohexane (MCH) to form unique supramolecular polymers *(15—18)*. For example, when naphthalene

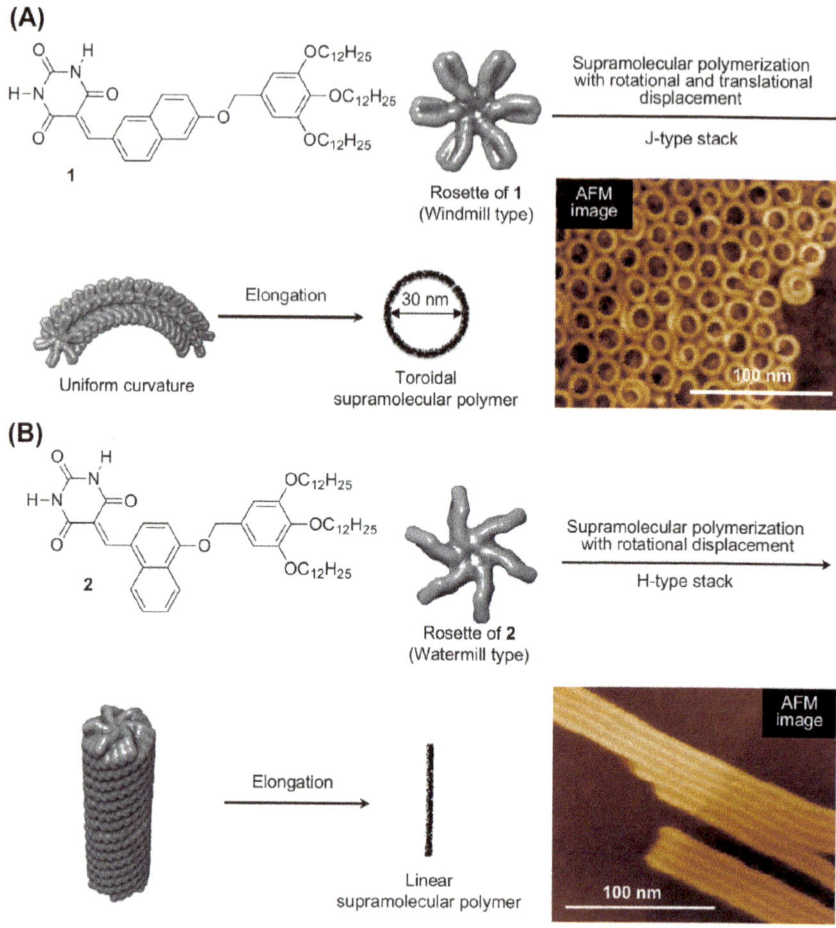

Figure 11.2 Supramolecular polymerization of naphthalene-based barbituric acid compounds (A) **1** and (B) **2** through the formation of cyclic rosettes.

derivative **1** was dissolved in MCH by heating to obtain a molecular dispersion (monomer) and then subsequently cooled to room temperature, toroidal supramolecular polymers (nanorings) were obtained in yields exceeding 90% based on monomer (Fig. 11.2A). On the other hand, a regioisomeric naphthalene derivative **2** afforded a linearly elongated supramolecular polymer (Fig. 11.2B). It was surprising to us that dramatically different topologies of supramolecular polymers were shown by **1** and **2**. Analysis of the supramolecular polymerization processes of these monomers by temperature-dependent absorption spectra suggested cooperative aggregation behavior consisting of nucleation and elongation processes *(18,19)*.

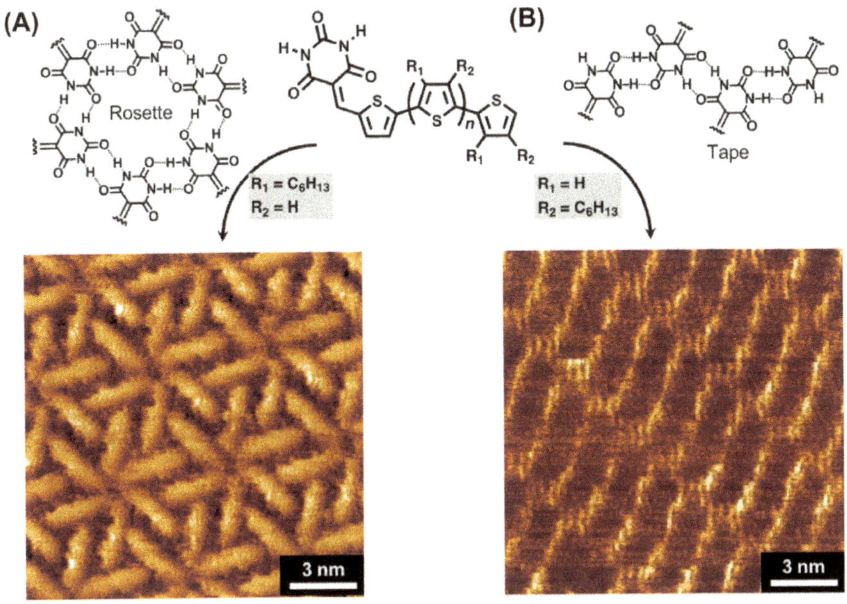

Figure 11.3 Scanning tunneling microscopy images of (A) discrete rosette and (B) infinite tapelike molecular arrangements of oligo(alkylthiophene)-conjugated barbituric acids at solid–liquid interface.

How do compounds **1** and **2** organize into supramolecular polymers with strikingly different topologies? A hint was obtained in another project on organic solar cells in our research group. We found that oligo(alkylthiophene)-conjugated barbituric acids formed linear one-dimensional nanostructures and showed outstanding solar cell performance (power conversion efficiency ~ 4.5%) as hydrogen-bonding materials on mixing with soluble fullerene derivatives *(20–22)*. Investigation of hydrogen bond–mediated molecular arrangements of these oligo(alkylthiophene)-conjugated barbituric acids by scanning tunneling microscopy at the solid–liquid interface revealed that they form either cyclic hexamers or tapelike polymers (Fig. 11.3). Although the latter infinite molecular arrangement is a supramolecular structure itself, it cannot be a basic structure of supramolecular polymers of **1** and **2** due to several structural reasons as well as its high crystallinity *(15)*.

We thus structurally optimized rosettes of **1** and **2** by molecular mechanics calculation and found a critical difference between them that can rationally explain the topological difference of supramolecular polymers. The rosette of **1** adopted a windmill-like geometry with lower degree of twisting of the naphthalene rings (Fig. 11.2A), whereas the rosette of **2** was optimized to a watermill-like geometry wherein naphthalene rings are considerably twisted due to intramolecular steric hindrance (Fig. 11.2B). Taking into account these distinct geometries, we have proposed following structural models for the formation of linear and toroidal supramolecular polymers. The planar rosette of **1** can stack not only with rotational but also with translational displacement to produce a spontaneous curvature (Fig. 11.2A). In contrast, the twisted rosette of **2** can stack only with rotational displacement to provide a rigid columnar structure (Fig. 11.2B). These stacking models are also consistent with spectroscopic observations which indicate that naphthalene moieties of **1** and **2** are arranged as J- and H-type stacking modes on supramolecular polymerization, respectively *(17)*. Later on, these structural models were supported by a simultaneous measurement of small-angle neutron scattering and small-angle X-ray scattering (SAXS) in solution for **1** *(19)* and a wide angle X-ray diffraction measurement of the oriented thin film for **2** *(18)*.

3. KINETIC CONSIDERATION ON THE FORMATION OF TOROIDAL STRUCTURE

The unique stacking mode of the rosette of **1** to produce spontaneous curvature can be used for the design of supramolecular polymers with well-defined topology. If the curvature continuously is generated in the same turning direction during the elongation process of supramolecular polymerization, a helically folded structure would be formed (Fig. 11.4). In contrast, if the curvature is generated randomly by changing the direction, a randomly coiled structure would be the outcome (Fig. 11.4). The diameter of toroidal supramolecular polymers of **1** is ca. 30 nm, which suggests the degree of polymerization is ca. 1000 based on monomer and ca. 170 based on rosette. When the formation process of toroids was monitored by absorption spectroscopy and the process was analyzed by a nucleation-elongation model, the number of molecules involved in the nucleus is estimated to be 20. Therefore, the nucleus consists of stacks of only a few rosettes (ca. four rosettes), and the formation of toroids seems to occur in the subsequent elongation process in which supramolecular polymer extended significantly. A closer look of AFM image of **1** (Fig. 11.2A) shows that there are also a few open-ended curved supramolecular polymer nanofibers, which should survive from cyclization. Probably, there would be two pathways in the elongation process, one is cyclization to form toroidal supramolecular polymers and the other is extension to form curved

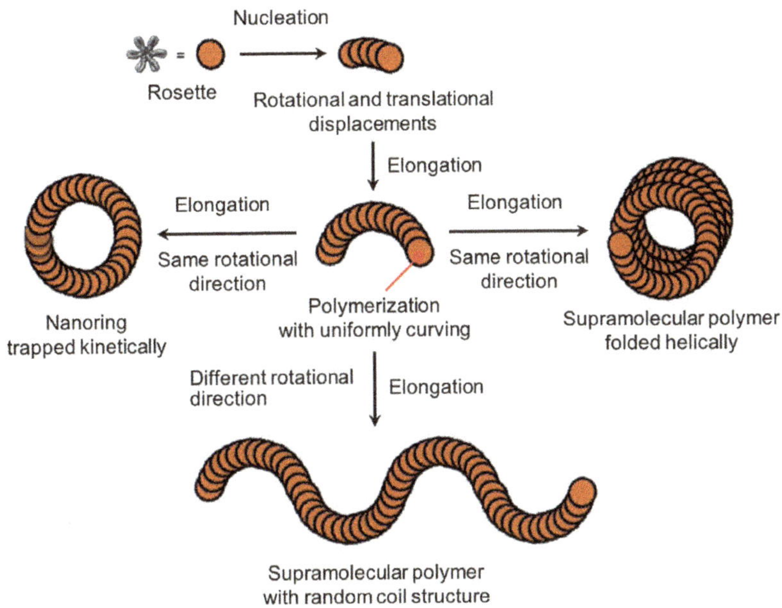

Figure 11.4 Aggregation pathways of curvature-forming rosettes toward the formation of various supramolecular polymers.

open-ended supramolecular polymers (Fig. 11.4). If the supramolecular polymerization process is reversible, toroidal supramolecular polymers may be able to go back to another pathway to produce open-ended supramolecular polymers if they are thermodynamically more stable (vide infra). However, under a kinetic polymerization condition wherein a hot solution was simply cooled at room temperature, the kinetically formed toroidal structures would be stabilized due to a rapid decrease of the solution temperature ("kinetic trap"). Accordingly, if supramolecular polymerization is carried out under such a condition that is capable of realizing equilibrium, kinetically formed toroidal structure could open up to elongate further, leading to thermodynamically stable species such as helically folded supramolecular polymers wherein interactions between helically folded loops are expected to stabilize them similar to the stabilization mechanism of protein α helices. Foldamers, synthetic oligomers that can adopt folded structures, are also stabilized by the similar mechanism *(23)*. Based on these assumptions, we cooled a hot solution of **1** at a rate of 0.1°C/min as a very thermodynamic condition. However, toroidal supramolecular polymers were again obtained as the main product, presumably suggesting that these closed structures were thermodynamically stable products for **1**.

How could we shallow the kinetic trap of the toroidal supramolecular polymers and dominate the path to the extended (helically folded) supramolecular polymers? For practical reasons, we usually prepared aggregates at around hundreds μM scale. In this concentration regime, the degree of polymerization of **1**, estimated from the temperature-dependent absorption experiment, is ca. 800, which is consistent well with the value estimated from the average circumference of toroids visualized by AFM. This result thus demonstrates that molecule **1** intrinsically does not have a high association ability to achieve high degrees of polymerization that ensure the formation of elongated supramolecular polymers. Applying higher monomer concentrations can theoretically increase the degree of polymerization, but the frequency of nucleation could also be increased at the same time, which would ultimately cancel the concentration effect. Thereby, the degree of polymerization of individual supramolecular polymer does not necessarily increase in proportion to the monomer concentration (indeed, toroidal structures have been predominantly observed even at millimolar regime).

An alternative approach can be the modification of monomer structure that shows stronger π–π stacking interaction without compromising the structural unit responsible for generating curvature. When we synthesized molecules possessing additional π-conjugated unit between the naphthalene and terminal aliphatic moieties, some of them indeed afforded a mixture of randomly coiled and helically folded supramolecular polymers. In the latter half of this chapter, we review the results of helically folded supramolecular polymers that can be unfolded with light in the viewpoint of kinetics.

4. HELICALLY FOLDED SUPRAMOLECULAR POLYMER

We have developed a variety of photoresponsive supramolecular assemblies using photochromic moieties such as azobenzene *(24–29)*, stilbene *(30)*, and diarylethene *(31–33)*. As the structure or the conformational flexibility of photochromic molecules changes significantly through photoisomerization, it is possible to reversibly control their supramolecular structures or aggregated states noninvasively. As an extension of the toroidal supramolecular polymer, we initially expected that if this closed supramolecular polymer could be constructed from compound **3** incorporating azobenzene unit into the parent molecule **1** (Fig. 11.5A), the photoinduced deterioration of curvature of the main chain should lead to a "photoinduced ring-opening" at the supramolecular polymer level. However, when MCH

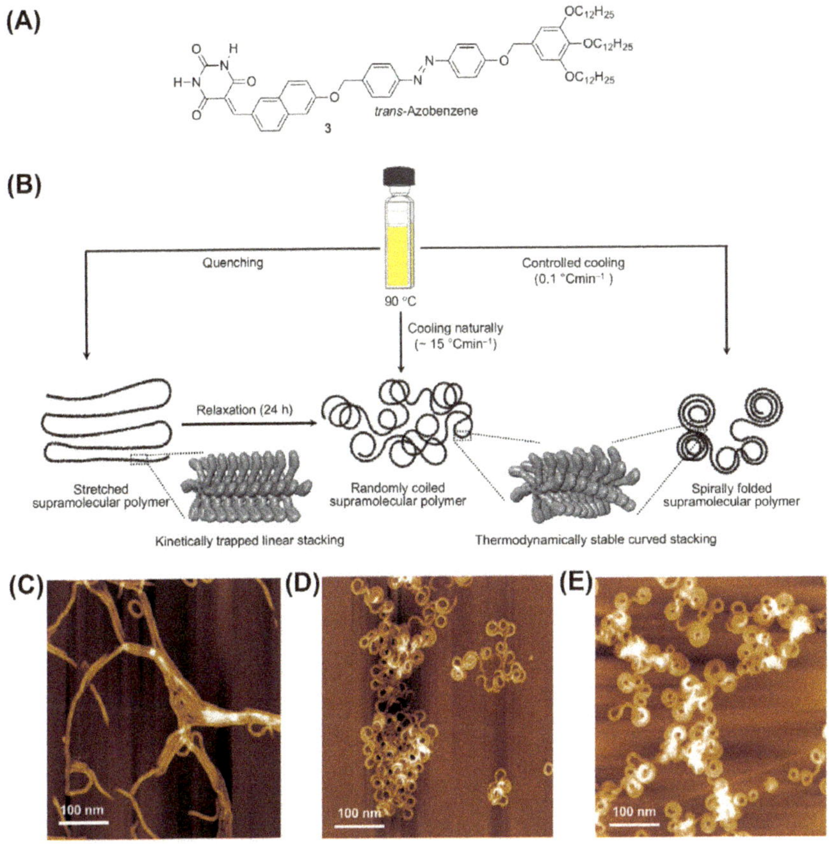

Figure 11.5 (A) Chemical structure of azobenzene-incorporated molecule **3**. (B) Schematic representation for kinetically controlled preparation of supramolecular polymers of **3** in MCH by various cooling rate. (C–E) AFM images of (C) stretched, (D) randomly coiled, and (E) spirally folded supramolecular polymers of **3**.

solution of **3** was cooled from 100°C to room temperature naturally (Fig. 11.5B, center), a random coiled supramolecular polymer having a constant curvature was obtained unexpectedly (Fig. 11.5D). The degree of curvature is almost the same as that of **1**, indicating that the introduction of azobenzene does not affect the specific stacking arrangement of rosettes. We then investigated the supramolecular polymerization process of **3** by temperature-dependent absorption measurements and obtained an elongation enthalpy (ΔH_e) of −108 kJ/mol, which is remarkably higher than that of **1** ($\Delta H_e = -83.4$ kJ/mol). This result indicates that the introduction of azobenzene unit reinforces the intermolecular interaction, which becomes more pronounced when it forms rosette *(34)*.

Why have not helically folded structures been obtained in the above experiment by natural cooling? In the randomly coiled supramolecular polymers visualized by AFM (Fig. 11.5D), many of the adjacent curvature domains turn in opposite directions, in other words, the main chains are meandering. The direction of the curvature of our supramolecular polymers is originally determined by the turning direction of the "rotational displacement" between rosettes. If the turning direction can be maintained continuously on elongation, it should result in a helically folded structure ultimately (Fig. 11.4). In such a case, the intermolecular interactions between the rosettes should be maximized due to uniform molecular packing over a long distance, leading to thermodynamically stable state. Moreover, the helically folded structure is additionally stabilized by intermolecular forces working along the axial direction of the helix ("interloop" interactions). However, during the supramolecular polymerization driven by natural cooling, the rate of the association of rosette is too fast, so that the supramolecular polymer is formed without considering the thermodynamic stability of products. In other words, the supramolecular polymerization is kinetically controlled. This was supported by an experiment employing an extremely kinetic condition achieved by a fast quenching of the solution from 90 to 20°C (>100°C/min, Fig. 11.5B, left) using an ice bath. As it can be seen in the AFM image of Fig. 11.5C, this recipe yielded stretched supramolecular polymers without spontaneous curvature. These stretched fibers share the same cross-sectional width (9.7 ± 0.3 nm) and thickness (2.7 ± 0.2 nm) of their main chains, suggesting that they organize via an identical self-assembly process of **3** via rosettes, but only with the translational displacement (Fig. 11.5B, left). Reflecting their very kinetic nature, storing the quenched solution for 24 h at 20°C led to the almost complete conversion of stretched fibers into randomly coiled supramolecular polymers, which demonstrates the spontaneous recovery of unique internal stacking order of rosettes.

In order to obtain thermodynamically stable molecular packing, rosettes must overcome a high energy barrier (activation energy) (Fig. 11.6). Under kinetic conditions, rosettes cannot overcome this high energy barrier and they polymerize kinetically by frequently changing the direction of rotation, leading to randomly coiled supramolecular polymers. Because the interaction between rosettes of **3** is very strong, these kinetically formed (metastable) supramolecular polymers do not convert into thermodynamically stable structures (kinetically trapped). Accordingly, for obtaining thermodynamically stable helically folded

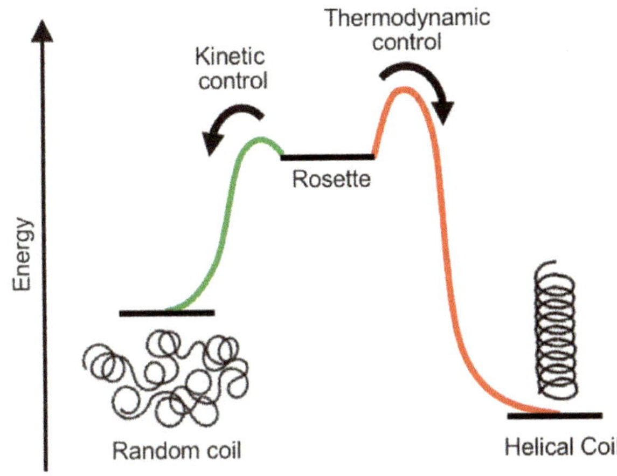

Figure 11.6 Energy landscape of supramolecular polymerization of rosettes under kinetic (left) and thermodynamic control (right).

supramolecular polymers, it is necessary to maintain the direction of rotation of rosettes over a long range by applying a thermodynamic polymerization condition in which energetically unfavorable stacking could be repaired. The simplest method would be slowing down the cooling rate *(35–37)*. It is now possible to equip the spectrometer with a Peltier device capable of precisely controlling temperature change. We thus cooled a hot MCH solution of **3** by the cooling rate of 0.1°C/min (Fig. 11.5B, right). As shown in Fig. 11.5E, the turning direction of curvatures is maintained for much longer distance compared with the natural cooling, and we obtained supramolecular polymers with spirally folded domains (Fig. 11.5C). Although the topology had steadily approached to helix, we needed a more thermodynamic condition to extend such domains with the same turning direction of curvatures. Because the application of further slow cooling (e.g., 0.01°C/min) is impractical from the viewpoint of solvent volatilization and the burden on the apparatus, we employed alternative method using chloroform, a good solvent for **3**. Because the presence of the good solvent improves the reversibility of supramolecular polymerization during cooling, we anticipated that the similar effect can also be achieved here to that of extremely slow cooling method mentioned above. We studied structures/morphologies of supramolecular polymers obtained by slow cooling (0.1°C/min) the hot MCH–chloroform solutions of **3** with varying chloroform content by the cooling rate of 0.1°C/min. When we used MCH:chloroform = 85:15 (vol ratio) mixture, we obtained helically

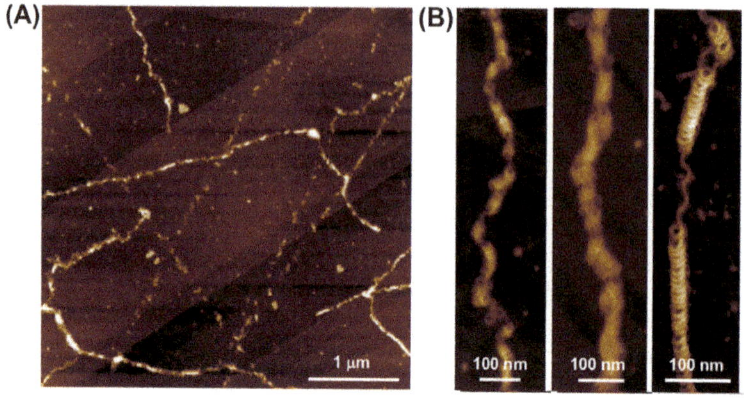

Figure 11.7 AFM images of helically folded supramolecular polymers of **3** prepared in the MCH:chloroform = 85:15 mixture by the cooling rate of 0.1°C/min. (A) A large area image (4 × 4 μm^2) showing several supramolecular polymers. (B) High-resolution images of a single supramolecular polymer.

folded supramolecular polymers as major product (Fig. 11.7). Although there is a biasness in helical handedness in a single supramolecular polymer nanofiber, no biasness in helical handedness was noticed for the overall system (fiber to fiber). This is not surprising because monomer **3** does not have any asymmetric center on its molecular structure.

5. PHOTO-UNFOLDING OF HELICALLY FOLDED SUPRAMOLECULAR POLYMER

On irradiation of UV light to a helically folded supramolecular polymer solution, AFM visualized unfolding of helically folded structures to randomly coiled structures when 20% of *trans*-azobenzene photoisomerized to *cis*-isomer, as confirmed by the absorption measurement (Fig. 11.8). Further UV irradiation resulted in a transformation into stretched supramolecular polymer fibers without curvature at 30% of the photoisomerization. Although the AFM observation of the topology change was performed after spin coating the MCH solution on HOPG substrate, the occurrence of the same topology change in solution state was confirmed by SAXS analysis in solution *(34)*.

When the above UV-irradiated solution containing fully unfolded supramolecular polymers was mixed with a nonirradiated solution containing intact helically folded supramolecular polymers, both folded and unfolded supramolecular polymers coexisted for a few days without averaging out their structure. This demonstrates that the monomer

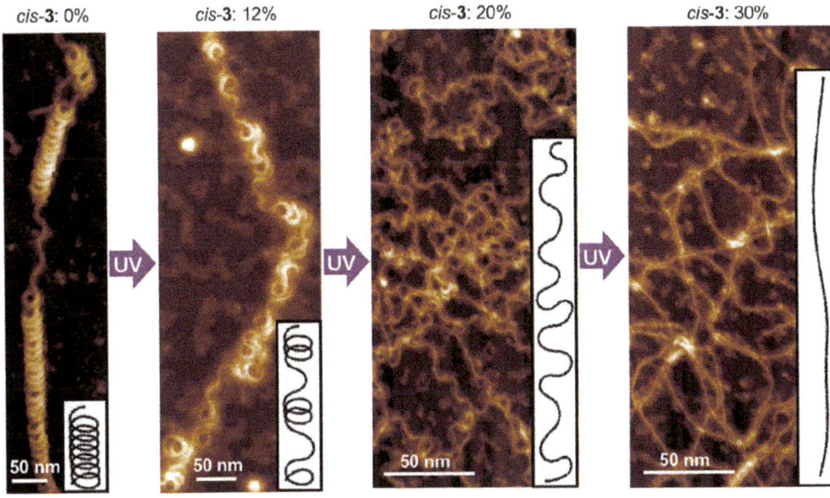

Figure 11.8 AFM images capturing the unfolding process of helically folded supramolecular polymers by UV light irradiation.

exchange between supramolecular polymers does not take place at all or the exchange is negligibly slow at a practical timescale. This, in turn, suggests that the UV-induced transformation of supramolecular polymers occurs directly via the change in the internal order of their main chains by the trans-to-cis isomerization of azobenzene and does not occur indirectly through the monomer exchange or reorganization, namely, isomerization-induced dissociation of supramolecular polymer into rosettes or monomers and subsequent reconstruction of new supramolecular polymers *(33)*.

Why is the curvature completely lost when 30% of *trans*-azobenzene isomerize to cis-form? The ratio of 30% implies that on an average only two azobenzenes out of the six photoisomerize in a rosette. These *cis*-azobenzene units are presumably the two which are along the axial direction of supramolecular polymers, in which the rosettes are arranged linearly (without rotational displacement) by the bent *cis*-arms that act as a "bulge" to align the rosettes with rotational displacement (Fig. 11.9). This was

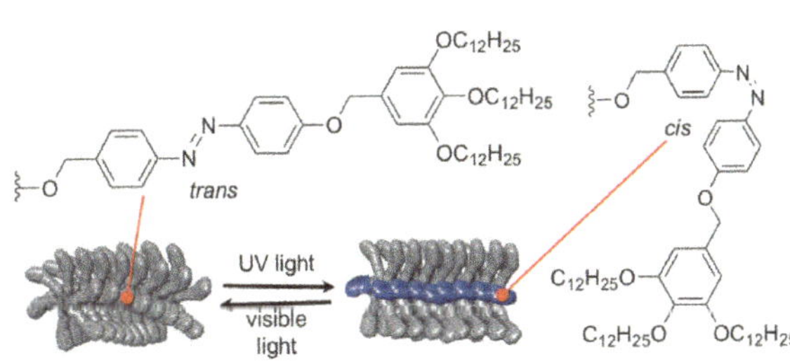

Figure 11.9 A proposed mechanism of photoinduced modulation of curvature.

supported by the fact that the photoisomerization of azobenzene was apparently saturated when it reached 30% cis, whereas further isomerization was only possible by irradiating with very strong UV light.

6. REFOLDING OF (PHOTO)UNFOLDED SUPRAMOLECULAR POLYMER

On irradiation of a weak visible light to the solution of the aforementioned extended supramolecular polymers (*cis*-isomer: 30%), we observed the *cis*-to-*trans* back isomerization of azobenzene, and a photostationary state (PSS) was reached at ca. 11%. At this state, AFM showed the reversion to randomly coiled structures (Fig. 11.10A,B), suggesting that almost recovery of spontaneous curvature on back isomerization (Fig. 11.9). When this PSS solution was aged to thermally revert the remaining *cis*-isomer at maximum (*cis*-isomer: 5%), however, we never observed the original helically folded supramolecular polymers. The randomly coiled supramolecular polymers are most likely the kinetically stable structure at the nanoscopic level and cannot fold back spontaneously into a helical structure.

In another experiment, we examined whether the recovery of spontaneous curvature could immediately occur if we abruptly reduce the *cis*-azobenzene unit using intense visible light. When the solution of the

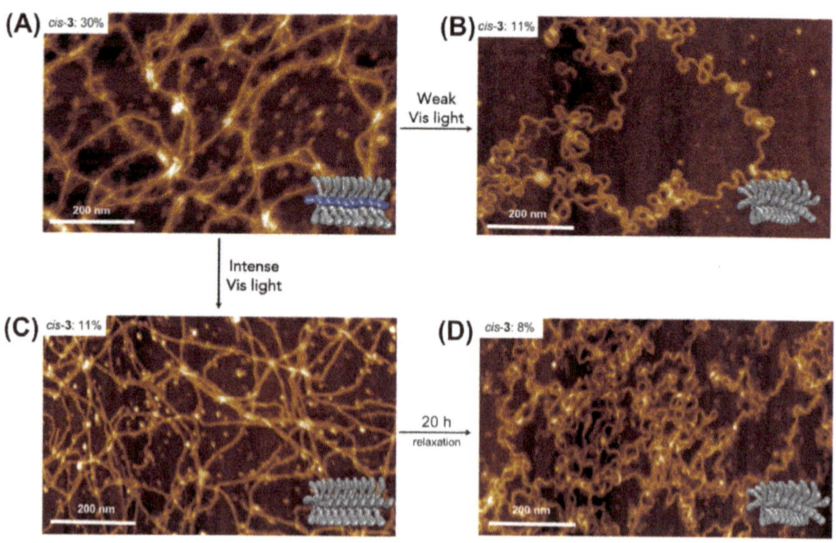

Figure 11.10 AFM images showing photo-refolding of (A) stretched supramolecular polymers to (B) randomly coiled state by weak visible light, to (C) kinetically stretched state by strong visible light, and to (D) its relaxation to randomly coiled state on aging.

extended supramolecular polymers was irradiated with intense visible light, surprisingly, we did not observe any apparent topological change (Fig. 11.10A–C) even though the *cis*-isomer was reduced to the same amount of 11%. This result suggests that the observed fibers are in the kinetically stretched state. In other words, the abruptly generated planar rosettes with five or six azobenzenes in *trans*-isomer (based on 11% of the *cis*-isomer) cannot form their thermodynamically preferred stacking mode with spontaneous curvature and kinetically maintain linear stacking mode for some time. When this solution was aged for 20 h, the stretched supramolecular polymers slowly converted into the randomly coiled ones (Fig. 11.10D).

7. SUMMARY AND FUTURE PROSPECTS

Although the primary structures of supramolecular polymer can be constructed infinitely by the molecular design of monomeric units, supramolecular polymers capable of controlling their higher-order structures, that is, topology, have hardly been reported. In this chapter, we have introduced our initial endeavor toward the construction of such supramolecular polymers with well-defined topological features including cylindrical fiber, ring (toroid), random coil, helix, and their stretched mutant. Although these sequential topological mutations have partially relied on the structural modification of π-conjugated core to produce spontaneous curvature and to assure high degree of polymerization, the topological control at higher order is achievable by understanding and controlling the kinetics of supramolecular polymerization. We have also successfully implemented controllability of topology by using light as an external stimulus by incorporating azobenzene unit. Based on the same rosette-forming molecular design leading to spontaneous curvature, our research group has already developed further exciting supramolecular polymers with many static and dynamic topological features. Some of them indeed exhibit unique properties and functions that are definitely originated from their unique topological features. We believe that, by understanding the structure–property relationship underpinning these novel phenomena, we would be able to demonstrate the importance of topology on the physical properties and hence functions of supramolecular polymers that have been overlooked so far. From materials viewpoint, topology-dependent physical properties of one-dimensional materials in the mesoscopic region are unknown and have the potential to explore a new direction in supramolecular polymer as well as polymer research.

REFERENCES

1. Rawlings, C. J.; Taylor, W. R.; Nyakairu, J.; Fox, J.; Sternberg, M. J. E. Reasoning About Protein Topology Using the Logic Programming Language PROLOG. *J. Mol. Graph.* **1985,** *3,* 151–157.
2. Tezuka, Y.; Oike, H. Topological Polymer Chemistry. *Prog. Polym. Sci.* **2002,** *27,* 1069–1122.
3. Takata, T.; Aoki, D. Topology-transformable Polymers: Linear–branched Polymer Structural Transformation via the Mechanical Linking of Polymer Chains. *Polym. J.* **2018,** *50,* 127–147.
4. Sijbesma, R. P.; Beijer, F. H.; Brunsveld, L.; Folmer, B. J. B.; Ky Hirschberg, J. H. K.; Lange, R. F. M.; Lowe, J. K. L.; Meijer, E. W. Reversible Polymers Formed from Self-complementary Monomers Using Quadruple Hydrogen Bonding. *Science* **1997,** *278,* 1601–1604.
5. Brunsveld, L.; Folmer, B. J. B.; Meijer, E. W.; Sijbesma, R. P. Supramolecular Polymers. *Chem. Rev.* **2001,** *101,* 4071–4097.
6. Hirschberg, J. H. K. K.; Brunsveid, L.; Ramzi, A.; Vekemans, J. A. J. M.; Sijbesma, R. P.; Meijer, E. W. Helical Self-assembled Polymers from Cooperative Stacking of Hydrogen-bonded Pairs. *Nature* **2000,** *407,* 167–170.
7. Aida, T.; Meijer, E. W.; Stupp, S. I. Functional Supramolecular Polymers. *Science* **2012,** *335,* 813–817.
8. de Greef, T. F. A.; Smulders, M. M. J.; Wolffs, M.; Schenning, A. P. H. J.; Sijbesma, R. P.; Meijer, E. W. Supramolecular Polymerization. *Chem. Rev.* **2009,** *109,* 5687–5754.
9. Whitesides, G. M.; Simanek, E. E.; Mathias, J. P.; Seto, C. T.; Chin, D.; Mammen, M.; Gordon, D. M. Noncovalent Synthesis: Using Physical-organic Chemistry to Make Aggregates. *Acc. Chem. Res.* **1995,** *28,* 37–44.
10. Adhikari, B.; Lin, X.; Yamauchi, M.; Ouchi, H.; Aratsu, K.; Yagai, S. Hydrogen-bonded Rosettes Comprising Π-conjugated Systems as Building Blocks for Functional One-dimensional Assemblies. *Chem. Commun.* **2017,** *53,* 9663–9683.
11. Ten Cate, A. T.; Sijbesma, R. P. Coils, Rods and Rings in Hydrogen-bonded Supramolecular Polymers. *Macromol. Rapid Commun.* **2002,** *23,* 1094–1112.
12. Ercolani, G. Physical Basis of Self-assembly Macrocyclizations. *J. Phys. Chem. B* **1998,** *102,* 5699–5703.
13. Kim, J. K.; Lee, E.; Huang, Z.; Lee, M. Nanorings from the Self-assembly of Amphiphilic Molecular Dumbbells. *J. Am. Chem. Soc.* **2006,** *128,* 14022–14023.
14. Yagai, S.; Mahesh, S.; Kikkawa, Y.; Unoike, K.; Karatsu, T.; Kitamura, A.; Ajayaghosh, A. Toroidal Nanoobjects from Rosette Assemblies of Melamine-linked Oligo(p-phenyleneethynylene)s and Cyanurates. *Angew. Chem. Int. Ed.* **2008,** *47,* 4691–4694.
15. Yagai, S.; Kubota, S.; Saito, H.; Unoike, K.; Karatsu, T.; Kitamura, A.; Ajayaghosh, A.; Kanesato, M.; Kikkawa, Y. Reversible Transformation between Rings and Coils in a Dynamic Hydrogen-bonded Self-assembly. *J. Am. Chem. Soc.* **2009,** *131,* 5408–5410.
16. Yagai, S.; Goto, Y.; Karatsu, T.; Kitamura, A.; Kikkawa, Y. Catenation of Self-assembled Nanorings. *Chem. Eur J.* **2011,** *17,* 13657–13687.
17. Yagai, S.; Goto, Y.; Lin, X.; Karatsu, T.; Kitamura, A.; Kuzuhara, D.; Yamada, H.; Kikkawa, Y.; Saeki, A.; Seki, S. Self-Organization of Hydrogen-Bonding Naphthalene Chromophores into J-type Nanorings and H-type Nanorods: Impact of Regioisomerism. *Angew. Chem. Int. Ed.* **2012,** *51,* 6643–6647.
18. Yamauchi, M.; Adhikari, B.; Prabhu, D. D.; Lin, X.; Karatsu, T.; Ohba, T.; Shimizu, N.; Takagi, H.; Haruki, R.; Adachi, S. I.; Kajitani, T.; Fukushima, T.; Yagai, S. Supramolecular Polymerization of Supermacrocycles: Effect of Molecular Conformations on Kinetics and Morphology. *Chem. Eur J.* **2017,** *23,* 5270–5280.

19. Hollamby, M. J.; Aratsu, K.; Pauw, B. R.; Rogers, S. E.; Smith, A. J.; Yamauchi, M.; Lin, X.; Yagai, S. Simultaneous SAXS and SANS Analysis for the Detection of Toroidal Supramolecular Polymers Composed of Noncovalent Supermacrocycles in Solution. *Angew. Chem. Int. Ed.* **2016,** *55,* 9890–9893.
20. Yagai, S.; Suzuki, M.; Lin, X.; Gushiken, M.; Noguchi, T.; Karatsu, T.; Kitamura, A.; Saeki, A.; Seki, S.; Kikkawa, Y.; Tani, Y.; Nakayama, K. I. Supramolecular Engineering of Oligothiophene Nanorods without Insulators: Hierarchical Association of Rosettes and Photovoltaic Properties. *Chem. Eur J.* **2014,** *20,* 16128–16137.
21. Ouchi, H.; Kizaki, T.; Yamato, M.; Lin, X.; Hoshi, N.; Silly, F.; Kajitani, T.; Fukushima, T.; Nakayama, K.; Yagai, S. Impact of Helical Organization on the Photovoltaic Properties of Oligothiophene Supramolecular Polymers. *Chem. Sci.* **2018,** *9,* 3638–3643.
22. Lin, X.; Suzuki, M.; Gushiken, M.; Yamauchi, M.; Karatsu, T.; Kizaki, T.; Tani, Y.; Nakayama, K. I.; Suzuki, M.; Yamada, H.; Kajitani, T.; Fukushima, T.; Kikkawa, Y.; Yagai, S. High-fidelity Self-assembly Pathways for Hydrogen-bonding Molecular Semiconductors. *Sci. Rep.* **2017,** *7,* 43098.
23. Hill, D. J.; Mio, M. J.; Prince, R. B.; Hughes, T. S.; Moore, J. S. A Field Guide to Foldamers. *Chem. Rev.* **2001,** *101,* 3893–4011.
24. Yagai, S.; Nakajima, T.; Karatsu, T.; Saitow, K. I.; Kitamura, A. Photo-triggered Self-assembly of Hydrogen-bonded Rosette. *J. Am. Chem. Soc.* **2004,** *126,* 11500–11508.
25. Yagai, S.; Nakajima, T.; Kishikawa, K.; Kohmoto, S.; Karatsu, T.; Kitamura, A. Hierarchical Organization of Photoresponsive Hydrogen-bonded Rosette. *J. Am. Chem. Soc.* **2005,** *127,* 11134–11139.
26. Yagai, S.; Karatsu, T.; Kitamura, A. Photocontrollable Self-assembly. *Chem. Eur J.* **2005,** *11,* 4054–4063.
27. Yagai, S.; Iwashima, T.; Kishikawa, K.; Nakahara, S.; Karatsu, T.; Kitamura, A. Photoresponsive Self-assembly and Self-organization of Hydrogen-bonded Supramolecular Tapes. *Chem. Eur J.* **2006,** *12,* 3984–3994.
28. Yagai, S.; Yamauchi, M.; Kobayashi, A.; Karatsu, T.; Kitamura, A.; Ohba, T.; Kikkawa, Y. Control over Hierarchy Levels in the Self-assembly of Stackable Nanotoroids. *J. Am. Chem. Soc.* **2012,** *134,* 18205–18208.
29. Yamauchi, M.; Kanao, N.; Adhikari, B.; Karatsu, T.; Yagai, S. Phototriggered Supramolecular Polymerization of Barbituric Acid Rosette. *Chem. Lett.* **2017,** *46,* 111–114.
30. Yamauchi, M.; Ohba, T.; Karatsu, T.; Yagai, S. Photoreactive Helical Nanoaggregates Exhibiting Morphology Transition on Thermal Reconstruction. *Nat. Commun.* **2015,** *6,* 8936.
31. Yagai, S.; Ohta, K.; Gushiken, M.; Iwai, K.; Asano, A.; Seki, S.; Kikkawa, Y.; Morimoto, M.; Kitamura, A.; Karatsu, T. Photoreversible Supramolecular Polymerization and Hierarchical Organization of Hydrogen-bonded Supramolecular Copolymers Composed of Diarylethenes and Oligothiophenes. *Chem. Eur J.* **2012,** *18,* 2244–2253.
32. Yagai, S.; Iwai, K.; Karatsu, T.; Kitamura, A. Photoswitchable Exciton Coupling in Merocyanine-diarylethene Multichromophore Hydrogen-bonded Complexes. *Angew. Chem. Int. Ed.* **2012,** *51,* 9679–9683.
33. Yagai, S.; Iwai, K.; Yamauchi, M.; Karatsu, T.; Kitamura, A.; Uemura, S.; Morimoto, M.; Wang, H.; Würthner, F. Photocontrol over Self-assembled Nanostructures of $\Pi-\pi$ Stacked Dyes Supported by the Parallel Conformer of Diarylethene. *Angew. Chem. Int. Ed.* **2014,** *53,* 2602–2606.
34. Adhikari, B.; Yamada, Y.; Yamauchi, M.; Wakita, K.; Lin, X.; Aratsu, K.; Ohba, T.; Karatsu, T.; Hollamby, M. J.; Shimizu, N.; Takagi, H.; Haruki, R.; Adachi, S.; Yagai, S. Light-induced Unfolding and Refolding of Supramolecular Polymer Nanofibres. *Nat. Commun.* **2017,** *8,* 15254.

35. Korevaar, P. A.; George, S. J.; Markvoort, A. J.; Smulders, M. M. J.; Hilbers, P. A. J.; Schenning, A. P. H. J.; de Greef, T. A. F.; Meijer, E. W. Pathway Complexity in Supramolecular Polymerization. *Nature* **2012,** *481,* 492–496.
36. Ogi, S.; Stepanenko, V.; Sugiyasu, K.; Takeuchi, M.; Würthner, F. Mechanism of Self-assembly Process and Seeded Supramolecular Polymerization of Perylene Bisimide Organogelator. *J. Am. Chem. Soc.* **2015,** *137,* 3300–3307.
37. Sorrenti, A.; Leira-Iglesias, J.; Markvoort, A. J.; de Greef, T. F. A.; Hermans, T. M. Non-equilibrium Supramolecular Polymerization. *Chem. Soc. Rev.* **2017,** *46,* 5476–5490.

CHAPTER 12

Kinetically Controlled Supramolecular Structures Created in Nonequilibrium Microflow Spaces

Munenori Numata
Department of Biomolecular Chemistry, Graduate School of Life and Environmental Sciences, Kyoto Prefectural University, Kyoto, Japan

Contents

1. INTRODUCTION

The kinetic self-assemblies of molecules and polymers are ubiquitous in both natural and artificial systems in comparison with their thermodynamic counterparts *(1a)*. The self-assembled structures created via kinetic processes are incredibly diverse in comparison with other thermodynamic structures, thereby providing a variety of structures and functionalities. Kinetic self-assembled structures can be classified according to their stabilities. Therefore, it is important to consider how these kinetic self-assembly processes are generated. Self-assembled structures found in nature feature a large number of components and are kinetically stabilized supramolecular structures. Such self assembly processes in nature are accomplished with the use of energy supplied from external systems such as chemical reactions. This results in the

Kinetic Control in Synthesis and Self-Assembly
ISBN 978-0-12-812126-9
https://doi.org/10.1016/B978-0-12-812126-9.00012-2

creation of kinetically stabilized supramolecular structures characterized by their molecular complexity and scalar hierarchy. In this sense, the kinetic processes of molecular self-assembly systems in nature are coupled with nonequilibrium phenomena *(1b)*. In conventional supramolecular systems, molecular self-assemblies occur under thermodynamic conditions. This process can result in discrete supramolecular structures composed of at most several tens of molecular units. This is in contrast to the fact that in nature, molecular self-assemblies containing more than 1000 components may arise and thus produce a variety of supramolecules. Despite the great effort being made by chemists, the precise control of kinetic self-assembly processes in a reproducible manner remains a challenge *(1c, 1d)* because of a lack of versatile strategies. The construction of a self-assembly system coupled with energy consumption can provide a potential solution for overcoming this difficulty; therefore this has been sought after in supramolecular chemistry.

Photoirradiation, thermal diffusion, pH change, mechanical stimuli, solvent diffusion, or chemical reactions can potentially provide a driving force for self-assembly under kinetic conditions. One of the main concerns for this process is the control of intermolecular interactions with the use of these stimuli. Photoirradiation and chemical reactions convert molecular structures via energy consumption. On the other hand, the precise regulation of chemical environments surrounding the molecules can also affect the self-assembling abilities of these molecules. Regulating self-assembly processes with environmental changes, which can be accomplished by utilizing energy, may provide an alternative way of forcing the self-assembly of molecules. For example, the use of solvent- or thermodiffusion can provide versatile ways for producing a nonequilibrium state at molecular levels and can potentially accelerate the self-assembly of solute molecules. In fact, mixing or heating solutions is a conventional method for initiating kinetic self-assembly *(1a)*. Here, I would like to highlight the importance of simultaneous environmental changes during these processes; that is, when thousands of component molecules in a restricted space are subjected to environmental changes at the same time, which can only be realized in a state of nonequilibrium, they can acquire unique self-assembly capabilities.

2. MICROFLOW: A NOVEL PLATFORM FOR MOLECULAR SELF-ASSEMBLY WITH ENERGY CONSUMPTION

The techniques of microfluidics have been applied in several disciplines, ranging from basic sciences such as physics and biology, to engineering disciplines. In chemistry, microfluidic techniques have seen a phenomenal growth in several areas, especially in the high-throughput analysis of bioactive molecules, including kinetic studies of these molecules *(2)* and their organic synthetic chemistry *(3)*. In material sciences, microflow systems have attracted great attention because of their potential use in the fabrication of materials with various shapes *(4,5)*. Despite the wide range of applications, the use of microfluidics in controlled self-assembly processes has been limited. Recently, the potential of microflow systems for the creation of supramolecular structures was reviewed by several research groups *(1d,5e)*. This research area has attracted much attention; a novel approach making use of these systems will represent a new frontier in nanomaterial sciences.

Microflow systems allow chemists to control mass transfer precisely at both nano- and micrometer scales. The injected molecules travel along the channel at a controlled rate. The molecular dynamics of the self-assembly are related to the travel distance of the molecules along the channel (Fig. 12.1A). Thereby, it is expected that the self-assembly process can be controlled by predicting the environmental changes along the channel in relation to time. Furthermore, the stream lines of the injected solutions in microflow never cross one another, even in the case of miscible solvents, such that laminar flow is maintained (Fig. 12.1B). Solution mixing in a microflow channel, therefore, proceeds only through molecular diffusion across the stream lines. From the viewpoint of nonequilibrium phenomena, laminar flow can be regarded as a typical self-organized liquid structure at a micrometer scale and can be maintained via energy expenditure. Although

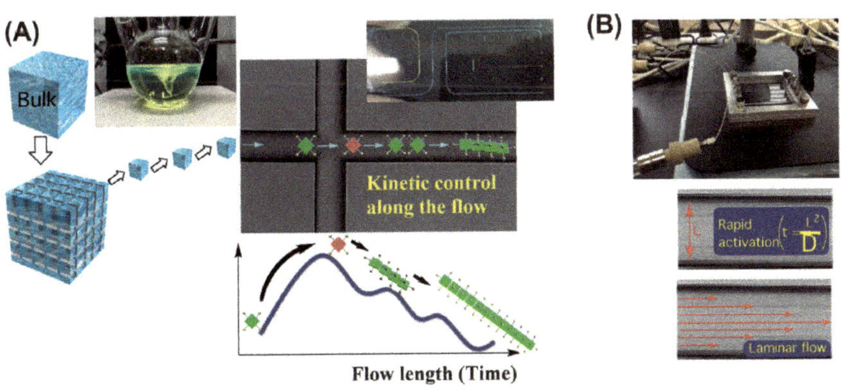

Figure 12.1 (A) Kinetic control in a flowing solution; microscopic image of the microflow channel and (B) photographic image of the microflow system and hydrodynamic nature of a flowing solution in a microflow channel.

the self-assembly behavior of individual molecules does not directly affect the laminar flow itself, which is in a nonequilibrium state at a micrometer scale, the homogenous solvent diffusion across the stream lines at micrometer range result in nonequilibrium environments at a molecular level. Thus, laminar flow provides dynamic homogeneous environments for self-assembling molecules, enabling simultaneous environmental changes for thousands of molecules. Hence, the application of microflow provides the following advantages for molecular self-assembly processes:

- The bulk solution is divided into picoliter-volume solutions. This dividing process lessens the number of molecules to be controlled compared with conventional vial-mixing processes, enabling the precise regulation of interactions (Fig. 12.1A).
- The chemical environments surrounding the assembling molecules, such as the solvent polarity, pH, and temperature, are controllable via rapid solvent diffusion in the picoliter-volume solutions. This enables precise kinetic control of the molecules via the rapid and uniform diffusion of the solvents, which results in a reliable and reproducible kinetic self-assembly process.
- The self-assembly events start simultaneously and end at the same spatial point. Therefore, in principle, the self-assembly dynamics can be controlled along a channel. For example, in the case of supramolecular polymerization, the self-assembly number (molecular weight) can be regulated by the flow distance (Fig. 12.1A).

3. KINETICALLY TRAPPED SUPRAMOLECULAR ARCHITECTURES CREATED IN A MICROFLOW SPACE

3.1 Kinetic Self-Assembly Coupled With Proton Diffusion

The diffusion of molecules or ions and thermal diffusion are typical nonequilibrium environments that can be present at a nanometer scale. A steep gradient of the concentrations of the components, pH, or temperature across an interface that creates a high energy state should be used to generate dynamic environments. If such dynamic environmental changes, such as the nonequilibrium state surrounding the assembling molecules, can be coupled with molecular self-assembly, then this can provide a pathway for creating novel supramolecular structures that cannot be fabricated through conventional strategies under other thermodynamic states. Considering the early era of the Earth, alkaline hydrothermal vents, which provided proton gradients, allowed for nonequilibrium conditions and were a driving force in the generation of life *(6)*. The topology of such alkaline vents are similar to that of cells, in which a pH gradient forms across the phospholipid membranes. Laminar flow generated in a microflow channel can potentially provide chemical environments similar to those observed in nature. In artificial systems, only microflow may provide a chance to generate such a steep gradient of concentrations over a molecular scale in a reproducible manner.

The diffusion constant of a proton is on the order of 10^{-9} m^2/s, meaning that the diffusion of a proton is complete within microseconds in a standard microflow channel. Taking the diffusion constant into consideration, protons are most suitable for diffusion-driven self-assembly in microflow. A steep gradient of the proton concentration across the liquid—liquid interface in laminar flow, akin to that across the membranes of a cell, can be a driving force for self-assembly in an energetically uphill fashion.

To demonstrate how intermolecular interactions can be affected by this steep gradient of protons, (i.e., the rapid diffusion of protons), we first used tetrakis(4-sulfonatophenyl)porphyrin (TPPS, with H_2TPPS and H_4TPPS representing its free and protonated forms, respectively) as an ideal model. H_2TPPS and H_4TPPS were under equilibrium in a vial, where pH values (via proton concentration) governed the equilibrium and the H_4TPPS concentration directly affected the subsequent self-assembly. Thus, the equilibrium shifted toward H_4TPPS on reduction of the pH and consequently accelerated the J-aggregate formation (Figs. 12.2 and 12.3).

Fig. 12.2A shows an illustration of the microflow channel. An aqueous H_2TPPS solution was injected into the central channel, aqueous HCl was injected into the side channels of this cross-type microflow, and the final pH was adjusted to a value in the range of 3—5. Generally, the diffusion time is proportional to the square of the diffusion distance. Fluorescence microscope imaging of the flowing TPPS solution revealed that the width of the central TPPS channel became narrower on increasing of the flow rates. Protons also diffused into the central TPPS solution more rapidly on increase of the flow rates, implying that the entirety of the injected TPPS solution can be protonated at the same time under higher flow rate conditions. UV—Vis spectroscopic images of the eluted solution collected at different flow rates clearly show that the conversion of H_2TPPS to the J-aggregate was dramatically enhanced by a decrease in the width of the layered solution. In other words, the formation of J-aggregates was dramatically accelerated on increase of the flow rates (i.e., the central

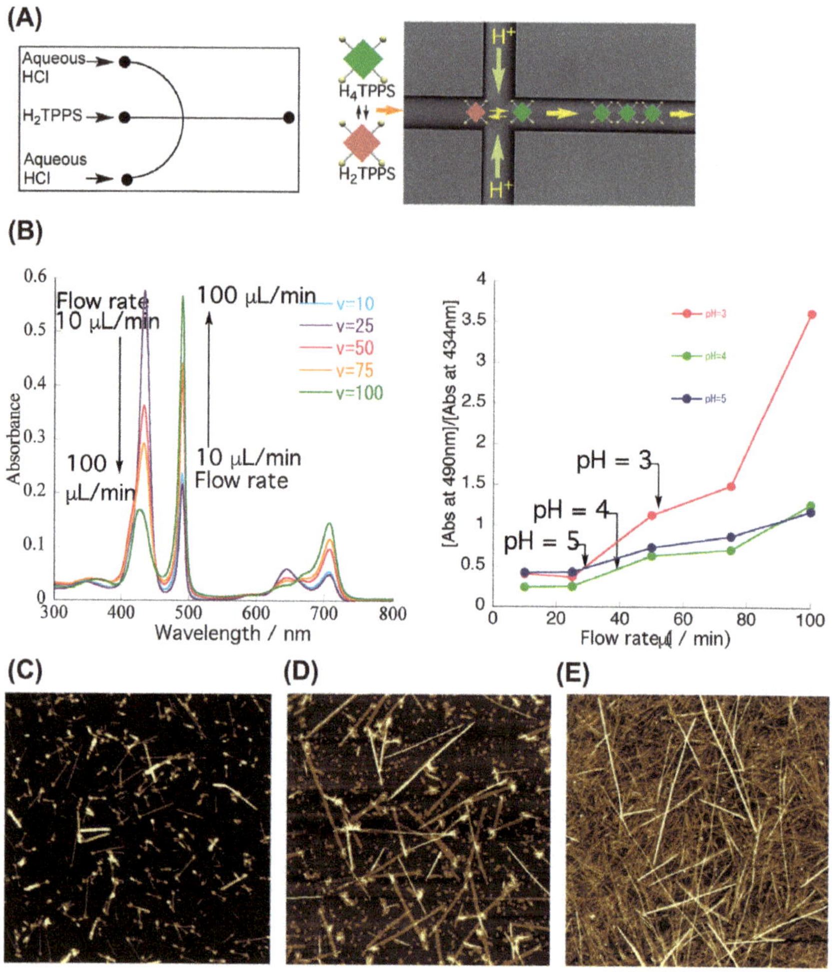

Figure 12.2 (A) Schematic representation of the microflow channel used for the synchronized activation of H_2TPPS. (B) UV–Vis spectra recorded at various flow rates and a plot of the A_{490}/A_{434} ratio with respect to flow rate (0.267 mM; 0.1 mm cell; r.t.). (C)–(E) AFM images of the J-aggregate prepared at pH 3.0; (C) in a vial and (D and E) in the microflow channel at flow rates of (D) 50 and (E) 100 μL/min (10 × 10 μm).

solution was squeezed; Fig. 12.2B). This result clearly indicates that not only do pH values govern the self-assembly of H_4TPPS, but the time required for protonation also plays a role *(7)*.

Along the same line, the average length of the J aggregates increased with the increase in flow rates, as revealed by atomic force microscopy (AFM) images (Fig. 12.2C–E). It could also be recognized from the AFM images that the microflow field produced supramolecular polymers of

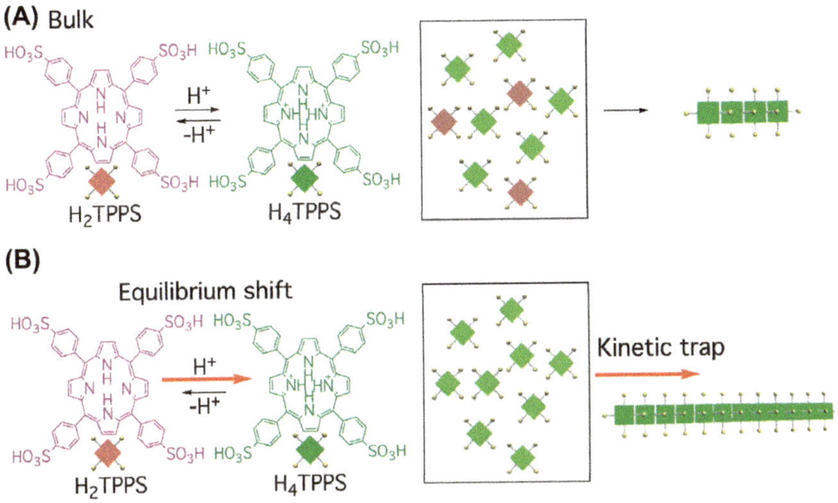

Figure 12.3 Chemical structure of TPPS and (A) its path toward self-assembly under thermodynamic control. (B) Temporal equilibrium shift toward H_4TPPS and the enhancement of intermolecular interactions.

almost uniform length. The J-aggregate fibers obtained were stable and did not undergo further structural changes. These results imply that the degree of supramolecular polymerization could be controlled by changing the time required for proton diffusion, i.e., by merely adjusting the flow rates. Uniform proton diffusion allowed all of the TPPS to polymerize in the same chemical environment. This allows for the possibility of long-range intermolecular interactions over a molecular scale.

We concluded from our experiments that the rapid protonation of H_2TPPS in microflow allowed for the entirety of the H_2TPPS to be protonated at the same time, thereby shifting the equilibrium and increasing the H_4TPPS concentration. Such a rapid proton diffusion resulted in a shift of the equilibrium and an increase in the H_4TPPS concentration. The temporal equilibrium shift generated an activated state of the molecules that facilitated self-assembly (Fig. 12.3). Kinetic self-assembly resulting from such a temporally activated state resulted in J-aggregate formation. Generally, the rapid and homogeneous diffusion of molecules or ions can change chemical environments and generate the activated states of a large number of self-assembling molecules across a wide range. This situation resembles that of an alkaline hydrothermal vent, which previously generated primordial supramolecules. Self-assembly in microflow channels would compensate for such a rapid solvent diffusion that may generate a large amount entropy in a restricted time (dS/dt) and space. Therefore, in principle, energetically uphill processes would be permitted under these conditions.

Using a similar strategy for controlled kinetic self-assembly, Sorrenti and coworkers have recently focused on the supramolecular chirality induced in microflow. They have also used the self-assembly of H_2TPPS with the assistance of a cationic surfactant (cetyltrimethylammonium bromide) as a model for effective chiral induction. In a conventional flask-mixing process, the self-assembly of H_2TPPS in the presence of a chiral inducer afforded chiral supramolecular structures in a sergeant-and-soldier manner. This effect of a chiral inducer was enhanced in a microflow channel even at a low concentration of the inducer *(8c)*.

Helmich and Meijer have demonstrated that a microflow system can control the thermodynamic state of a solution containing supramolecular polymers. To show that the thermodynamic equilibrium state of the supramolecular structures alters with solvent diffusion, they used an equilibrium mixture composed of a stacked porphyrin aggregate and its dimer as a model. On elution of this solution under equilibrium into pure methylcyclohexane by using an H-cell, equilibrium between the aggregate and dimers was achieved. After the separation of the two streams, each eluted solution reached an equilibrium state *(9)*.

The driving force for the self-assembly of H_4TPPS is hydrogen bonding in addition to the electrostatic and $\pi-\pi$ stacking interactions. In this case, the thinner TPPS layer helped us to precisely regulate the time required for all of the TPPS to be protonated, such that this occurred at the same time. Generally, hydrogen-bonding interactions are not created via an applied driving force for the creation of micrometer-sized supramolecular structures because of their fragility. Synchronized protonation under kinetic control might overcome this general difficulty. On the basis of this concept, we used 5,10,15,20-tetrakis(4-carboxyphenyl) porphyrin (TCPP) as a second model compound because of its potential ability to form two-dimensional hydrogen-bonding networks under kinetic conditions (Fig. 12.4A) *(10)*. According to the TPPS system, an alkaline solution of ZnTCPP was injected into the central flow channel and squeezed with aqueous HCl solutions to vary the pH (1–7). Although amorphous aggregates formed even below a pH of 4 on mixing of the ZnTCPP solution with aqueous HCl, scanning electron microscopy (SEM) images revealed that micrometer-sized sheet structures formed under a microflow pH of 4 (Fig. 12.4B). A fourfold rotationally symmetric hydrogen-bonded network could be reasonably proposed to have formed on the basis of the XRD

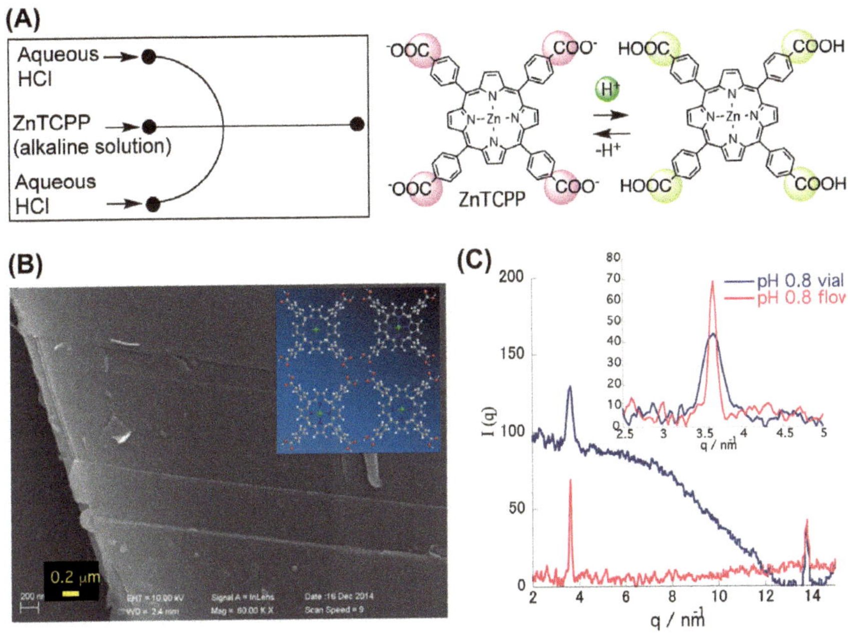

Figure 12.4 (A) Chemical structure of ZnTCPP and a schematic representation of the microflow channel used for the hydrogen-bonded network formation under rapid protonation. (B) SEM image of the microsheet structures created in a microflow channel (25 μM, pH 2.0) (inset: DFT calculations of the hydrogen-bonded networks with fourfold rotational symmetry). (C) Synchrotron XRD patterns of the microsheet (red line) and vial sample (blue line). The periodicity observed at 1.7 nm ($q = 3.64$) is assignable to the distance between the ZnTCPP molecules. The peak sharpness relative to that of the aggregate prepared in the vial is indicative of the formation of regular hydrogen-bonded networks (inset).

analysis of the sheets (Fig. 12.4C). In this network, the ZnTCPPs molecules were connected to each other in a regular manner on a scale of up to several tens of micrometers. The IR spectra also supported the presence of hydrogen-bonded networks in the sheets. Note that such an extended hydrogen-bonding network cannot be created under conventional proton diffusion conditions in a vial. The simultaneous protonation of all carboxylic acid (-COOH) groups in ZnTCPP enabled us to achieve such a large hydrogen-bonded supramolecular structure under kinetic control.

These results suggest that simultaneous protonation via very rapid proton diffusion allows numerous self-assembling molecules (i.e., TPPS and ZnTCPP) to behave as a kind of cluster during self-assembly. In principle, thousands of molecules can be activated (or protonated, as in the present case) toward self-assembly, and interactions among these molecules are formed at the same time. From these findings, we have proposed the possibility that molecules can be synchronized under very rapid environmental changes, such as a change in pH or solvent polarity. The synchronous behavior of the molecules can be defined as the state in which a large number of molecules are in the same chemical environment in a restricted space and may temporally behave as a type of cluster. The anomalous amplification of weak interactions through the synchronized behavior of the components of these clusters is ubiquitous in nature at any scale, ranging from elementary particles to a global scale. Such behavior of molecules has not seen much attention in chemistry. By using simple self-assembly models, we demonstrated the synchronous phenomena of the molecules, which may lead to the production of a novel strategy for creating kinetically controlled self-assembly processes.

3.2 Kinetic Self-Assembly in Laminar Flow

In the previous section, it was shown that intermolecular interactions could be promoted using rapid and uniform proton diffusion in microflow channels. Next, we applied this concept to a more general self-assembly system induced by a solution-mixing process. Specifically, the supramolecular polymerization of perylene bisimide (PBI) dyes is highlighted in this section as a model, where rapid and uniform solvent diffusion plays a crucial role in the self-assembly of PBIs.

Amide-appended PBI derivatives with various alkyl spacers ($C_n = 3$) were designed according to the literature *(11)*. These PBIs have the potential to form intramolecular hydrogen bonds (closed form). Remarkably, in the case of PBI ($C_n = 3$), the closed form is less active toward self-assembly because of the steric hindrance of the peripheral dodecyl groups, whereas the nonintramolecular–hydrogen-bonded (open) form is active in self-assembly (Fig. 12.5A). DFT calculations for the intramolecular hydrogen-bonded forms of PBI ($C_n = 3$) revealed that the looped alkyl spacer suppressed the $\pi-\pi$ stacking interactions. These two conformers of PBI ($C_n = 3$) are at equilibrium in solution.

We used a microflow channel with a crossing point. Dichloromethane (DCM) solutions of PBI (at concentrations of 10, 20, 30, 40, 50, and 60 μM) were injected into the central channel at a flow rate of 8 μL/min, and hexane was injected in the two lateral flow channels (at 6 μL/min each) to produce laminar flow (the concentration of the eluted solutions became 4, 8, 12, 16, 20, and 24 μM, respectively). With this experimental setup, hexane molecules are able to diffuse uniformly into the DCM solution (Fig. 12.5B).

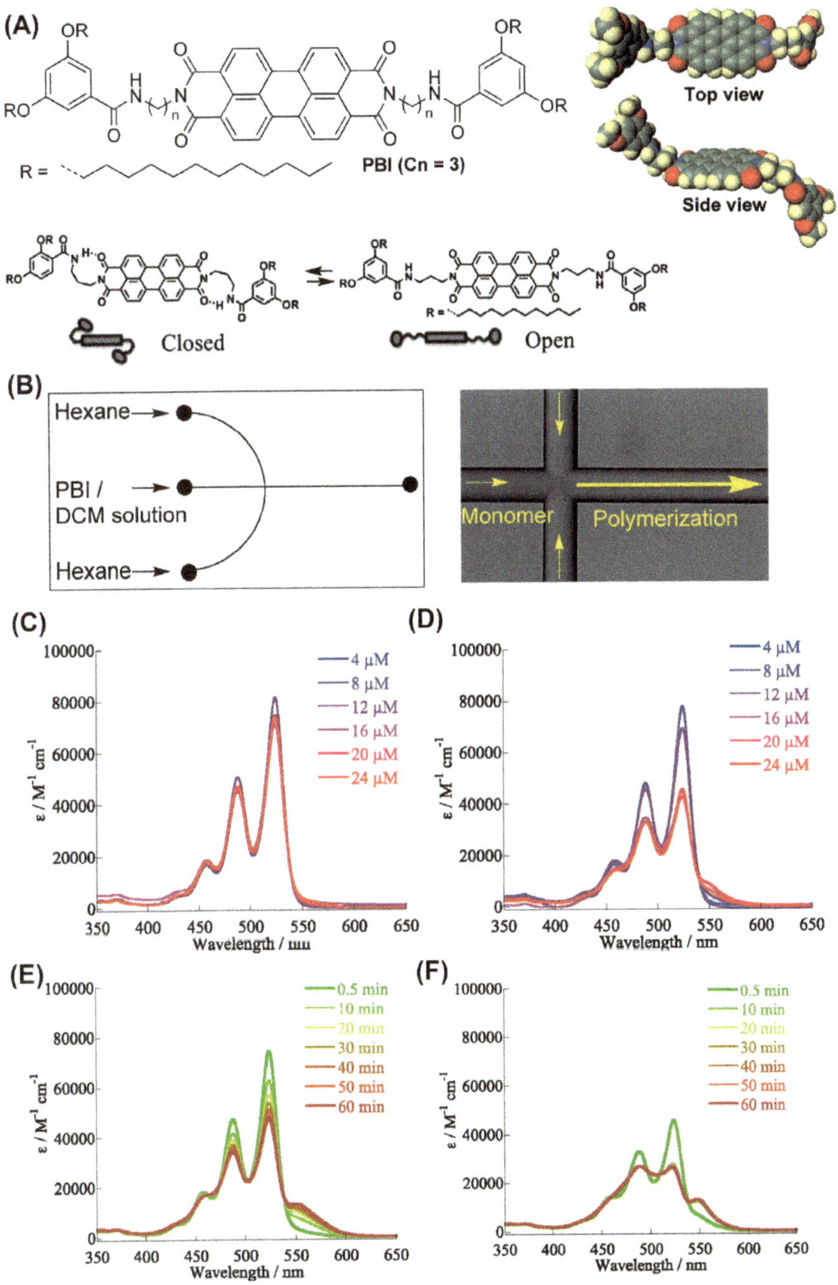

Figure 12.5 (A) The chemical structure of the PBI derivatives and the calculated models of PBI ($C_n = 3$). All dodecyl groups have been replaced by methyl groups for simplification. (B) Schematic representation of the microflow channel. Concentration dependent UV–Vis spectral changes of PBI ($C_n = 3$) prepared (C) in a vial and (D) after elusion from the microflow channel (r.t.). Time-dependent UV–Vis spectral changes of PBI ($C_n = 3$) prepared (E) in a vial and (F) after elusion from the microflow channel (r.t.).

The concentration-dependent UV–Vis spectra of the eluted PBI ($C_n = 3$) solutions revealed that the mole fraction of the aggregated dye (α_{agg}) tended to increase at higher concentrations. The critical lower concentration was estimated to be around 12 μM. α_{agg} reached its maximum value (c. 60%) in the range of 16–24 μM (Figs. 12.5C,D and 12.6A). On the other hand, there was no clear concentration dependence for the mixed vial sample throughout this concentration range, where the α_{agg} was at most 12%. The absorption spectra of PBI ($C_n = 3$) gradually changed with time after the DCM solutions were mixed with hexane. Time-dependent absorption changes accompanied by significant hypochromic effects were also observed for the flow samples (Fig. 12.5E and F). The plot displaying the changes of α_{agg} with respect to time clearly indicates that self-assemblies occurred even under lower concentrations in the microflow sample in comparison with the corresponding vial mixing sample. The final α_{agg} for PBI ($C_n = 3$) at 8 μM

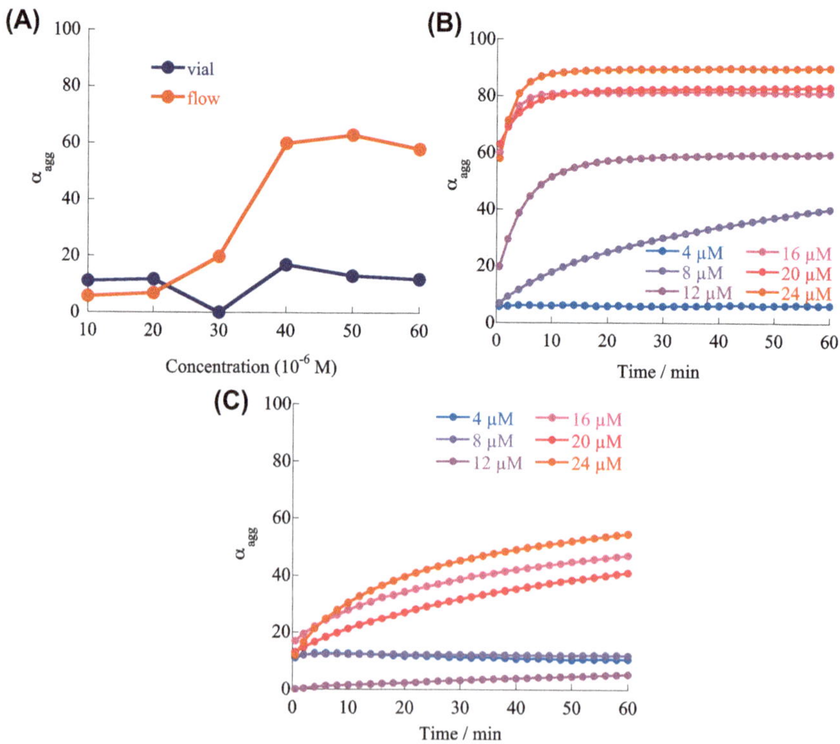

Figure 12.6 (A) Plot of the mole fraction of the aggregated dye (α_{agg}) against the concentration of the injected solutions (blue line: vial sample, red line: flow sample). Plots of the time-dependent changes of α_{agg} at various concentrations: (B) microflow samples and (C) vial sample.

reached 40% for the flow sample (Fig. 12.6B). This is in sharp contrast to the conventional vial-mixing process, where no self-assembly occurred at the same concentration (8 μM) (Fig. 12.6C). The final α_{agg} value at 24 μM increased up to 90% in the microflow field, whereas α_{agg} was ca. 55% for the corresponding vial samples. In addition to the final conversion from monomer to polymer, the self-assembly rates also accelerated in the micro-flow channel; that is, the aggregation rate was approximately 20 times higher in the microflow system. These results were likely responsible for temporally raising the local concentration of activated PBI ($C_n = 3$) and reducing the activation energy of nucleation in the presence of rapid and uniform solvent diffusion *(12)*. A spontaneous change in the polarity of the surrounding PBI molecules would promote the activated molecules toward self-assembly, in which a large number of PBIs were solvated uniformly.

It is well-known that the concentration of the reactant affects the reaction rates. In the self-assembly process, the reaction rates are also governed by the concentration of the self-assembling molecules in a solution. Under kinetic control, which is generally triggered by solution mixing, thermal diffusion, photoirradiation, etc., the effective concentration required for self-assembly is varied by the uniformity of the solution at a nanometer scale. We have proposed that overall, simultaneous microenvironmental changes that inherently occurred in the microflow channel can raise the effective concentration of the self-assembling species. The effective concentration of the self-assembling species, which are fabricated from molecules under the same chemical environment conditions, is directly correlated with the self-assembly rate rather than the concentration of the solution.

4. CREATION OF METASTABLE SUPRAMOLECULAR STRUCTURES IN A MICROFLOW SPACE

To gain insight into the self-assembled structures of PBI ($C_n = 3$), eluted DCM/hexane-mixed solutions (24 μM) were cast onto mica substrates. In the case of a vial-mixing sample containing PBI ($C_n = 3$), short nanofibers were observed by AFM imaging. In contrast, well-extended fibrous structures were observed in the microflow sample. Time-dependent AFM observations revealed that the nanofibers created under the microflow conditions were metastable by nature; that is, the nanofibers were not very stable after elution and broke up into short nanofibers within 60 min (Fig. 12.7). The rapid diffusion of the solvent induces spontaneous

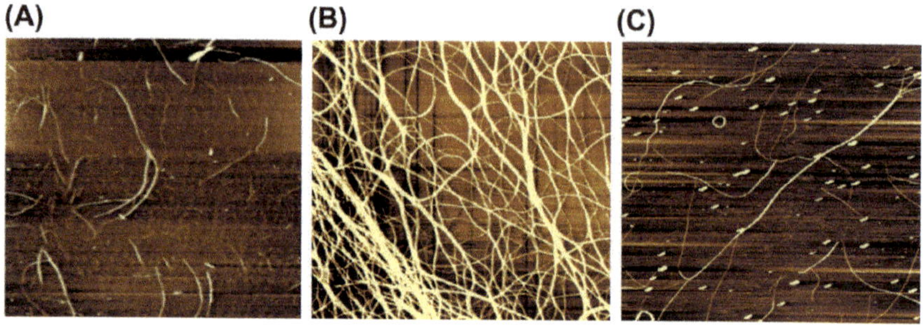

Figure 12.7 AFM images of the PBI ($C_n = 3$) nanofibers; the samples were prepared (A) in a vial (24 μM), (B) immediately after elusion from the microflow channel, and (C) 60 min after elution from the channel.

molecular assembly under kinetic control, where tight molecular packing may be suppressed. In the present case, it is likely that the somewhat loose intermolecular interactions in the nanofibers became tighter after elution from the channel. Such a rearrangement of intermolecular interactions would cause the degradation of the nanofibers *(12)*.

Next, to investigate whether kinetic self-assembly under rapid solvent diffusion conditions could occur generally, we designed an oligo(*p*-phenylenevinylene) (OPV) solution with two independent interactive moieties to undergo hydrogen bonding and π-stacking (Fig. 12.8A). A THF solution of OPV (60 μM) was injected into the central channel at a flow rate of 8 μL/min, and distilled water was injected into the two lateral flow channels (at 6 μL/min each). Time-dependent UV−Vis spectral changes of the eluting solution clearly revealed that the absorption maximum at 405 nm gradually increased with time and finally became constant after 30 min (Fig. 12.8B). The time-dependent absorption changes indicated that the π−π stacking became looser with time after elution from the channel. Time-dependent IR spectra revealed that the main intermolecular interaction switched from π-stacking to hydrogen bonding within 30 min. Time-dependent AFM images revealed the metastable nature of the formed nanofibers; that is, these nanofibers underwent drastic morphological changes from fibers to bundled structures, to fan-shaped sheets, and, finally, to dots (Fig. 12.8C). The metastable nanofibers created in the microflow channel decomposed within tens of minutes, producing thermodynamically stable dots. As also indicated by the results of spectroscopy, the change in intermolecular interactions from π-stacking to hydrogen bonding induced the decomposition of the fibers. Remarkably, this decomposition occurred at the end of each fiber at the same time, thereby leading to the formation of unique fan-shaped sheets.

The inherent metastable nature of the created fibers was due to the lack of tight π-stacking and hydrogen-bonding interactions.

From these results, we can propose that self-assembly in a microflow channel enabled us to fabricate energetically unstable supramolecular structures from molecules with nonoptimal interaction moieties, even if their interactions in a bulk solution were inactive. This situation is similar to that observed in the self-assembly of TPPS, which underwent rapid proton diffusion and generated a large amount of entropy in a restricted time and

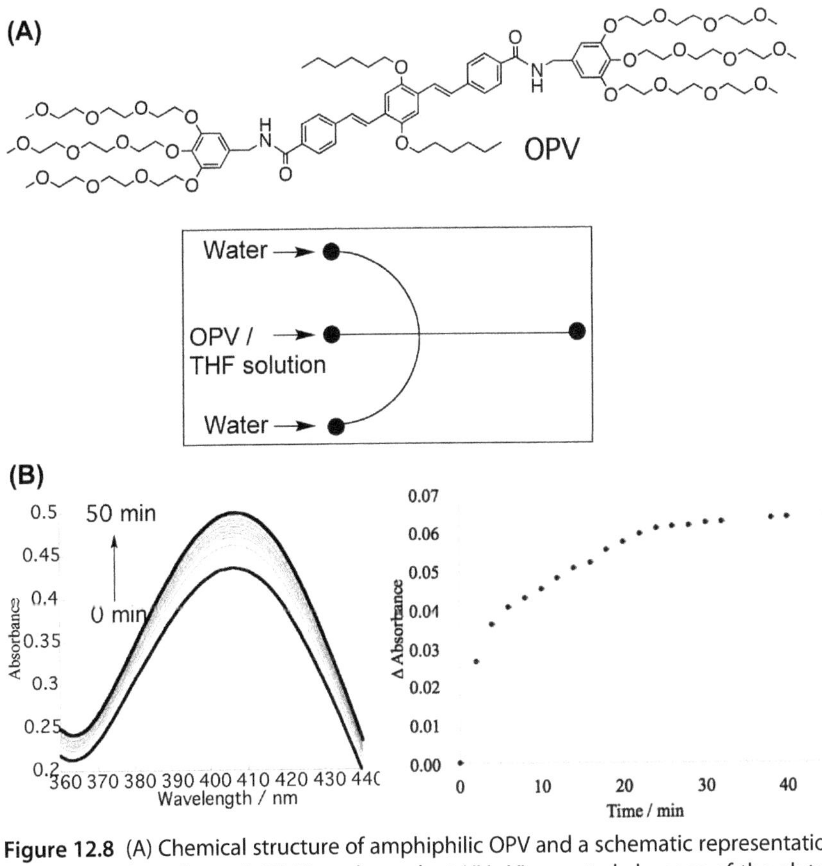

Figure 12.8 (A) Chemical structure of amphiphilic OPV and a schematic representation of the microflow channel. (B) Time-dependent UV–Vis spectral changes of the eluted solution and plot of ΔAbs. at 405 nm with respect to time. (C) Time-dependent AFM images of the eluted solution. The nanofibers created in the microflow channel were metastable; after elution, they demonstrated various kinetic structures in a cascade manner before reaching the thermodynamically stable dot structure. (D) Time-delayed action of the metastable supramolecular fibers; illustration of a series of morphological transitions over time.

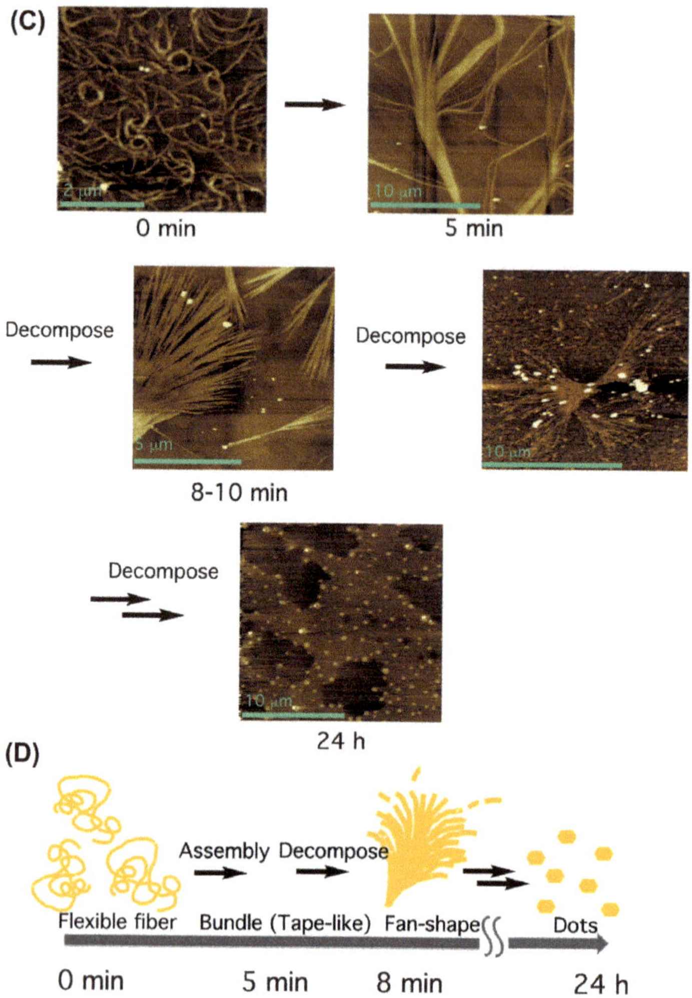

Figure 12.8 Cont'd

space, allowing for the energetically uphill process. It should be emphasized that the field of microflow self-assembly can provide new opportunities for creating various novel supramolecular structures through precisely regulated kinetic pathways that are inaccessible through bulk mixing *(13)*.

5. SUMMARY

Nonequilibrium processes in the self-assembly of molecules may be classified roughly into two groups on the basis of driving forces: one involves the activation of the initial stage (i.e., the activation of the self-assembling molecules) and the other involves the stabilization of the final stage (i.e., stabilizing self-assemblies). Both of these forces can lead to an equilibrium shift resulting in self-assembly. Self-assembly in microflow channels is always initiated through the activation of the initial stage, in which external energy is consumed to generate the activated state of the self-assembling molecules under dynamic homogeneous environments. Laminar flow, a micrometer-scale nonequilibrium phenomenon, is a driving force for molecular self-assembly through diffusion phenomena. In addition to the diffusion phenomena, thermal diffusion, pressure, and photoirradiation can be integrated into the same channel to allow for the design of a programmed self-assembly field akin to an assembly line of molecular building blocks (i.e., a nanometer-sized plant for molecular self-assembly).

Microflow channels are versatile self-assembly platforms that are independent of their molecular nature. Self-assembly based on molecular design is still practical, but it has less of an effect on the final structure. Balancing the molecular and flow-field designs will govern the final structures of the self-assemblies. As found in this study, self-assembly in a designed microflow space provides a new means for creating diverse supramolecular structures. Combining microflow channels with various shapes specialized for separation, molecular synthesis, and analysis may provide an opportunity for the automated production of desired nanomaterials from a molecular source via a single process.

ACKNOWLEDGMENTS

This research was funded by JSPS KAKENHI (grant number 15H03532) and the Asahi Glass Foundation.

REFERENCES

1. (a) Jackson, K. A. *Kinetic Processes, Crystal Growth, Diffusion, and Phase Transitions in Materials;* Wiley: Weinheim, **2010**.
(b) Nicolis, G.; Prigogine, I. *Self-Organization in Nonequilibrium Systems: From Dissipative Structures to Order through Functions;* Wiley: Hoboken, **1977**. For recent reviews on non-equilibrium self-assembly.
(c) Mattia, E.; Otto, S. Supramolecular Systems Chemistry. *Nat. Nanotechnol.* **2015,** *10,* 111−119.
(d) Sorrenti, A.; Leira-Iglesias, J.; Markvoort, A. J.; de Greef, T. F. A.; Hermans, T. M. Non-Equilibrium Supramolecular Polymerization. *Chem. Soc. Rev.* **2017,** *46,* 5476−5490.

2. (a) Bernard, A.; Michel, B.; Delamarche, E. Micromosaic Immunoassays. *Anal. Chem.* **2001,** *73,* 8–12.
(b) Lipman, E. A.; Schulter, B.; Bakajin, O.; Eaton, W. A. Single-Molecular Measurement of Protein Folding Kinetics. *Science* **2003,** *301,* 1233–1235.
(c) Hertzog, D. E.; Michalet, X.; Jäger, M.; Kong, X.; Santiago, J. G.; Weiss, S.; Bakajin, O. Femtomole Mixer for Microsecond Kinetic Studies of Protein Folding. *Anal. Chem.* **2004,** *76,* 7169–7178.
(d) Goto, M.; Sato, K.; Murakami, A.; Tokeshi, M.; Kitamori, T. Development of a Microchip-Based Bioassay System Using Cultured Cells. *Anal. Chem.* **2005,** *77,* 2125–2131.
3. (a) Kobayashi, J.; Mori, Y.; Kobayashi, S. Multiphase Organic Synthesis in Microchannel Reactors. *Chem. Asian J.* **2006,** *1–2,* 22–35.
(b) In *Microreactors in Organic Synthesis and Catalysis;* Wirth, T., Ed.; Wiley-VCH: Weinheim, **2008**.
(c) Yoshida, J. *Basics of Flow Microreactor Synthesis;* Springer: Berlin, **2015**.
4. (a) Kenis, P. J. A.; Ismagilov, R. F.; Whitesides, G. M. Microfabrication Inside Capillaries Using Multiphase Laminar Flow Patterning. *Science* **1999,** *285,* 83–85.
(b) Takayama, S.; Otsuni, E.; Qian, X.; McDonald, J. C.; Jiang, X.; LeDuc, P.; Wu, M.-H.; Ingber, D. E.; Whitesides, G.,M. Topographical Micropatterning of Poly(dimethylsiloxane) Using Laminar Flows of Liquids in Capillaryes. *Adv. Mater.* **2001,** *13,* 570–574.
(c) Puigmatí-Luis, J.; Rubo-Martínez, M.; Hartfelder, U.; Imaz, I.; Maspoch, D.; Dittrich, P. S. Coodination Polymer Nanofibers Generated by Microfluidic Synthesis. *J. Am. Chem. Soc.* **2011,** *133,* 4216–4219.
(d) Kiriya, D.; Ikeda, M.; Onoe, H.; Takinoue, M.; Komatsu, H.; Shimoyama, Y.; Hamachi, I.; Takeuchi, S. Meter-long and Robust Supramolecular Strands Encapsulated in Hydrogel Jackets. *Angew. Chem. Int. Ed.* **2012,** *51,* 1553–1557.
(e) Zang, H.-Y.; Ruiz de la Oliva, A.; Miras, H. N.; Long, D.; McBurney, R. T.; Cronin, L. Discovery of Gigantic Moleculer Nanostructures Using a Flow Reaction Array as a Search Engine. *Nat. Commun.* **2014**. https://doi.org/10.1038/ncomms4715.
(f) Rodrigez-San-Miguel, D.; Abrishamkar, A.; Novarro, J. A. R.; Rodriguez-Trujillo, R.; Amabilino, D. B.; Mas-Balleste, R.; Zamora, F.; Puigmartí-Luis, J. Crystalline Fibers of a Covalent Organic Framework through Bottom-Up Microfluidic Synthesis. *Chem. Commun.* **2016,** *52,* 9212–9215.
5. For recent reviews and a book, see:
(a) Song, Y.; Hormes, J.; Kumar, C. Microfluidic Synthesis of Nanomaterials. *Small* **2008,** *4,* 698–711.
(b) In *Microfluidic Devices in Nanotechnology, Fundamental Concepts;* Kumar, C. S., Ed.; Wiley: Hoboken, New Jersey, **2010**.
(c) Pumera, M. Nanomaterials Meet Microfluidics. *Chem. Commun.* **2011,** *47,* 5671–5680.
(d) Puigmartí-Luis, J. Microfluidic Platforms: A Mainstream Technology for the Preparation of Crystals. *Chem. Soc. Rev.* **2014,** *43,* 2253–2271.
(e) Sevim, S.; Sorrenti, A.; Franco, C.; Furukawa, S.; Pane, S.; deMello, A. J.; Puigmartí-Luis, J. Self-Assembled Materials and Supramolecular Chemistry within Microfluidic Environments: From Common Thermodynamic States to Non-equilibrium Structures. *Chem. Soc. Rev.* **2018,** *47,* 3788–3803.
6. Möller, F. M.; Krigel, F.; Kieβ, M.; Sojo, V.; Braun, D. Steep PH Gradients and Directed Colloid Transport in a Microfluidic Alkaline Hydrothermal Pore. *Angew. Chem. Int. Ed.* **2017,** *56,* 2340–2344.
7. Numata, M.; Sakai, R. Kinetically Controllable Supramolecular Polymerization through Synchronized Activation of Monomers. *Bull. Chem. Soc. Jpn.* **2014,** *87,* 858–862.

8. (a) Arteaga, O.; Canillas, A.; Crusats, J.; El-Hachemi, Z.; Llorens, J.; Sacristan, E.; Ribo, J. M. Emergence of Supramolecular Chirality by Flows. *ChemPhysChem* **2010,** *11,* 3511–3516.
(b) Sorrenti, A.; El-Hachemi, Z.; Crusats, J.; Ribo, J. M. Effects of Flow-Selectivity on Self-Assembly and Auto-Organization Processes: An Example. *Chem. Commun.* **2011,** *47,* 8551–8553.
(c) Sorrenti, A.; Rodriguez-Trujillo, R.; Amabilino, D. B.; Puigmartí-Luis, J. *J. Am. Chem. Soc.* **2016,** *138,* 6920–6923.
9. Helmich, F.; Meijer, E. W. Controlled Perturbation of the Thermodynamic Equilibrium by Microfluidic Separation of Porphyrin-Based Aggregates in a Multi-Component Self-Assembling System. *Chem. Commun.* **2013,** *49,* 1796–1798.
10. Numata, M.; Nishino, Y.; Sanada, Y.; Sakurai, K. Creation of Kinetically Stabilized Porphyrin Microfilms through Synchronized Hydrogen-Bonding Interactions in Microflow. *Chem. Lett.* **2015,** *44,* 861–863.
11. Ghosh, S.; Li, X.-Q.; Stepanenko, V.; Würthner, F. Control of H- and J-type π Stacking by Peripheral Alkyl Chains and Self-Sorting Phenomena in Perylene Bisimide homo- and Heteroaggregates. *Chem. Eur. J.* **2008,** *14,* 11343–11357.
12. Numata, M.; Nogami, R.; Kitamura, A. Enhanced Self-Assembly Abilities Coupled with Nano- and Microscale Non-Equilibrium Phenomena in Flowing Microfields. *ChemNanoMat* **2018,** *4,* 175–182.
13. Numata, M.; Sato, A.; Nogami, R. Energy-dissipative Self-assembly Driven in Microflow: A Time-programmed Self-organization and Decomposition of Metastable Nanofibers. *Chem. Lett.* **2015,** *44,* 995–997.

CHAPTER 13

Development of Polymorphic Control Technology for Pharmaceutical Compounds

Mihoko Maruyama[1,2] **Masashi Yoshimura**[1,3] **Kazufumi Takano**[2], **Yusuke Mori**[1,4]
[1]Graduate School of Engineering, Osaka University, Suita, Osaka, Japan; [2]Graduate School of Life and Environmental Sciences, Kyoto Prefectural University, Kyoto, Japan; [3]Institute of Laser Engineering, Osaka University, Suita, Osaka, Japan; [4]Osaka University, Suita, Osaka, Japan

Contents

1. INTRODUCTION

Many of the organic low molecular weight compounds used as pharmaceuticals have crystal polymorphism. Polymorphism is very important in the pharmaceutical industry because different polymorphs of the same active ingredient exhibit different physical properties such as stability, solubility, high or low melting point, processability, or bioavailability. In the process of drug development, each pharmaceutical compound undergoes polymorphism screening *(1)*. The search for crystal polymorphism is

Kinetic Control in Synthesis and Self-Assembly
ISBN 978-0-12-812126-9
https://doi.org/10.1016/B978-0-12-812126-9.00013-4

important from two viewpoints. First, the identification of crystal polymorphism is important for selection of the most thermodynamically stable polymorph. In general, the most thermodynamically stable polymorph is used in commercial formulations, as changes in drug properties associated with transformation from one polymorph to another are highly undesirable *(2)*. However, it is not known which polymorph is the most thermodynamically stable, and it is extremely difficult to identify all polymorphs. Ritonavir (Abbott) is a famous example of polymorphism and temporary market withdrawal. After form I of Ritonavir was launched, the more stable form II, which is less soluble, appeared. Abbott withdrew the original, less stable capsules from the market *(3)*. This example underscores the importance of complete polymorphism identification, as well as physical and chemical evaluations. The second reason that identification of crystal polymorphism is important is that it contributes to the development of more effective pharmaceuticals. As already mentioned, the most stable phase of a compound is generally selected for the commercial formulation. A stable phase is the most insoluble phase; in other words, its bioavailability is lower than that of other metastable phases. Recently, much attention has been focused on using less stable polymorphs (metastable phases) to improve properties such as solubility and final bioavailability *(4—7)*. However, it is very difficult to control metastable phases precisely due to, for example, the difficulty of crystallizing the metastable phase only. When two or more polymorphs are crystallized, the less stable phase transforms into the stable phase as a result of the difference in solubility between the polymorphs, which makes it difficult to control the metastable phase. For both these reasons, then, crystallization techniques for metastable phases with high stability are urgently needed.

2. A STRATEGY FOR METASTABLE PHASE CRYSTALLIZATION

We must consider three processes in the effort to control the polymorphism of pharmaceutical compounds: nucleation, crystal growth, and phase transition. Which polymorphism appears at the time of nucleation is determined by the kinetic factors of the crystal polymorphism *(8)*. Nucleation occurs via the formation of small embryos of the new phase inside the large volume of the old phase when the solid phase is thermodynamically more stable than the solution phase (supersaturation of the solution is larger than equilibrium). The embryos grow by repeated molecular collision, coalescence, and disassembly. The change of the Gibbs free energy is then

$$\Delta G(r) = -\frac{4\pi r^3}{3v}\Delta\mu + 4\pi r^2\gamma, \tag{13.1}$$

where r is the radius of the embryo, γ is the interfacial free energy between the solution and the solid phase, $\Delta\mu$ is the chemical potential, and v is the volume per molecule. ΔG consists of two terms: the volume term $\frac{4\pi r^3}{3v}\Delta\mu$ and the surface term $4\pi r^2\gamma$. The minus before the volume term indicates that energy is gained when the solid phase is thermodynamically stable. $\Delta\mu$ of the solution is written as

$$\Delta\mu = \mathrm{k}\,T\ln\frac{C}{C_e} = kT\ln(1+\sigma), \quad \sigma = \frac{C - C_e}{C_e}, \tag{13.2}$$

where C is the concentration of the solution and C_e is the equilibrium concentration. As shown in Fig. 13.1, ΔG displays a maximum at some critical size r_c given by

$$r_c = \frac{2v\gamma}{kT\ln(1+\sigma)}, \tag{13.3}$$

when $r = r_c$, the $\Delta G(r_c)$ is shown as

$$\Delta G(r_c) = \frac{16\pi v^2\gamma^3}{3\{kT\ln(1+\sigma)\}^2}. \tag{13.4}$$

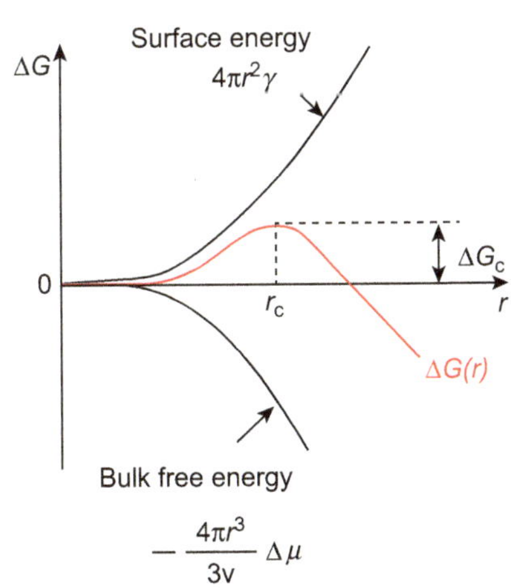

Figure 13.1 Dependence of the crystal radius of the Gibbs free energy change ΔG connected with the formation of nuclei from a supersaturated solution phase.

We will consider a compound with two polymorphs (form I and form II). The crystal structure is maintained by bonding such as hydrogen bonding and van der Waals force, and if the crystal polymorphisms are different, the force with which they interact is also different, so that they take different energy states. Usually, a solvent exists in the crystallization operation of organic low molecular weight compounds. Considering the two-component system of solute and solvent, solubility is lower in the stable phase than in the metastable phase. When the solubility of form II is higher than that of form I in all temperature ranges, form I is always stable and form II becomes metastable (Fig. 13.2A, a monotropic system). As shown in Fig. 13.2B, the system in which the solubility curves intersect is called an enantiotropic system. Because the solubility curves of forms I and II cross at a transition temperature T_p, form I is stable below this temperature and form II becomes stable above this temperature. As a simple example of nuclear generation, we will consider a monotropic system. The nucleation rate J can be expressed by Eqs. (13.5) and (13.6) for form I (stable form) and form II (metastable form). Here, A is a proportional constant.

$$J_\mathrm{I} = \mathrm{A}_\mathrm{I}\exp\left[-16\pi\gamma_\mathrm{I}^3 v_\mathrm{I}^2/\left\{3(kT)^3(\ln\sigma_\mathrm{I})^2\right\}\right] \quad (13.5)$$

$$J_\mathrm{II} = \mathrm{A}_\mathrm{II}\exp\left[-16\pi\gamma_\mathrm{II}^3 v_\mathrm{II}^2/\left\{3(kT)^3(\ln\sigma_\mathrm{II})^2\right\}\right] \quad (13.6)$$

These equations show that the nucleation rate is determined by a competition between the degree of supersaturation σ and surface energy γ of each phase. In the case of a monotropic system, $\sigma_\mathrm{I} > \sigma_\mathrm{II}$ is always established, which is advantageous for form I (stable phase). Although the

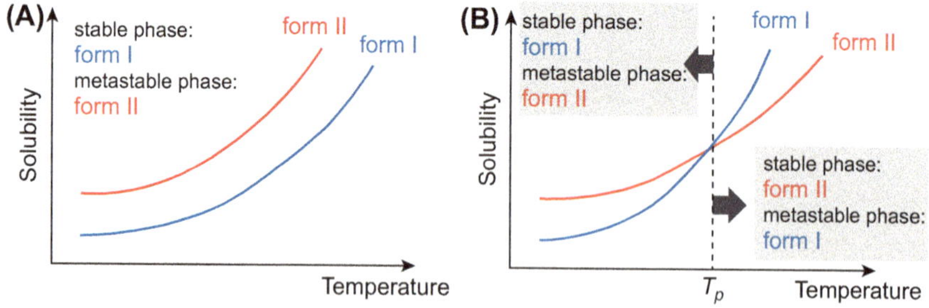

Figure 13.2 (A) Solubility curves without intersection (a monotropic system). Form I crystal is stable phase in this temperature condition. (B) Solubility curves with intersection (an enantiotropic system). Form I crystal is the stable phase under T_p, and form II crystal is the stable phase over T_p.

interfacial energy γ is a property that cannot be easily measured, the metastable phase with larger solubility generally tends to have a smaller γ, i.e., $\gamma_{I} > \gamma_{II}$ *(9,10)*. If the degree of supersaturation is sufficiently high for both forms I and II, nucleation of form II (metastable phase) is expected to prove advantageous, because the difference in γ works with the cube. On the other hand, if the supersaturation is low and the supersaturation of the metastable phase is very small, nucleation of form I (stable phase) will be advantageously promoted.

After nucleation occurs, crystal growth progresses. In the case where stable and metastable phases are mixed, the stable phase is always advantageous in terms of supersaturation, so stable phase growth proceeds preferentially. As the crystal growth progresses and the supersaturation of the solution decreases, the condition reaches a stage that is supersaturated for the stable phase but unsaturated for the metastable phase. In this case, dissolution of the metastable phase progresses, and the stable phase grows in such a way that it absorbs the solute, finally becoming the stable phase only. This is the phase transition process that progresses with crystal growth. In many systems, as described above, even if there is a metastable phase at the beginning, the stable phase becomes predominant over time *(10)*. So, how can a metastable phase be obtained?

To answer this question, we refer back to Fig. 13.2A. The first strategy for realizing the metastable form is to forcibly generate a large amount of crystal nuclei at a stage where adequate supersaturation is achieved; in other words, where nuclear generation of the metastable form is advantageous ($\gamma_{I} > \gamma_{II}$). As a result, the supersaturation is lowered rapidly so that the stable phase does not nucleate *(11)*. Nucleation requires an appropriate induction time. If the solution reaches a metastable state (supersaturated but not capable of nucleation) before the incubation time of the stable phase passes, the environment achieves a pure metastable phase. We are currently using such a strategy to develop technology to forcibly induce crystal nucleation. Another strategy is to provide a template that is suitable for metastable phase nucleation. For example, polymers have specific functional groups, and it is expected that small organic molecules will be arranged using those sequences as templates. In addition, considering the energy required for crystal nucleation, heterogeneous nucleation on the surface of something (e.g., a container, particles in a solution) is easier than nucleation without any surface (homogeneous nucleation) *(12,13)*. In the next section, we introduce the technology that promotes crystallization in detail.

3. TECHNOLOGY TO PROMOTE CRYSTAL NUCLEATION

3.1 Crystal Nucleation Technique by Femtosecond Laser Irradiation

Laser irradiation of a solution is an effective way to generate crystal nuclei *(14–17)*. The different groups employing this method have achieved crystal nucleation with different physical phenomena, and the details are best investigated in the reports themselves. The laser technology developed by our group induces crystal nucleation by generating cavitation bubbles in solution with femtosecond laser irradiation. Originally, this technology was developed to crystallize proteins for 3D structural analysis *(15,18)*. We applied the technique to polymorphic control of indomethacin (an antiphlogistic analgesic), which is a model substance of organic low molecules.

Indomethacin has two polymorphs: γ-form (stable phase, a platelike morphology) and α-form (metastable phase, a needlelike morphology) (Fig. 13.3) *(19–21)*. Fig. 13.4 is the experimental setup. The detailed laser conditions were shown in Ref *(21)*. In this experiment, laser focal points were important parameter. Laser focal points were set at three deferent points: the center of the solution (Fig. 13.4A), approximately 100 μm from the sidewall of the glass vial (Fig. 13.4B) and approximately 100 μm from air at the air–liquid interface (Fig. 13.4C). In contrast, a Teflon-coated magnetic stirring bar was added to the solution to attempt crystallization by the

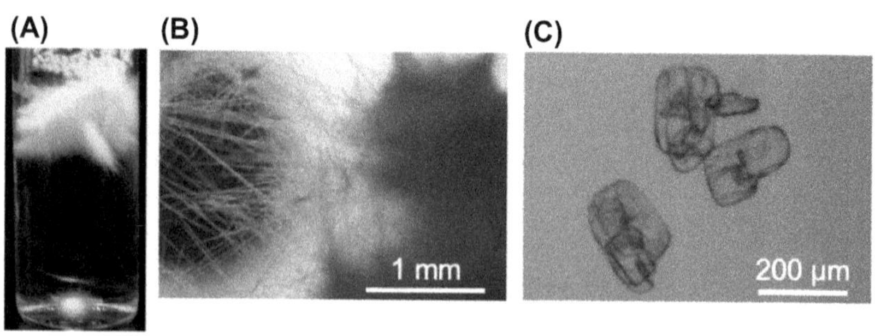

Figure 13.3 (A) Indomethacin a-form crystals in a 1 mL glass vial. Bright field images of (B) typical a-form and (C) g-form. *(Reprinted with permission from the Japan Society of Applied Physics, Ref. Ikeda, K.; Maruyama, M.; Takahashi, Y.; Mori, Y.; Yoshikawa, H. Y.; Okada, S.; Adachi, H.; Sugiyama, S.; Takano, K.; Murakami, S.; Matsumura, H.; Inoue, T.; Yoshimura, M.; Mori, Y. Selective Crystallization of the Metastable Phase of Indomethacin at the Interface of Liquid/Air Bubble Induced by Femtosecond Laser Irradiation.* Appl. Phys. Express ***2015,*** 8 *(4), 045501-1-4.)*

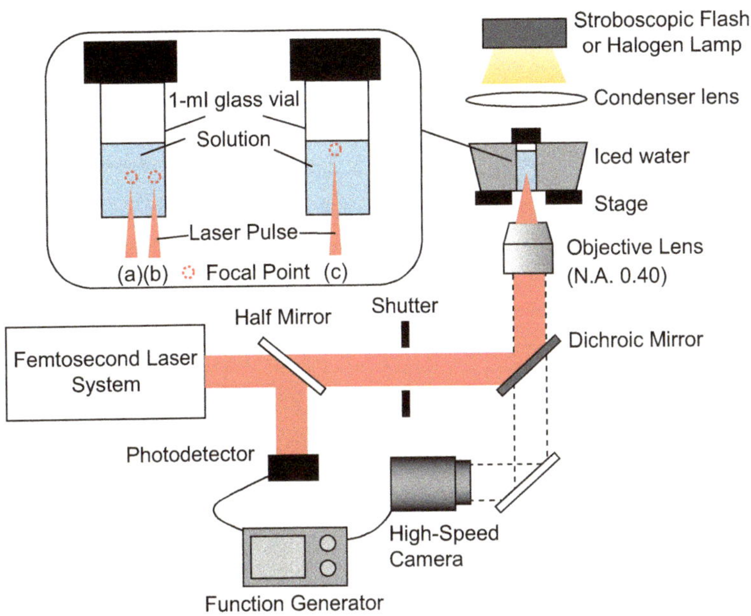

Figure 13.4 An experimental setup for femtosecond laser irradiation and fast imaging of solution movement. *(Reprinted with permission from the Japan Society of Applied Physics, Ref. Ikeda, K.; Maruyama, M.; Takahashi, Y.; Mori, Y.; Yoshikawa, H. Y.; Okada, S.; Adachi, H.; Sugiyama, S.; Takano, K.; Murakami, S.; Matsumura, H.; Inoue, T.; Yoshimura, M.; Mori, Y. Selective Crystallization of the Metastable Phase of Indomethacin at the Interface of Liquid/Air Bubble Induced by Femtosecond Laser Irradiation.* Appl. Phys. Express ***2015,*** 8 *(4), 045501-1-4.)*

impact of magnetic stirring. Crystallization probability was calculated as $(n_{cry} = n_{total}) \times 100$, where n_{cry} and n_{total} were the number of crystallized samples and number of total samples, respectively. Without laser irradiation and stirring, nucleation did not occur in the supersaturated solutions after a period of 2 d at 0°C ($\sigma_\gamma = 3.5$). Laser irradiation showed different results depending on the focal points. When the laser irradiation was focused at the center of the solution (Fig. 13.5A), α-form nucleation occurred within 24 h, but within another 24 h, the resulting crystals turned into γ-form. With laser irradiation near the sidewall of the glass vial or air–liquid interface (Fig. 13.5B and C), nucleation occurred also within 24 h. The resulting crystals were α-form. The crystals showed high temporal stability, and it was at least 8 months in air at room temperature and in solutions at 0°C. In the case of stirring, nucleation occurred within 1 h, and the resulting crystals were also α-form. However, all of these crystals were transformed into α-form within about 24 h after crystallization;

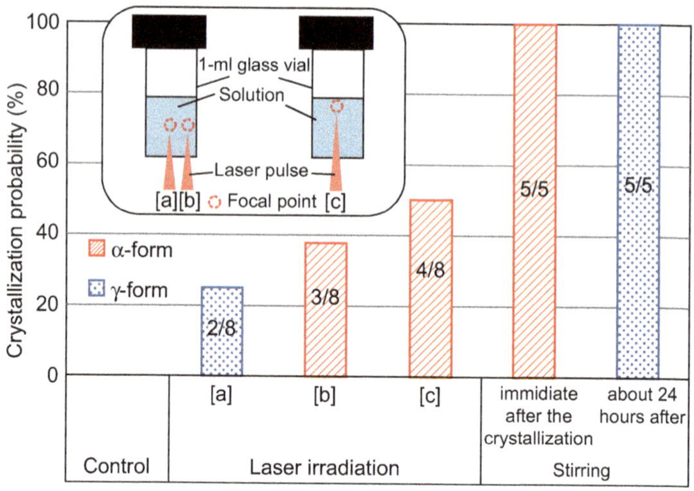

Figure 13.5 Probability of crystallization for each case. Without forcible nucleation techniques (control), with laser irradiation (at the center of the solution [a], near the sidewall of the glass vial [b], and at the air–liquid interface [c]), and with stirring (immediately after crystallization and about 24 h after), [a], [b], and [c] correspond to the positions indicated in inserted picture. *(Reprinted with permission from the Japan Society of Applied Physics, Ref. Ikeda, K.; Maruyama, M.; Takahashi, Y.; Mori, Y.; Yoshikawa, H. Y.; Okada, S.; Adachi, H.; Sugiyama, S.; Takano, K.; Murakami, S.; Matsumura, H.; Inoue, T.; Yoshimura, M.; Mori, Y. Selective Crystallization of the Metastable Phase of Indomethacin at the Interface of Liquid/Air Bubble Induced by Femtosecond Laser Irradiation.* Appl. Phys. Express ***2015,*** 8 *(4), 045501-1-4.)*

they followed a similar tendency of crystals obtained by laser irradiation focused at the center of the solution.

Cavitation bubbles occur when a solution is irradiated with a femtosecond laser. Because this cavitation bubble repeatedly expands and contracts in a short period of time (in the order of microseconds), the local concentration of solute is increased around the bubble. This localized high-concentration environment induces crystal nucleation. Magnetic stirring also induces crystal nucleation. However, in the case of polymorphic control, merely inducing nucleation does not result in a highly stable metastable form. As shown in Section 2 of this chapter, the main point is to promptly promote metastable crystallization and lower supersaturation to the extent of metastable concentration. The α-form crystals obtained by laser irradiation near the sidewall of the glass vial or air–liquid interface showed high temporal stability. This is because many crystals nucleated rapidly at the conditions comparing to other conditions. In both the cases, a lot of bubbles appeared and stayed along the meniscus of the solution.

The nucleation of the α-form occurred on these lasting bubbles. The environment promoted nucleation of α-form effectively, then the supersaturation rapidly decreased to the extent of metastable concentration; γ-form could not nucleate because of the rapid change of supersaturation. As a result, the laser irradiation technique realized metastable crystallization and improved stability *(21,22)*. Another model case using acetaminophen (an antipyretic analgesic) was also shown in Ref *(22)*.

3.2 Crystal Nucleation by Ultrasonic Irradiation

There is a long history of crystallization by ultrasonic irradiation, dating back to the initial report of Richards and Loomis in 1927 *(23)*. Throughout this body of literature, ultrasonic irradiation has often been used in the nucleation stage of the crystallization process. Various effects such as induction of nucleation, narrowing of the width of the metastable region, shortening of the induction time to crystallization, narrowing of the dispersion of crystal size distribution, and selection of crystal polymorphism have been reported *(24–30)*. Although the mechanisms underlying these effects have not been completely elucidated, many researchers consider that the phenomena occur due to the cavitation bubbles generated by irradiation of the solutions with ultrasonic waves *(31–33)*. As mentioned in Section 3.1, we succeeded in crystallizing metastable forms of indomethacin and acetaminophen by generating cavitation bubbles by femtosecond laser irradiation of the solution. Therefore, we focused on ultrasonic waves as a method for generating more cavitation bubbles and worked on metastable crystallization using the technique.

Various polymorphs of acetaminophen have been reported: form I (stable phase *(34)*), form II (metastable phase *(35)*), form III (metastable phase *(36)*), and three hydrates (monohydrate *(37)*, dehydrate *(38)*, and trihydrate *(39)*). Form I is used in commercial formulations because it is easily crystallized by various methods *(40,41)*, and its properties do not deteriorate easily. The experimental setup for ultrasonic irradiation is shown in Fig. 13.6. The supersaturated acetaminophen solution was dispensed in 0.8 mL aliquots into 1 mL glass vials placed in a block incubator at 60°C. These samples were then placed in an incubator at 55°C, cooled to 0°C at a constant cooling rate of 3°C/h, and maintained at that temperature for about 1 day *(42)*.

We confirmed the absence of spontaneous crystallization before ultrasonic irradiation. An ultrasonic cleaner (as one VS-100III) capable of generating frequencies of 28, 45, and 100 kHz was used as the

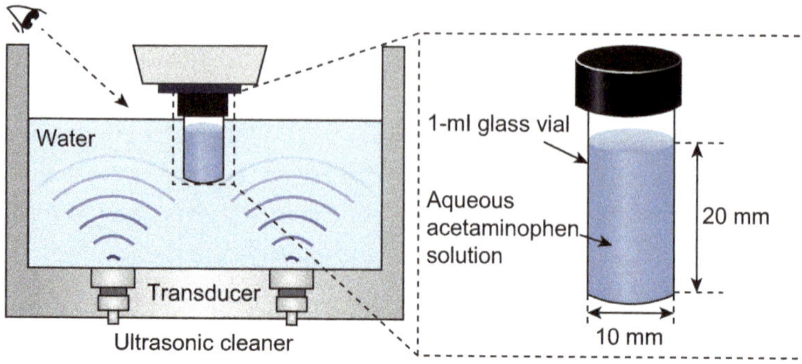

Figure 13.6 Experimental setup for ultrasonic irradiation. Glass vial containing aqueous acetaminophen solution was located at the position between transducers. *(Reprinted with permission from the Japan Society of Applied Physics, Ref. Mori, Y.; Maruyama, M.; Takahashi, Y.; Ikeda, K.; Fukukita, S.; Yoshikawa, Y. H.; Okada, S.; Adachi, H.; Sugiyama, S.; Takano, K.; Murakami, S.; Matsumura, H.; Inoue, T.; Yoshimura, M.; Mori, Y. Selective Crystallization of Metastable Phase of Acetaminophen by Ultrasonic Irradiation.* Appl. Phys. Express ***2015,*** 8 *(6), 065501-1-4.)*

ultrasonic generator. Periodic ultrasonic irradiation (irradiation for 10 s with 10 s of rest) at frequencies of 28, 45, and 100 kHz was applied to the solution. Periodic irradiation was used to maintain the water temperature at 0°C during the experiments. It was continued for up to 800 s and stopped immediately when crystallization was confirmed by the naked eye through the space in the apparatus for fixing the glass vial and ultrasonic cleaner. A stirring experiment was also conducted in an incubator at 0°C for comparison.

Crystallization probability was calculated as $(n_{cry} = n_{total}) \times 100$, where n_{cry} and n_{total} were the number of crystallized samples and number of total samples, respectively. Fig. 13.7A shows the crystallization probability for each condition. These results indicate that stirring and ultrasonic irradiation with low frequencies of 28 and 45 kHz is effective for the promotion of nucleation. This stirring condition realized form I crystallization with 100% probability (Fig. 13.7A and B). On the other hand, we revealed that ultrasonic irradiation with low frequencies of 28 and 45 kHz was effective for crystallization of form II (Fig. 13.7A and C). When ultrasonic irradiation at 100 kHz was applied, no crystallization occurred. The differences between the stirring and ultrasonic irradiated conditions made it clear that the presence of cavitation bubbles was necessary for form II crystallization.

In the case of ultrasonic irradiation, a comparison of the results for 28, 45, and 100 kHz indicated that differences in cavitation behavior caused by frequency differences affected nucleation. Ultrasound creates cavitation bubbles in liquid media. It is generally known that the formation of

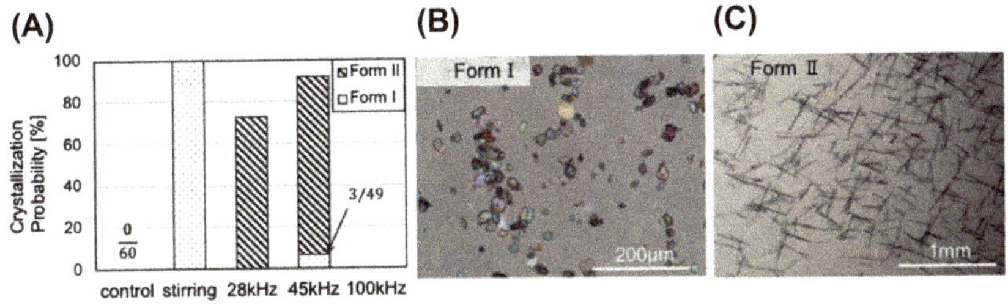

Figure 13.7 (A) Crystallization probability for cases with stirring and ultrasonic irradiation (28, 45, and 100 kHz), and without forcible nucleation techniques (control). (B) Platelike crystals crystallized by solution stirring. Identified as form I. (C) Needlelike crystals crystallized by ultrasonic irradiation. Identified as form II. *(Reprinted with permission from the Japan Society of Applied Physics, Ref. Mori, Y.; Maruyama, M.; Takahashi, Y.; Ikeda, K.; Fukukita, S.; Yoshikawa, Y. H.; Okada, S.; Adachi, H.; Sugiyama, S.; Takano, K.; Murakami, S.; Matsumura, H.; Inoue, T.; Yoshimura, M.; Mori, Y. Selective Crystallization of Metastable Phase of Acetaminophen by Ultrasonic Irradiation.* Appl. Phys. Express ***2015,*** 8 *(6), 065501-1-4.)*

cavitation bubbles becomes more difficult, and bubble size smaller, when the ultrasound frequency is increased *(43)*. Given these facts, low-frequency ultrasound should be effective for creating local high-concentration regions because the size of cavitation bubbles generated by low-frequency ultrasound will be greater than that of cavitation bubbles generated by high-frequency ultrasound. At present, we consider that the principle of polymorphic control by ultrasonic application proceeds as follows. Nucleation occurs in the solution immediately because ultrasonic irradiation can generate many cavitation bubbles in the solution. Under the highly supersaturated conditions that can initiate the nucleation of both the stable and metastable phases, the metastable phase alone crystallizes if we can shorten the waiting time for nucleation (details are presented in Section 2). Under the condition of 100 kHz, the diameter of the cavitation bubble is small, and the concentration and pressure of the bubble during expansion and contraction are moderately weakened; consequently nucleation does not occur.

Other polymorphic studies using ultrasonic irradiation have proposed different ideas. For example, Gracin et al. used ultrasonic irradiation to crystallize p-Aminobenzoic acid (with α- and β-forms) and succeeded in crystallizing the β-form under a condition in which the α-form normally crystallizes *(29)*. They offered the following rationale: p-Aminobenzoic acid has a carboxyl group, and in solution, carboxyl groups join hydrogen bonds to form a dimer. Because the α-form has a crystal structure based on this dimer, it crystallizes with the usual crystallization method. On the other hand, when ultrasonic irradiation is performed, the β-form tends to crystallize as a result of the weakening of the hydrogen bond of the dimer, and progresses with the separation of the dimer at high temperature/high pressure in a process caused by cavitation. Kurotani et al. investigated the effect of ultrasound irradiation on the crystallization of sulfamerazine and reported that under conditions where only the metastable form crystallizes, the stable form could be nucleated at the same time by ultrasonic irradiation *(30)*. They suggested that the stable form nucleates because the ultrasonic irradiation provides energy exceeding the free energy barrier for stable nucleation. We still cannot definitively state which mechanism is correct. However, by carefully considering various physical phenomenon generated by ultrasonic irradiation, we will be able to develop polymorphic control techniques by ultrasonic irradiation more effectively and with better versatility.

3.3 Polymer-Induced Heteronucleation

In the crystal nucleation process, the template plays important roles. When the template surface and the solution containing the target solute have a high affinity, the template becomes a site where heterogeneous nucleation occurs and promotes nucleation *(44,45)*. There are templates of various materials with different surface conditions *(46–48)*. Among them, the crystallization technique using a polymer as a template is called polymer-induced heteronucleation (PIHn), and it has been applied to the crystallization of many materials *(49,50)*. It has been reported that the polymorphism depends on the kind of polymer, so PIHn has attracted attention as a very effective technique for the discovery of new polymorphism and selective polymorphic crystallization. Lang et al. tested 84 kinds of polymers by applying PIHn to the crystallization of acetaminophen. They found that, depending on the type of polymer, metastable form II could be selectively obtained *(46)*. Although the detailed mechanism has not yet been clarified, it is believed that organic molecules are oriented specifically at the polymer interface *(51,52)*.

In the course of an experiment, we accidentally discovered a polymorphic control method using the PIHn method. A doctoral course student joined two kinds of containers by applying an adhesive, and then used it in ultrasonic irradiation experiments. When he dispensed the acetaminophen solution to the junction container and allowed it to stand, he noticed that crystals with characteristic shape had crystallized. The crystal shape was first observed in our laboratory at that time. After careful observation and reproduction, it was revealed that acetaminophen

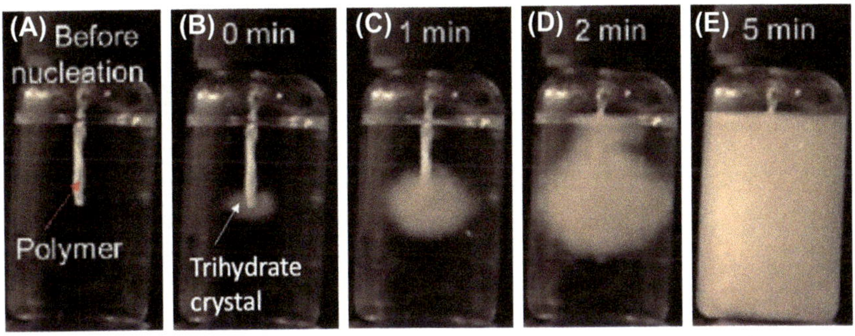

Figure 13.8 In situation observation of nucleation and growth of trihydrate. (A) A stick-shaped polymer (white body) was immersed in aqueous acetaminophen solution before nucleation. (B) The initiation of nucleation was defined as 0 min.Trihydrate crystals nucleated on the surface of polymer. (C)-(E) The trihydrate crystals gradually grew as time passed. *(Reprinted with permission from the Japan Society of Applied Physics, Ref. Mori, Y.; Maruyama, M.; Takahashi, Y.; Yoshikawa, Y. H.; Okada, S.; Adachi, H.; Sugiyama, S.; Takano, K.; Murakami, S.; Matsumura, H.; Inoue, T.; Yoshimura, M.; Mori, Y. Metastable Crystal Growth of Acetaminophen Using Solution-Mediated Phase Transformation.* Appl. Phys. Express ***2017,*** 10 *(1), 015501.)*

trihydrate crystals crystallized from the surface of the adhesive (one kind of polymer) used for bonding the containers (Fig. 13.8) *(53)*.

Although the existence of trihydrate has been reported, it is a very unstable phase, easily dehydrated within 1 min in air. Even in solution, the trihydrate rearranged to form I by solution-mediated phase transition at a temperature condition of 5 °C or higher *(39,54)*. However, the trihydrate obtained by our method had high stability in solution and maintained the crystal form of trihydrate even under changing temperature. Utilizing this high stability, we have succeeded in obtaining the solubility curve of acetaminophen trihydrate for the first time *(53)*.

3.4 Crystallization by Plastic Ball—Assisted Ultrasonic Irradiation

We have introduced femtosecond laser technology, the ultrasonic irradiation technique, and polymer technology as effective methods for obtaining a metastable form selectively. We combined the ultrasonic irradiation with the polymer to create a more efficient and effective method. Here, we again use acetaminophen as a model system.

The normal ultrasonic irradiation setup is shown in Fig. 13.6 in Section 3.2. In this new setup, a 3-mm diameter plastic ball made of polypropylene (PP) was set as shown in Fig. 13.9. Note that the PP ball

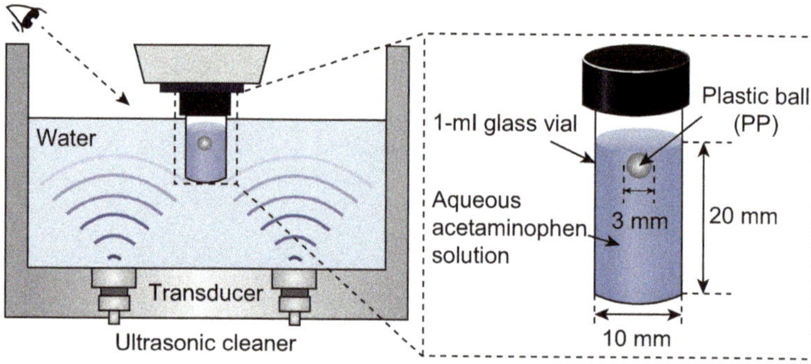

Figure 13.9 Experimental setup for ultrasonic irradiation. Glass vial containing aqueous acetaminophen solution was located at the position between transducers. It should be noted that polypropylene (PP) ball floats on the solution. *(Reprinted with permission from the Japan Society of Applied Physics, Ref. Mori, Y.; Maruyama, M.; Takahashi, Y.; Yoshikawa, Y. H.; Okada, S.; Adachi, H.; Sugiyama, S.; Takano, K.; Murakami, S.; Matsumura, H.; Inoue, T.; Yoshimura, M.; Mori, Y. Metastable Crystal Growth of Acetaminophen Using Solution-Mediated Phase Transformation.* Appl. Phys. Express ***2017,*** 10 *(1), 015501.)*

floats on the solution. PP ball characteristics include a simple structure and the fact that they barely dissolve in acetaminophen solution. A PP ball improves the nucleation probability of acetaminophen form II (Fig. 13.10A). When supersaturation was very low ($\sigma_{II} = 0.4$), crystallization did not occur whether the PP ball was present or not. At $\sigma_{II} = 0.9$, the probability of form II crystallization increased to 100% when the PP ball was present and was 0% without the PP ball. At higher supersaturation values ($\sigma_{II} = 1.4$–3.3), form II crystallization was 100% in the presence of a PP ball, and without the ball it increased gradually with an increase in supersaturation, finally reaching 100% at $\sigma_{II} = 3.3$. Fig. 13.10B shows the induction time for form II crystallization in the absence and presence of a PP ball as a function of supersaturation. In both cases, the induction time decreased with increased supersaturation. The presence of a PP ball significantly reduced the variation in induction time. At $\sigma_{II} = 1.9$–3.3, there was a significant difference between the induction times for samples obtained in the absence and presence of a PP ball. Thus, the presence of a plastic ball in aqueous acetaminophen solution undergoing ultrasonic irradiation effectively increases the nucleation probability and reduces the induction time of form II nucleation.

As to why the PP spheres increase the efficiency of crystallization of the metastable form, two suggestions can be offered. The first is that the

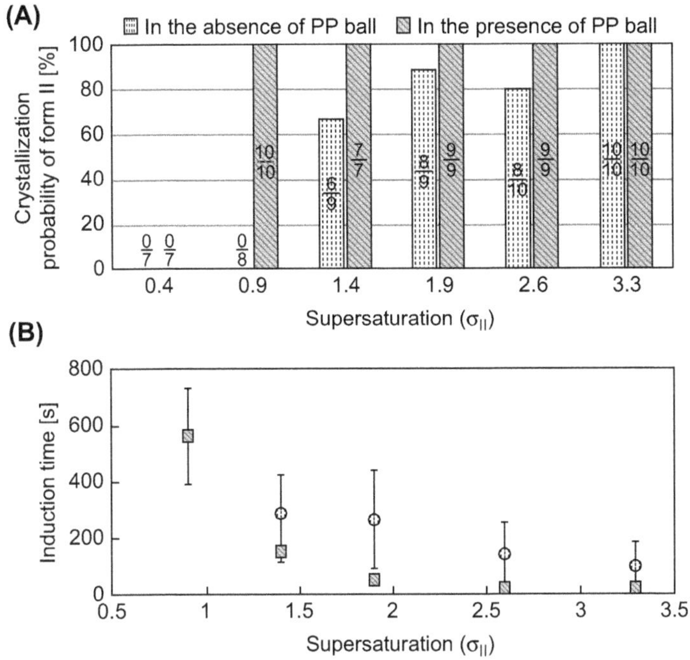

Figure 13.10 (A) Crystallization probability of acetaminophen form II by ultrasonic irradiation in the absence and presence of polypropylene ball as a function of the supersaturation. (B) Induction time for crystallization of acetaminophen form II in the absence and presence of polypropylene ball as a function of the supersaturation. Each symbol is the average of the crystallized samples. Error bars represent SD. *(Reprinted with permission from the Japan Society of Applied Physics, Ref. Mori, Y.; Maruyama, M.; Takahashi, Y.; Yoshikawa, Y. H.; Okada, S.; Adachi, H.; Sugiyama, S.; Takano, K.; Murakami, S.; Matsumura, H.; Inoue, T.; Yoshimura, M.; Mori, Y. Metastable Crystal Growth of Acetaminophen Using Solution-Mediated Phase Transformation.* Appl. Phys. Express ***2017,*** 10 *(1), 015501.)*

surface of the PP ball has an effect on form II crystallization, as well as on the PIHn as described above (Section 3.3). Lang et al. reported that PP promotes form II crystallization *(46)*. In addition to the nucleation-inducing effect of ultrasonic irradiation, this polymer surface effect likely made the form II crystallization highly efficient. The second possibility is that the ball enhances cavitation. The addition of solid particles into liquid under ultrasonic irradiation results in a decrease of the cavitation threshold. It is reasonable to think that form II nucleation would be promoted by an increase in cavitation bubbles. Actually, both of these mechanisms may have contributed to the observed results.

4. TECHNOLOGY FOR IMPROVING CRYSTAL QUALITY AND STABILITY

4.1 What Determines the Stability of Crystals?

The metastable crystal obtained by the introduced techniques (femtosecond laser irradiation, ultrasonic irradiation, PIHn, etc.) has higher stability than the crystal obtained by conventional methods *(42,55–58)*. The former can be stably stored for more than 1 year in a solution at normal temperature and pressure *(21,42)*. In this section, we will consider what determines the stability of metastable crystals.

At least two steps are necessary for the metastable form to exist stably in solution. As a first step, it is necessary that only the metastable form crystallizes and does not mix with the stable form. This is because the solvent-mediated phase transition progresses due to the difference in solubility when stable forms coexist, and all stable form crystals are formed in a very short period (depending on the conditions, for example, several seconds to several days) *(10)*.

The next important point is for the crystal to have as few defects that trigger the phase transition as possible. Once the single crystal is removed from the solution, we may think that the crystal is free from solvent-mediated phase transition, but that is incorrect. Actually, the solution is often incorporated inside the crystals as defects, and solvent-mediated phase transition may continue to progress. Solid phase rearrangement may also occur. Solid phase transition will also occur from the crystal defects, and the transition will then spread throughout the crystals *(59)*. Therefore, it is important to reduce defects inside the crystal as much as possible to obtain high-quality metastable crystals.

4.2 Conditions Suitable for High-Quality Crystal Growth

What kinds of conditions can reduce defects inside the crystals? From the viewpoint of crystal growth, slow growth under low supersaturation can reduce defects. When growing crystals at high speed under high supersaturation, the growth rate increases and the concentration gradient of the solution near crystal interfaces becomes large (the Berg effect) (Fig. 13.11A). Such a concentration gradient causes hopper growth of crystals; the edges of the crystal grow preferentially, whereas growth at the center of the crystal surface is delayed, causing the crystal to become a "skeletal" crystal (i.e., one in which voids remain) *(60)*. Hopper growth tends to be resolved by a slower crystal growth rate, so that defects close as

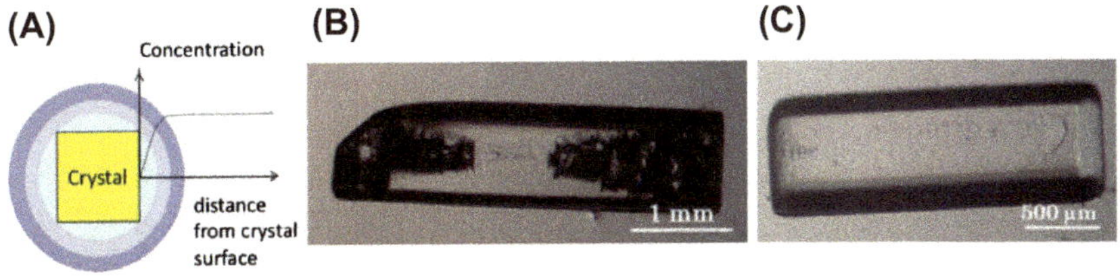

Figure 13.11 (A) A concentration gradient around the interface of a crystal. (B) Optical microscope images of acetaminophen form II crystals with defects inside the crystal (B) and without any serious defects inside the crystal (C). *(Reprinted with permission from the Japan Society of Applied Physics, Ref. Nii, K.; Maruyama, M.; Okada, S.; Adachi, H.; Takano, K.; Murakami, S.; Yoshikawa, Y. H.; Matsumura, H.; Inoue, T.; Imanishi, M.; Tsukamoto, K.; Yoshimura, M.; Mori, Y. Improvement of Metastable Crystal of Acetaminophen via Control of Crystal Growth Rate.* Appl. Phys. Express ***2018,*** 11 *(3), 035501-1-4.)*

the crystal grows. The growing solution is often incorporated in the crystal as inclusion *(61)*. Fig. 13.11B shows a form II crystal of acetaminophen grown under a relatively high supersaturation condition, revealing many defects inside the crystal. In contrast, form II crystals grown under low supersaturation have few internal defects and overall high transparency (Fig. 13.11C). We evaluated the stability of crystals by grouping these two types of crystals and concluded that crystals with fewer defects have higher stability *(61)*.

4.3 Control of Supersaturation

There are a variety of ways to achieve supersaturation control. The easiest way is to change the amount of dissolved solute. Because the substance has inherent solubility as explained in Section 2, operations such as raising the temperature are necessary to dissolve the solute. If the dissolution process is performed at a sufficiently high temperature and then cooled, higher supersaturation is realized. If low supersaturation is desired, the amount of solute can be reduced. However, there is a dilemma; crystal nucleation does not occur (or does not occur readily) under low supersaturation, so a high supersaturation is necessary to realize nucleation. Even with the ultrasonic application method, there is a threshold supersaturation condition required to realize nucleation generation of form II crystals. In other words, when considering the growth process for high-quality crystals, we must consider the nucleation and crystal growth processes separately.

We separated these processes by seeding form II crystals in a low supersaturation solution (metastable conditions where nucleation usually does not occur). This approach realized high-quality and stable form II crystals.

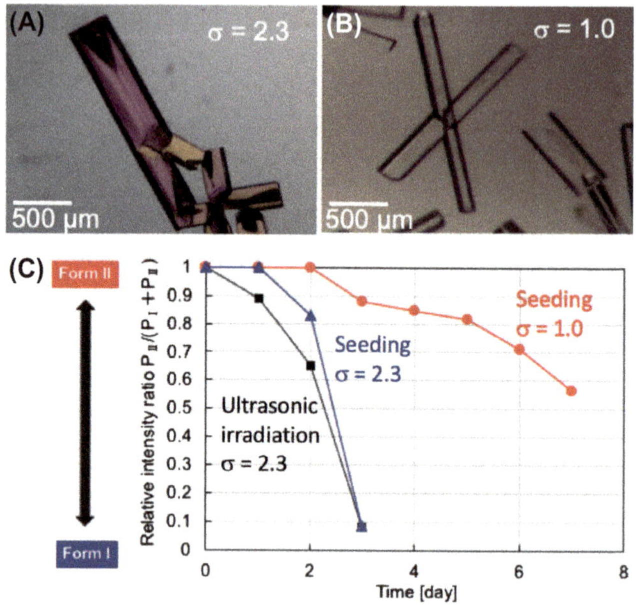

Figure 13.12 Acetaminophen form II crystals grown in a high supersaturated solution ($\sigma = 2.3$) (A) and grown in a low supersaturated solution ($\sigma = 1.0$) (B). (C) Temporal stability of acetaminophen form II obtained by ultrasonic irradiation $\sigma = 2.3$ (square) and seed crystal growth in a high supersaturated solution $\sigma = 2.3$ (triangle) and a low supersaturated solution $\sigma = 1.0$ (circle). *(Reprinted with permission from the Japan Society of Applied Physics, ref. Nii, K.; Maruyama, M.; Okada, S.; Adachi, H.; Takano, K.; Murakami, S.; Yoshikawa, Y. H.; Matsumura, H.; Inoue, T.; Imanishi, M.; Tsukamoto, K.; Yoshimura, M.; Mori, Y. Improvement of Metastable Crystal of Acetaminophen via Control of Crystal Growth Rate.* Appl. Phys. Express ***2018,*** 11 *(3), 035501-1-4.)*

The form II crystals used for seeding were prepared by the ultrasonic irradiation method. Fig. 13.12A is an image of a typical acetaminophen form II grown under relatively high supersaturation ($\sigma = 2.3$), and Fig. 13.12B is an image of a typical form II grown under relatively low supersaturation ($\sigma = 1.0$). The crystals obtained at $\sigma = 1.0$ were transparent with almost no visible defects. Fig. 13.12C shows the time history of the crystal stability of form II. PXRD was employed to distinguish between form I and II crystals. Note that the samples were allowed to remain at a high temperature (41.3°C ± 0.7°C) and high humidity (72% RH ± 6%) to accelerate the phase transformation *(53)*. This is because the form II crystals grown previously by our group were stable for a long time (at least 11 months) in air at room temperature, such that evaluation of the stability in the short term was difficult.

The relative intensity ratio $[P_{II} = (P_I + P_{II})]$ was calculated according to the characteristic intensity of the diffraction signal of form I (P_I, around 26.5 degrees) and form II (P_{II}, around 24 degrees). The relative intensity indicates the stability of form II. The form II crystals obtained by applied ultrasonic irradiation began to transform into form I within a day. In contrast, form II crystals obtained by seeding growth had better stability. Moreover, the temporal stability of high-quality crystals obtained under low supersaturation ($\sigma = 1.0$) was longer than that of the defective crystals obtained under high supersaturation ($\sigma = 2.3$) *(61)*.

Several final points should be kept in mind. First, it is a good idea to grow crystals only after improving the quality of the seed crystals. In addition, the idea that supersaturation in the vicinity of individual crystal surfaces (the surface supersaturation degree) affects the crystal growth must be regarded. When many crystals are growing at the same time, it is almost impossible to control the degree of surface supersaturation of each crystal. However, by increasing the number of seed crystals of uniform size, it is possible to keep individual surface supersaturation low and, as a result, to improve the quality of individual crystals *(61)*.

5. FUTURE PROSPECTS

In this chapter, basic strategies for obtaining metastable crystals and techniques for promoting crystal nucleation were introduced. Examples such as indomethacin and acetaminophen, selected as models of polymorphic control, were shown. Generalization of technologies is anticipated, so that technology can be applied to many drug candidate compounds other than model substances. In addition, because many drug candidate compounds are poorly water soluble, techniques applicable to solution growth systems using organic solvents and melt growth systems are also needed. Polymorphic control is characterized by numerous complex elements, and thus it tends to be discussed only in case studies for individual substances. However, the end to which the principles are generalized is also important. We hope to improve the lives of people suffering from intractable diseases or high medical expenses through the development of useful crystallization technologies.

ACKNOWLEDGMENTS

We gratefully acknowledge Research Fellowships of JSPS and Shiseido Female Researcher Science Grant for MM. We also acknowledge funding under a Grant-in-Aid for Scientific Research (B) No. 26286042 and Grant-in-Aid for Challenging Exploratory Research 15K13380 and 17H02774 for YM.

REFERENCES

1. Lee, E. H. A Practical Guide to Pharmaceutical Polymorph Screening & Selection. *Asian J. Pharm. Sci.* **2014,** *9* (4), 163–175.
2. Müller, M.; Meier, U.; Wieckhusen, D.; Beck, R.; Pfeffer-Hennig, S.; Schneeberger, R. Process Development Strategy to Ascertain Reproducible API Polymorph Manufacture. *Cryst. Growth Des.* **2006,** *6* (4), 946–954.
3. Takata, N.; Terada, K. In *Advanced Technology, Application and Development of Polymorphic Crystals;* Matsuoka, M., Ed.; 2005; pp 105–116.
4. Yokoyama, T.; Umeda, T.; Kuroda, K.; Nagafuku, T.; Yamamoto, T.; Asada, S. Studies on Drug Nonequivalence. IX. Relationship between Polymorphism and Rectal Absorption of Indomethacin. *Yakugaku Zasshi* **1979,** *99* (8), 837–842.
5. Aguiar, A. J.; Krc, J., Jr.; Kinkel, A. W.; Samyn, J. C. Effect of Polymorphism on the Absorption of Chloramphenicol from Chloramphenicol Palmitate. *J. Pharm. Sci.* **1967,** *56* (7), 847–853.
6. Kato, Y.; Okamoto, Y.; Nagasawa, S.; Ishihara, I. New Polymorphic Forms of Phenobarbital. *Chem. Pharm. Bull.* **1984,** *32* (10), 4170–4174.
7. Salole, E. G.; Al-Sarraj, F. A. Spironolactone Crystal Forms. *Drug Dev. Ind. Pharm.* **1985,** *11* (4), 855–864.
8. Ostwald, W. Studien über die Bildung und Umwandlung fester Körper: *Z. Phys. Chem.* **1897,** *22,* 289–330.
9. Nielsen, A. E.; Sohnel, O. Interfacial Tensions Electrolyte Crystal-Aqueous Solution, from Nucleation Data. *J. Cryst. Growth* **1971,** *11* (3), 233–242.
10. Kitamura, M. In *Advanced Technology, Application and Development of Polymorphic Crystals;* Matsuoka, M., Ed.; CMC Publication: Tokyo, 2005; pp 39–50.
11. Maruyama, M.; Yoshikawa, H. Y. Active Control of Growth Mode and Polymorph Nucleation of Organic Molecules by Femtosecond Laser. *J. Jpn. Assoc. Cryst. Growth* **2017,** *44* (3), 1–12.
12. Liu, X. Y. Interfacial Process of Nucleation and Molecular Nucleation Templator. *Appl. Phys. Lett.* **2001,** *79* (1), 39–41.
13. Markov, I. V. *Cryatal Growth for Beginners, Fundamentals of Nucleation, Crystal Growth and Epitaxy,* 2 ed.; World Scientific Publishing Co. Pte. Ltd.: Singapore, 2003.
14. Garetz, B. A.; Aber, J. E.; Goddard, N. L.; Young, R. G.; Myerson, A. S. *Phys. Rev. Lett.* **1996,** *77,* 3475–3476.
15. Adachi, H.; Takano, K.; Hosokawa, Y.; Inoue, T.; Mori, Y.; Matsumura, H.; Yoshimura, M.; Tsunaka, Y.; Morikawa, M.; Kanaya, S.; Masuhara, H.; Kai, Y.; Sasaki, T. Laser Irradiated Growth of Protein Crystal. *Jpn. J. Appl. Phys.* **2003,** *42* (Part 2, No. 7B)), L798–L800.
16. Hosokawa, Y.; Adachi, H.; Yoshimura, M.; Mori, Y.; Sasaki, T.; Masuhara, H. Femtosecond Laser-Induced Crystallization of 4-(Dimethylamino)-N-methyl-4-stilbazolium Tosylate. *Cryst. Growth Des.* **2005,** *5* (3), 861–863.
17. Okutsu, T.; Furuta, K.; Terao, M.; Hiratsuka, H.; Yamano, A.; Ferté, N.; Veesler, S. Light-Induced Nucleation of Metastable Hen Egg-White Lysozyme Solutions. *Cryst. Growth Des.* **2005,** *5* (4), 1393–1398.
18. Yoshikawa, H. Y.; Murai, R.; Adachi, H.; Sugiyama, S.; Maruyama, M.; Takahashi, Y.; Takano, K.; Matsumura, H.; Inoue, T.; Murakami, S.; Masuhara, H.; Mori, Y. Laser Ablation for Protein Crystal Nucleation and Seeding. *Chem. Soc. Rev.* **2014,** *43* (7), 2147–2158.
19. Borka, L. The Polymorphism of Indomethacine. New Modifications, Their Melting Behavior and Solubility. *Acta Pharm. Suec.* **1974,** *11* (3), 295–303.
20. Kaneniwa, N.; Otsuka, M.; Hayashi, T. *Pharm. Factory* **1985,** *5,* 738.

21. Ikeda, K.; Maruyama, M.; Takahashi, Y.; Mori, Y.; Yoshikawa, H. Y.; Okada, S.; Adachi, H.; Sugiyama, S.; Takano, K.; Murakami, S.; Matsumura, H.; Inoue, T.; Yoshimura, M.; Mori, Y. Selective Crystallization of the Metastable Phase of Indomethacin at the Interface of Liquid/Air Bubble Induced by Femtosecond Laser Irradiation. *Appl. Phys. Express* **2015,** *8* (4), 045501-1-4.
22. Ikeda, K. *Control of Crystal Polymorphism of Drug Candidate Compounds by Gas-Liquid Interface Control;* Osaka University, 2015.
23. Richards, W. T.; Loomis, A. L. The Chemical Effects of High Frequency Sound Waves i. A Preliminary Survey. *J. Am. Chem. Soc.* **1927,** *49* (12), 3086−3100.
24. Amara, N.; Ratsimba, B.; Wilhelm, A. M.; Delmas, H. Crystallization of Potash Alum: Effect of Power Ultrasound. *Ultrason. Sonochem.* **2001,** *8* (3), 265−270.
25. Lyczko, N.; Espitalier, F.; Louisnard, O.; Schwartzentruber, J. Effect of Ultrasound on the Induction Time and the Metastable Zone Widths of Potassium Sulphate. *Chem. Eng. J.* **2002,** *86* (3), 233−241.
26. Ruecroft, G.; Hipkiss, D.; Ly, T.; Maxted, N.; Cains, P. W. Sonocrystallization: The Use of Ultrasound for Improved Industrial Crystallization. *Org. Process Res. Dev.* **2005,** *9* (6), 923−932.
27. Miyasaka, E.; Kato, Y.; Hagisawa, M.; Hirasawa, I. Effect of Ultrasonic Irradiation on the Number of Acetylsalicylic Acid Crystals Produced under the Supersaturated Condition and the Ability of Controlling the Final Crystal Size via Primary Nucleation. *J. Cryst. Growth* **2006,** *289* (1), 324−330.
28. Ueno, S.; Ristic, R. I.; Higaki, K.; Sato, K. In Situ Studies of Ultrasound-Stimulated Fat Crystallization Using Synchrotron Radiation. *J. Phys. Chem. B* **2003,** *107* (21), 4927−4935.
29. Gracin, S.; Uusi-Penttilä, M.; Rasmuson, Å. C. Influence of Ultrasound on the Nucleation of Polymorphs of p-Aminobenzoic Acid. *Cryst. Growth Des.* **2005,** *5* (5), 1787−1794.
30. Kurotani, M.; Hirasawa, I. Polymorph Control of Sulfamerazine by Ultrasonic Irradiation. *J. Cryst. Growth* **2008,** *310* (21), 4576−4580.
31. Nishida, I. Precipitation of Calcium Carbonate by Ultrasonic Irradiation. *Ultrason. Sonochem.* **2004,** *11* (6), 423−428.
32. Harzali, H.; Espitalier, F.; Louisnard, O.; Mgaidi, A. Sono-Crystallization of $ZnSO_4 \cdot 7H_2O$. *Phys. Proc.* **2010,** *3* (1), 965−970.
33. Nalajala, V. S.; Moholkar, V. S. Investigations in the Physical Mechanism of Sonocrystallization. *Ultrason. Sonochem.* **2011,** *18* (1), 345−355.
34. Haisa, M.; Kashino, S.; Kawai, R.; Maeda, H. The Monoclinic Form of p-Hydroxyacetanilide. *Acta Crystallogr. B* **1976,** *32* (4), 1283−1285.
35. Haisa, M.; Kashino, S.; Maeda, H. The Orthorhombic Form of p-hydroxyacetanilide. *Acta Crystallogr. B* **1974,** *30* (10), 2510−2512.
36. Perrin, M.-A.; Neumann, M. A.; Elmaleh, H.; Zaske, L. Crystal Structure Determination of the Elusive Paracetamol Form III. *Chem. Commun.* **2009,** *22,* 3181−3183.
37. Parkin, A.; Parsons, S.; Pulham, C. R. Paracetamol Monohydrate at 150 K. *Acta Crystallogr. E* **2002,** *58* (12), o1345−o1347.
38. Fabbiani, F. P. A.; Allan, D. R.; David, W. I. F.; Moggach, S. A.; Parsons, S.; Pulham, C. R. High-Pressure Recrystallisation - A Route to New Polymorphs and Solvates. *CrystEngComm* **2004,** *6,* 504−511.
39. McGregor, P. A.; Allan, D. R.; Parsons, S.; Pulham, C. R. Preparation and Crystal Structure of a Trihydrate of Paracetamol. *J. Pharm. Sci.* **2002,** *91* (5), 1308−1311.
40. Burger, A. Interpretation of Polymorphism Studies. *Acta Pharm. Technol.* **1982,** *28,* 1−20.

41. Sudha, C.; Nandhini, R.; Srinivasan, K. Polymer-Induced Selective Nucleation of Mono or Ortho Polymorphs of Paracetamol through Swift Cooling of Boiled Aqueous Solution. *Cryst. Growth Des.* **2014,** *14* (2), 705–715.
42. Mori, Y.; Maruyama, M.; Takahashi, Y.; Ikeda, K.; Fukukita, S.; Yoshikawa, Y. H.; Okada, S.; Adachi, H.; Sugiyama, S.; Takano, K.; Murakami, S.; Matsumura, H.; Inoue, T.; Yoshimura, M.; Mori, Y. Selective Crystallization of Metastable Phase of Acetaminophen by Ultrasonic Irradiation. *Appl. Phys. Express* **2015,** *8* (6), 065501-1-4.
43. Ashokkumar, M.; Lee, J.; Iida, Y.; Yasui, K.; Kozuka, T.; Tuziuti, T.; Towata, A. Spatial Distribution of Acoustic Cavitation Bubbles at Different Ultrasound Frequencies. *ChemPhysChem* **2010,** *11* (8), 1680–1684.
44. Turnbull, D. Kinetics of Heterogeneous Nucleation. *J. Chem. Phys.* **1950,** *18* (2), 198–203.
45. Chayen, N. E.; Saridakis, E.; Sear, R. P. Experiment and Theory for Heterogeneous Nucleation of Protein Crystals in a Porous Medium. *Proc. Natl. Acad. Sci. U.S.A.* **2006,** *103* (3), 597–601.
46. Lang, M.; Grzesiak, A. L.; Matzger, A. J. The Use of Polymer Heteronuclei for Crystalline Polymorph Selection. *J. Am. Chem. Soc.* **2002,** *124* (50), 14834–14835.
47. Hiremath, R.; Varney, S. W.; Swift, J. A. Selective Growth of a Less Stable Polymorph of 2-iodo-4-Nitroaniline on a Self-Assembled Monolayer Template. *Chem. Commun.* **2004,** *23,* 2676–2677.
48. Hiremath, R.; Basile, J. A.; Varney, S. W.; Swift, J. A. Controlling Molecular Crystal Polymorphism with Self-Assembled Monolayer Templates. *J. Am. Chem. Soc.* **2005,** *127* (51), 18321–18327.
49. Price, C. P.; Grzesiak, A. L.; Matzger, A. J. Crystalline Polymorph Selection and Discovery with Polymer Heteronuclei. *J. Am. Chem. Soc.* **2005,** *127* (15), 5512–5517.
50. Grzesiak, A. L.; Uribe, F. J.; Ockwig, N. W.; Yaghi, O. M.; Matzger, A. J. Polymer-Induced Heteronucleation for the Discovery of New Extended Solids. *Angew. Chem. Int. Ed.* **2006,** *45* (16), 2553–2556.
51. López-Mejías, V.; Knight, J. L.; Brooks, C. L.; Matzger, A. J. On the Mechanism of Crystalline Polymorph Selection by Polymer Heteronuclei. *Langmuir* **2011,** *27* (12), 7575–7579.
52. Curcio, E.; López-Mejías, V.; Di Profio, G.; Fontananova, E.; Drioli, E.; Trout, B. L.; Myerson, A. S. Regulating Nucleation Kinetics through Molecular Interactions at the Polymer–Solute Interface. *Cryst. Growth Des.* **2014,** 14 (2), 678–686.
53. Mori, Y.; Maruyama, M.; Takahashi, Y.; Yoshikawa, Y. H.; Okada, S.; Adachi, H.; Sugiyama, S.; Takano, K.; Murakami, S.; Matsumura, H.; Inoue, T.; Yoshimura, M.; Mori, Y. Metastable Crystal Growth of Acetaminophen Using Solution-Mediated Phase Transformation. *Appl. Phys. Express* **2017,** *10* (1), 015501.
54. Peterson, M. L.; McIlroy, D.; Shaw, P.; Mustonen, J. P.; Oliveira, M.; Almarsson, Ö. Crystallization and Transformation of Acetaminophen Trihydrate. *Cryst. Growth Des.* **2003,** *3* (5), 761–765.
55. Di Martino, P.; Conflant, P.; Drache, M.; Huvenne, J.-P.; Guyot-Hermann, A.-M. Preparation and Physical Characterization of Forms II and III of Paracetamol. *J. Therm. Anal.* **1997,** *48* (3), 447–458.
56. Nichols, G.; Frampton, C. S. Physicochemical Characterization of the Orthorhombic Polymorph of Paracetamol Crystallized from Solution. *J. Pharm. Sci.* **1998,** *87* (6), 684–693.
57. Mikhailenko, M. A. Growth of Large Single Crystals of the Orthorhombic Paracetamol. *J. Cryst. Growth* **2004,** *265* (3–4), 616–618.
58. Sudha, C.; Srinivasan, K. Nucleation Control and Separation of Paracetamol Polymorphs through Swift Cooling Crystallization Process. *J. Cryst. Growth* **2014,** *401,* 248–251.

59. Yasui, M. Phase Transition of Organic Crystals (Japanese). *J. Crystallogr. Soc. Jpn.* **1995,** *37* (1), 69–75.
60. Berg, W. L. Crystal Growth from Solutions. *Proc. Roy. Soc. Lond. Ser. A Math. Phys. Sci.* **1938,** *164* (916), 79.
61. Nii, K.; Maruyama, M.; Okada, S.; Adachi, H.; Takano, K.; Murakami, S.; Yoshikawa, Y. H.; Matsumura, H.; Inoue, T.; Imanishi, M.; Tsukamoto, K.; Yoshimura, M.; Mori, Y. Improvement of Metastable Crystal of Acetaminophen via Control of Crystal Growth Rate. *Appl. Phys. Express* **2018,** *11* (3), 035501-1-4.

INDEX

'*Note:* Page numbers followed by "f" indicate figures, "t" indicate tables.'

D

Q

R

S

CPI Antony Rowe
Chippenham, UK
2018-12-04 16:32